ADVANCES IN POTATO PEST MANAGEMENT

ADVANCES IN POTATO PEST MANAGEMENT

Edited by
JAMES H. LASHOMB
Rutgers—The State University

and

RICHARD CASAGRANDE
University of Rhode Island

Hutchinson Ross Publishing Company

Stroudsburg, Pennsylvania

Library of Congress Catalog Card Number: 81-6442
ISBN: 0-87933-407-X

84 83 82 1 2 3 4 5
Manufactured in the United States of America.

Library of Congress Cataloging in Publication Data
Main entry under title:
Advances in potato pest management.
"Proceedings of a conference . . . held at Rutgers University, January 23–25, 1980"—Pref.
Includes index.
1. Potatoes—Diseases and pests—Congresses.
2. Colorado potato beetle—Control—Congresses.
3. Pest control—Congresses. I. Lashomb, James H.
II. Casagrande, Richard.
SB608.P8A38 635'.2196 81-6442
ISBN 0-87933-407-X AACR2

Distributed world wide by Academic Press,
a subsidiary of Harcourt Brace Jovanovich,
Publishers.

CONTENTS

PREFACE

"Recent Advances in Potato Pest Management" was the theme of a conference held at Rutgers University, January 23–25, 1980. The conference was part of a series of activities held to honor the first one hundred years of the New Jersey Agricultural Experiment Station. Originally established as part of the Rutgers Scientific School, the Experiment Station is now part of Rutgers's Cook College, named after George Cook, first director of the station. In publishing these edited proceedings of the conference, we wish to celebrate this centennial and acclaim the achievements of this and other agricultural experiment stations.

Although pest management must still be considered an infant science, the concepts and practices of this science have been maturing rapidly during the past decade. The problem of managing pests is now generally recognized as a subset of the problems of design and management of ecosystems in general. Ten years ago this acknowledgment would have been little more than rhetoric. During the 1970s, however, it became increasingly apparent that no part of any ecosystem can be dealt with in isolation. Coevolving with this awareness was the new application of systems science to environmental management problems. Thus as we began to recognize a new hierarchy of needs, we also grew familiar with new and better methods for dealing with those needs. Even today, however, pest management in most crops is more of a philosophy than a reality because as yet we have had insufficient time to study the major components and interactions of crop ecosystems.

"Recent Advances in Potato Pest Management" was a milestone in the evolution of the science of pest management. The conference chronicled the deceptively successful history of potato crop management. As with other crops, we have experienced tremendous progress in crop production technology, most notably in the areas of variety selection, fertilization, harvesting, and storage. Our progress in crop protection has been less spectacular. Historically the approach to crop protection has been to develop remedial measures on a pest-by-pest basis. In many cases this has led to several undesirable effects including pesticide resistance, secondary pests, and environmental contamination.

The Colorado potato beetle provides one of the most dramatic examples of the outcome of shortsighted responses to a long term problem. Due to a lack of plant resistance and biological and cultural controls, we have relied exclusively on chemicals to control this pest for over a century. It has responded by developing insecticide resistance, and it continues to be a devastating pest, not only in the northeastern United States but throughout the world. The preponderance of conference papers devoted to this insect reflects the global magnitude of this pest.

This conference was intended to provide a broad survey of the state of the art of potato pest management. It stressed the multidisciplinary aspects of potato ecosystem management and highlighted the interregional, indeed international scope of common concerns. In reviewing recent developments in disease and insect forecasting schemes, the conference showed promise for the future of potato pest management. It also made clear the long and arduous task that challenges us for the future.

Many people were responsible for making the conference a success. We thank the authors for a thorough representation of their research in these papers. We thank Herbert Streu, Stuart Race, Loren Hiddleson, and Melvin Henninger for their help in organizing the conference. We also acknowledge and deeply appreciate the efforts of Phyllis Hornlein for her dedication while typing this manuscript; of Joanne Bellanca who filled many important roles during the course of this project; and of Leslie May who assisted with the proofreading of the final manuscript. Special thanks are extended to the following institutions for providing grants to help offset conference costs and to promote the distribution of the book: New Jersey Agricultural Experiment Station, American Cyanamid, BASF-Wyandotte, Chevron, Diamond Shamrock, FMC: ICI Americas, Mobay, Penick, Rohm and Haas, Sandoz, Shell, Stauffer, and Union Carbide.

JAMES H. LASHOMB
RICHARD CASAGRANDE

CONTRIBUTORS

T. E. ANDERSON, Department of Entomology, North Carolina State University, Raleigh, North Carolina

GEORGE W. BIRD, Department of Nematology, Michigan State University, East Lansing, Michigan

J. DE WILDE, Department of Entomology, Agricultural University, Wageningen, The Netherlands

ELMER E. EWING, Department of Vegetable Crops, Cornell University, Ithaca, New York

WILLIAM E. FRY, Department of Plant Pathology, Cornell University, Ithaca, New York

ANDREW J. FORGASH, Department of Entomology and Economic Zoology, Rutgers—The State University, New Brunswick, New Jersey

NORMAN L. GAUTHIER, Agway, Inc., Syracuse, New York

RICHARD N. HOFMASTER, Virginia Truck and Ornamental Station, Eastern Shore Branch, Painter, Virginia

T. HSIAO, Department of Biology, Utah State University, Logan Utah

GEORGE G. KENNEDY, Department of Entomology, North Carolina State University, Raleigh, North Carolina

ROGER A. LeBRUN, Department of Plant Pathology and Entomology, University of Rhode Island, Kingston, Rhode Island

PATRICK A. LOGAN, Department of Plant Pathology and Entomology, University of Rhode Island, Kingston, Rhode Island

D. R. MacKENZIE, Plant Pathologist, The Pennsylvania State University University Park, Pennsylvania

MICHAEL L. MAY, Department of Entomology and Economic Zoology, Rutgers—The State University, New Brunswick, New Jersey

DONALD W. ROBERTS, Insect Pathology Resource Center, Boyce Thompson Institute, Cornell University, Ithaca, New York

MAURIE SEMEL, Department of Entomology, Cornell University, Long Island Horticultural Research Laboratory, Riverhead, Long Island, New York

ZANE SMILOWITZ, Pesticide Research Laboratory and Graduate Studies Center, Department of Entomology, The Pennsylvania State University, University Park, Pennsylvania

RICHARD S. SOPER, USDA-SEA-AR, Insect Pathology Research Unit, Boyce Thompson Institute, Cornell University, Ithaca, New York

RICHARD H. STORCH, Department of Entomology, University of Maine, Orono, Maine

GEORGE TAMAKI, Yakima Agricultural Research Laboratory, USDA-SEA-AR, Yakima, Washington

WARD M. TINGEY, Department of Entomology, Cornell University, Ithaca, New York

ADVANCES IN POTATO PEST MANAGEMENT

Overview of Potato Physiology Related to Integrated Pest Management

Elmer E. Ewing

The annual farm value of potatoes in the United States in 1978 was $1.2 billion, far exceeding the value of any other vegetable (USDA, 1979). Few if any major crops surpass the potato in terms of potential yields of calories or of protein per hectare if calculated per day of growing season, and it ranks among the top half-dozen or so food crops in total world production of both protein and energy (derived from data in Adams, 1975; FAO, 1977).

The leading producer of potatoes in the world is the Soviet Union, with about 30 percent of total world production. Second is Poland, which has about 17 percent of the total, and third is the People's Republic of China. Potato production in China reportedly is increasing very rapidly, and by now might exceed even that of Poland (R. L. Sawyer, pers. comm.). The United States ranks in fourth place, with about 5 percent of the world production, followed by East Germany, India, West Germany, France, the United Kingdom, and the Netherlands. Almost all European countries have substantial potato production. The potato is also of major significance in many parts of Asia, Africa, and South America. India, for example, has more hectares of potatoes than does the United States, although total production is less due to lower yields. Bangladesh grows twice as many hectares as Ireland, South Korea more than Minnesota, and Japan surpasses Canada in both hectares and yields. Turkey, Colombia, Brazil, Argentina, Peru, Egypt, Australia, and Algeria all produce substantial quantities of potatoes. Clearly we must revise our notion that the potato is an important crop only in North America and Europe. It has gained major acceptance in many parts of the third world, including countries commonly thought to be rice eating (FAO, 1977; USDA, 1980).

This chapter provides a framework for the others in this book by outlining important features of potato physiology as these may relate to integrated pest management. I shall touch on several of the most important aspects of potato physiology, but my major emphasis will be on those facets with which my own research program is most involved. Most of this chapter therefore centers on the physiology of tuberization and the implications this has for integrated pest management. A number of excellent reference books on potato physiology may be consulted for further information and supporting references on other phases of the subject (Ivins and Milthorpe, 1963; Burton, 1966; Talburt and Smith, 1975; Smith, 1977; Harris, 1978).

PROPAGATION

The potato's clonal reproduction has several consequences for those who work with it. First, the development of new cultivars is affected greatly because it is not necessary to obtain a line of seeds that will breed true. Only a single seed is required to be the progenitor of a new cultivar. If there are a hundred seeds in a potato berry, we can expect a hundred recognizably different clones to arise from sowing them. The potato is notoriously heterozygous not only because of its asexual propagation but because it is tetraploid. The combination of heterozygosity and tetraploidy complicates inheritance studies, and those who wish to perform genetic analyses on the sources of pest resistance are confronted with far greater complexity than they would encounter in other crops.

A second consequence of asexual propagation is disease transmission. Bacterial, fungal, and viral diseases and the potato cyst nematode are readily spread from generation to generation through the planting of tubers. So serious is this problem that until the recent development of meristem culturing techniques (see E. G. Cutter in Harris, 1978, chap. 3), the entire stock of some cultivars was affected with one or more diseases. Even with recent advances in eliminating disease organisms from seed stocks, transmission of disease by seed tubers remains a major problem in potato production.

Apart from questions of disease, the physiological condition of seed tubers can affect their yielding potential in a manner distinctly different from sexually propagated crops (see D. C. E. Wurr in Harris, 1978, chap. 8; Smith, 1977, chap. 7). The effect of the physiological age of seed tubers is a major research topic in Europe at present. Proportionately it has received less attention in the United States and Canada, although some studies have been carried out here. Factors that contribute to the physiological aging of tubers include longer storage, storage at higher temperatures, earlier planting of the seed crop, warmer temperatures during the growing season, and varieties possessing shorter dormancy periods. Characteristics of physiologically old tubers are extensive sprouting, sprouting at many eyes, and relatively small, early-maturing haulms. Physiologically older seed tubers tend to give higher yields during short growing seasons but not from later harvests. Excessive physiological age of seed tubers leads to the production of sprout tubers rather than normal sprouts. In this disorder, a stolon and tuber or even a sessile tuber may develop from the sprouted eyes of the seed tuber. Typically this happens only after several sets of normal sprouts have been produced and lost due to removal during handling or to attack by Rhizoctonia and other pests. Any haulm developing from a seed tuber that has sprout tubers will produce little or no marketable yield. The problem is more common than many growers realize because missing hills are often blamed upon poor planter performance or seed piece decay. Greensprouting tubers before planting--exposure to diffuse light while sprouts are developing--serves to toughen sprouts and shorten internodes. This practice, common in Europe, probably would reduce sprout-tuber problems in America, but more hand labor is required to accomplish it than American growers are willing to expend.

Another difference between the handling of seed tubers in Europe and America is that whole tubers are planted universally in Europe, whereas economic considerations force American growers to cut much of their seed (see Smith, 1977, chap. 7). Most studies indicate that seed pieces, either whole or cut, should weigh about 50 grams (g). Cut tubers can produce just as well as can whole seed, but they present extra challenges for disease control. Cutting can spread disease from infected to healthy tubers, and the cut surface is more easily invaded by pathogens present in the soil.

Missing potato hills, whether from sprout tubers, seed piece decay, or planter failure, can be expected to reduce total yields and contribute to a lack of uniformity in tuber size. The optimal planting rates vary with cultivar and many other conditions, but increasing the quantity of seed planted will increase yields until very high rates are attained (see E. J. Allen in Harris, 1978, chap. 7; Smith, 1977, chap. 9). Population studies with potatoes are complicated by the ability of each seed piece to produce more than one stem. The number of stems per unit area may show a better correlation with tuber yield than does number of seed pieces per unit area. Stem number tends to increase with size of seed piece but is also affected by cultivar, by whether seed is whole or cut, and by storage temperature.

The quantity of seed tubers planted per hectare is of great importance, but within rather broad limits the planting arrangement is not. By arrangement I mean the distance between seed pieces within the row and the distance between rows for any given quantity of seed tubers per hectare. For example, seed tubers spaced in a square arrangement, 44 centimeters (cm) apart in 44-cm rows, can be expected to give no better yields than are obtained from the more conventional spacing of 22 cm between seed tubers in rows 88 cm apart. Even wide variations in the uniformity of arrangement within a row have little effect on yield or tuber size if the total quantity of seed tubers per row is not changed (Sieczka and Ewing, unpubl.).

Recently interest has grown in propagation of potatoes by seeds rather than seed tubers. Only 30 to 40g of seeds are required to plant 1 hectare, compared to more than 2.5 metric tons of tubers, and seeds are much less perishable. In contrast to the many diseases that can be carried in seed tubers, only spindle tuber is known to be transmitted through potato seeds. These advantages of seeds make them seem attractive for some developing countries that have difficulty obtaining healthy seed tubers, even though plants grown from seeds can be expected to lack uniformity, yield less, and mature later than comparable plants grown from healthy seed tubers.

It does not seem likely that seeds will replace seed tubers for commercial potato production in the United States in the foreseeable future. If seeds are to be a practical alternative even in third world countries, a new technology must be developed. It can be expected that fertilization, spacing, soil preparation, and other cultural practices will be different for potatoes grown from seeds than from seed tubers. Experiments will be needed to decide whether plants should be direct seeded or transplanted. Modifications will be needed in potato breeding requirements, including attention to homozygosity. Techniques of seed production must be worked out. Finally, it can be predicted safely that a different set of weed, insect, and disease problems will be encountered.

CONTROL OF TUBERIZATION

Tuber formation in the potato may be initiated soon after, or even before, plant emergence. Contrary to popular belief, it is not dependent upon flowering, although frequently the two processes occur at about the same time.

In the morphological sense a tuber is an underground stem (Figure 1). The "eyes" are nodes--buds in the axils of rudimentary leaves that form the "eyebrows." As in an ordinary stem, the buds at the apex exert dominance over the lateral buds. Exposure to light leads to the production of chlorophyll (and also increases the concentration of toxic glycoalkaloids). The vascular system and arrangement of tissues is basically that of an ordinary stem. As the tuber enlarges, the epidermis sloughs off and is replaced by a periderm, equivalent to the bark of a woody stem.

Tuber initiation is controlled by a hypothetical tuberization stimulus (Gregory, 1965). Formation of the stimulus is favored by photoperiods shorter than a

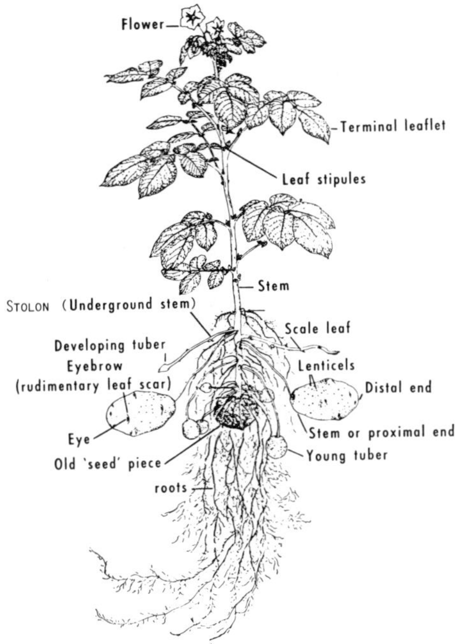

Figure 1
Potato plant (drawn with slight alteration from Kehr et al., 1964, Figure 1).

genetically determined critical photoperiod and by cool temperatures. Increasing temperature effectively shortens the critical photoperiod, reducing tuberization. High light intensity appears to favor the stimulus under greenhouse conditions, although it is not clear what level of light is optimal in the field. Under greenhouse conditions low rates of nitrogen fertilization also favor the stimulus, but more information is needed to determine how important nitrogen rate is in influencing the tuberization stimulus under field conditions. Physiologically old seed tubers seem to produce plants with higher levels of the stimulus.

The tuberization stimulus is probably initiated in the leaves. The chemical nature of the stimulus is not known (see E. G. Cutter and J. Moorby in Harris, 1978, chaps. 3, 4). It may be a single compound or result from changes in the concentrations of a group of compounds. Gibberellins reduce the expression of the stimulus, and there is considerable evidence that cytokinin concentration is well correlated with its presence. There is some inclination to conclude that the level of the stimulus depends on the ratio of cytokinin(s) to gibberellin(s), although other classes of compounds may be involved also, and the participation of a specific, still unidentified hormone that orchestrates the other hormonal changes cannot be ruled out. Whatever the chemical nature of the stimulus, it is transported from the leaves throughout the plant.

The entire morphology of the plant, not just tuber initiation, is controlled by the level of the tuberization stimulus. Under conditions that strongly favor the stimulus--short, cool days and genotypes with long critical photoperiods--leaves are large and form wide angles with the stem, internodes are short, stems are thick, above-ground axillary branching is suppressed, root growth is restricted, and tubers form sessilely or on very short stolons (Figure 2). The main shoot apex develops slowly and then stops growing altogether. Thus high levels of the tuberization stimulus produce a very high ratio of tuber weight to total plant weight. The restrictions in haulm and root development do not appear to be merely the consequence of photosynthate partitioning to tubers. These changes tend to occur even when tuber initiation is prevented artifically.

Until the chemical nature of the stimulus is elucidated, some technique other than chemical analysis is needed to follow changes in its relative level or intensity and to study its movement in the plant. In view of these needs, it is fortunate that tuber development is not limited to stolon tips; any bud on the potato plant can tuberize if placed in the dark, if given a high relative humidity, and if the tuberization stimulus is present (Gregory, 1965). Hence stem cuttings that contain at least one leaf and one bud afford a

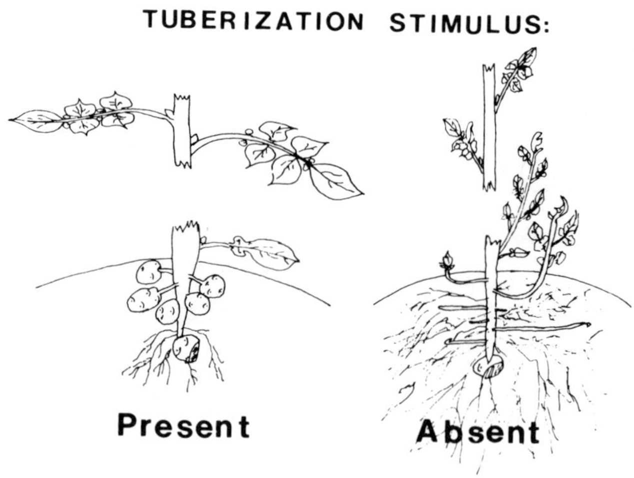

Figure 2
Differences between a potato plant that has not been induced to tuberize (right) and one that has been strongly induced (left).

convenient means of determining whether the tuberization stimulus was present in the leaf at the time the cutting was taken. The cutting is placed in a mist chamber to prevent wilting of the leaf, and the bud is covered with soil. The photoperiod in the mist chamber is kept long so that conditions after cutting will not promote formation of the tuberization stimulus. If the stimulus was present in the leaf at the time of cutting, the bud will form a sessile tuber under the soil within about ten days. In the absence of the tuberization stimulus, the bud will remain dormant or will develop into a leafy shoot.

Figure 3 shows how apical cuttings and one-node cuttings are taken, and Figure 4 depicts the range of cutting responses elicited by varying levels of induction prior to making the cuttings. If plants have been exposed only to days longer than their critical photoperiods before apical cuttings are taken, then the axillary bud inserted into the soil does not grow. Exposure to a few short days, but not enough to induce tubers, leads to the growth of a shoot or stolon from the underground bud of the cutting, suggesting that stolon formation is not the opposite of tuberization, but rather an intermediate response between bud dormancy and tuberization. Very strong induction before cutting produces growth of a sessile tuber at the underground bud (Ewing and Wareing, 1978).

We have found cuttings to be a useful tool for screening breeding populations for their response to

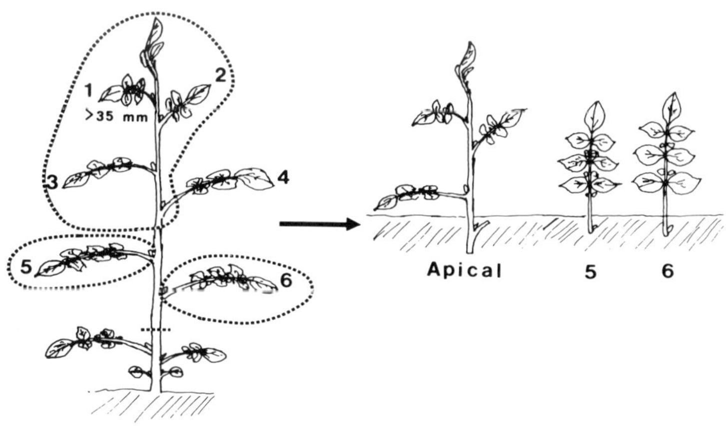

Figure 3
Diagram shows how apical and one-leaf cuttings are taken. Dotted lines indicate where stem is severed. Cuttings are inserted into a rooting medium with one bud underground and are placed in an ordinary horticultural mist bench. Photoperiods are extended to at least twenty hours. Cuttings may be examined in ten to twelve days.

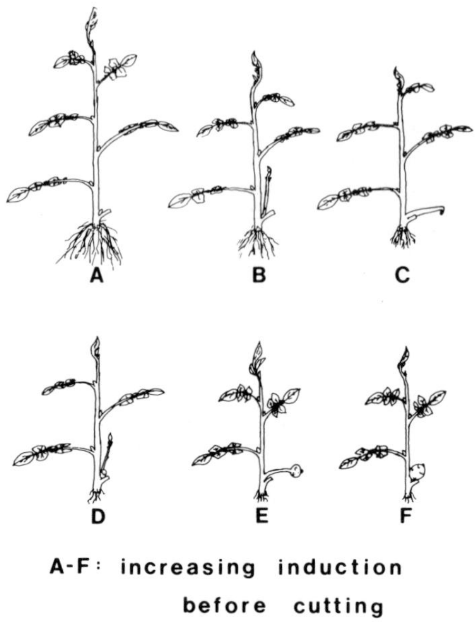

Figure 4
Cuttings after twelve days under long photoperiods in a mist bench. Note differences in growth of lowest bud, which was covered by soil after cutting. One-leaf cuttings exhibit a similar range of responses. (A) Cutting from a plant exposed to days much longer than its critical photoperiod. Note growth of shoot apex and abundant rooting but dormancy of underground bud. (B) Typical response from exposure to a few short photoperiods prior to cutting but not enough to induce tuberization. Underground bud has grown upward as a leafy shoot. (C) Slightly stronger induction produces diageotropic outgrowth as a stolon from the underground bud. (D) The central bud at the underground node produced a leafy shoot, but one or both ancillary buds at the same node were induced to tuberize. A variation of this response would be a stolon rather than a leafy shoot at the central bud. (E) A stolon terminating in a tuber, indicating a fairly high level of induction. (F) Exposure of plant to ten to fifteen photoperiods well below its critical photoperiod induces a sessile tuber at the underground bud. Growth of roots and the apical bud is very restricted on a strongly induced cutting such as this.

photoperiod. The potato as found in the Andes of Peru and Bolivia, its original home, has a critical photoperiod of only twelve to thirteen hours when tested at cool temperatures. If such clones, referred to as Andigena types, are grown at high latitudes, they will not tuberize until late in the fall when days are short and nights are cool. The Tuberosum type potato, by contrast, has been selected for its ability to tuberize under the long summer days of northern Europe and North America. Andigena populations carry valuable genetic traits, including resistance to various pests, that are not known in Tuberosum, but to take advantage of Andigena germ plasm, it is necessary to select for long critical photoperiod. Cuttings provide a convenient, rapid screening method to make such selections (Ewing, 1978).

Even standard Tuberosum cultivars will vary in critical photoperiod, though the range is not so extreme as the difference between typical Andigena and Tuberosum. Earliest maturing cultivars tend to have the longest critical photoperiods (Lazin et al., 1979). Consequently they are able to manufacture the tuberization stimulus even under the longest summer days. Because the stimulus limits haulm size, the ratio of tubers to total plant weight is higher in the early-maturing cultivars than in late ones, but total tuber yields are less.

Another way in which cuttings have been useful has been to study the movement of the tuberization stimulus. They have been used to show that the stimulus moves across grafts (Chapman, 1958; Gregory, 1956; Kumar and Wareing, 1973; Ewing and Wareing, 1978), that movement is not affected by gravity (Kahn and Ewing, unpubl.), and that the stimulus tends to move most strongly to those underground nodes farthest from light (Kahn and Ewing, unpubl.). Especially significant is the fact that the more leaves present on a cutting, the stronger is the shift from above-ground to below-ground bud growth and the tendency to tuberize (Kahn and Ewing, unpubl.). This is consistent with the hypothesis that the more leaves there are to export the stimulus, the stronger will be the tuberization response. It explains why the more leaves that have developed on a plant under a given set of conditions, the more likely the plant is to initiate tubers.

What has been said for tuber initiation also applies to tuber enlargement. Invariably, conditions that favor the first seem to favor the second. It is reasonable to conclude that the tuberization stimulus must be present for both initiation and growth of tubers.

IMPLICATIONS FOR PEST MANAGEMENT

Factors leading to high levels of the tuberization stimulus early in the growing season tend to favor early yield, but at the expense of the eventual total yield. For example, an early-maturing Tuberosum cultivar grown under cool, short days may tuberize quickly, but the rest of the plant will be so stunted by the high level of tuberization stimulus that photosynthate available for export to the tubers will be severely limited and actual yields will be low. Similar tendencies, though less pronounced, may be observed by the use of physiologically old seed and perhaps even by low rates of nitrogen fertilizer.

Low levels of the tuberization stimulus throughout the growing season lead to excessive root and haulm production and little or no tuber formation. Highest yields are obtained when the level of stimulus is low enough early in the season to permit vigorous root and haulm growth if this is followed by conditions that give high levels of the stimulus and consequent partitioning of photosynthate to the tubers. (For detailed consideration of relationships between haulm size or leaf area and tuber yield, see Harris, 1978, pp. 177-183, 678-696.) Theoretically, it would seem that the more leaf area produced during the long days early in the season, the greater the tuberization stimulus should be when days become short at the end of the season, and the higher the yields to be expected. The implication is that any loss of foliage would reduce tuber yields. In practice such is not necessarily the case. A suprisingly large amount of leaf area has been removed in some experiments with only minor effects on eventual tuber yields, especially if removed before full bloom (for example, Beresford, 1967, and other work cited therein). Caution must be exercised in extrapolating these data to other conditions, however. It is reasonable to expect that the degree of defoliation that can be tolerated will depend upon the intricate array of factors that control the level of the tuberization stimulus throughout the season. Differences in day length (due to latitude or time of planting), cultivar, temperature, physiological age of seed, and nitrogen fertilization might all influence the results.

Also, it may be expected that these factors would influence the threshold of damage that can be sustained from insects with piercing-sucking mouth parts or from nematodes or various pathogens of the root system. Through control of the size of the root system, the same factors could affect the ability of the plant to take up systemic pesticides applied to the soil. Effects on the haulm growth in turn would affect weed populations, as well as the severity of diseases (such as late blight and aerial blackleg) that are responsive to relative humidity.

Finally, the relation of the tuberization stimulus to maturity should be considered. Whether examining differences in pest tolerance among cultivars or comparing effects of cultural practices on pests, apparent differences in pest damage might be indirect effects of maturity differences brought about by changes in the level of the tuberization stimulus. For example, a cultivar with a short critical photoperiod will have lower levels of the stimulus and mature later in the season. The later maturity might well complicate the evaluation of its susceptibility to leafhoppers, verticillium wilt, early blight, late blight, nematodes, and other pests.

PERIDERM FORMATION

Another important aspect of maturity is its relation to periderm formation on tubers. The advent of modern insecticides after World War II made profitable the use of much higher rates of nitrogen fertilizer and resulted in much later haulm maturity. As a consequence, potato "skins" or periderms in many parts of the country are still thin at harvest unless a chemical top killer is employed. These certainly hasten periderm formation to a certain extent, but the chemical killing of very immature, green haulms will not induce as good a periderm as would be obtained from natural maturity. On the other hand, growers cannot afford to wait for natural haulm maturity if this means harvesting when soil temperatures are too low. Aside from the danger of field frost or the undesirable effects of chilling on tuber processing quality, potato tubers become much more subject to harvesting damage as the soil temperature is lowered.

Probably more potatoes are lost from cuts and bruises in the United States than from any pest, perhaps even more than from all pests combined. Studies in almost every major production area have shown that commonly 15 to 40 percent of tubers are damaged significantly during harvest. Severe cuts and bruises can ruin a tuber directly, but even less serious damage leads to problems. Water evaporates more rapidly from the cuts, increasing weight loss and shriveling, and producing increased susceptibility to the serious storage disorders, pressure bruise and blackspot. Moreover, cuts and bruises provide entry for various decay organisms. Evaluation of pest management practices therefore should assess any possible effects of the practices on maturity, as well as on such obvious parameters as tuber size and number.

CONCLUSIONS

The economic importance of the potato, the enormous number of pests that attack it, and the complicated physiological responses of the potato plant to changes in its environment make it an ideal subject for studies in integrated pest management.

REFERENCES

Adams, C. F., 1975, Nutritive Value of American Foods, Agr. Handbk. 456, Agr. Res. Serv., U.S. Department of Agriculture, 291p.

Beresford, B. C., 1967, Effect of Simulated Hail Damage on Yield and Quality of Potatoes, Amer. Potato J. 44:347-354.

Burton, W. G., 1966, The Potato, 2d ed., Veenman and Zonen N.V., Wageningen, 382p.

Chapman, H. W., 1958, Tuberization in the Potato Plant, Physiol. Plant 11:215-224.

Ewing, E. E., 1978, Critical Photoperiods for Tuberization: A Screening Technique with Potato Cuttings, Amer. Potato J. 55:43-53.

Ewing, E. E., and P. F. Wareing, 1978, Shoot, Stolon, and Tuber Formation on Potato (Solanum tuberosum L.) Cuttings in Response to Photoperiod, Plant Physiol. 61:348-353.

Food and Agriculture Organization of the United Nations, 1977, FAO Production Yearbook, vol. 31, FAO Statistics Series No. 15, 291p.

Gregory, L. E., 1956, Some Factors for Tuberization in the Potato Plant, Amer. J. Bot. 43:281-288.

Gregory, L. E., 1965, Physiology of Tuberization in Plants. (Tubers and tuberous roots.), Encyclopedia Plant Physiol. 15:1328-1354.

Harris, P. M., ed., 1978, The Potato Crop, Chapman and Hall, London, 730p.

Ivins, J. D., and F. L. Milthorpe, eds., 1963, The Growth of the Potato, Butterworths, London, 328p.

Kehr, A. E., R. V. Akeley, and G. V. C. Houghland, 1964, Commercial Potato Production, U.S. Department of Agriculture Handbook No. 267, 59p.

Kumar, D., and P. F. Wareing, 1973, Studies on Tuberization in Solanum andigena, I: Evidence for the Existence and Movement of a Specific Tuberization Stimulus, New Phytol. 72:283-287.

Lazin, M. B., I. J. Sung, E. E. Ewing, and J. B. Sieczka, 1979, Relation between Maturity of Varieties and Critical Photoperiod for Tuberization as Determined by the Stem Cutting Technique, Amer. Potato J. 56:166 (Abstr.).

Smith, O., 1977, Potatoes: Production, Storing, Processing, 2d ed., Avi, Westport, Conn., 776p.

Talburt, W. F., and O. Smith, 1975, Potato Processing, 3d ed., Avi, Westport, conn., 705p.

U.S. Department of Agriculture, Crop Reporting Board, Economics, Statistics, and Cooperative Services, 1979, Potatoes and Sweetpotatoes, 1977-78 Production, Disposition, Value, Stocks, and Utilization.

U.S. Department of Agriculture, Foreign Agric. Service, 1980, Foreign Agriculture Circular, Fresh and Processed Vegetables, Washington, D. C., FVEG2-80.

History of Colorado Potato Beetle Control

Norman L. Gauthier, Richard N. Hofmaster, and *Maurie Semel*

The most destructive insect pest of potato in the northeastern United States is the Colorado potato beetle, Leptinotarsa decemlineata (Say). The beetle, first found on the eastern slopes of the Rockies in 1824, where its feeding habits were restricted to native weeds, remained obscure until settlers pushing westward began growing potatoes. By 1874, the Coloradc potato beetle had spread to the Atlantic coast and has since invaded many other parts of the world.

This one insect probably has had a greater effect on developments in crop protection methods and application equipment than any other pest. The early spread of the beetle increased unchecked throughout its range until Paris green, the first arsenical insecticide, was discovered. According to Bronson and Anderson (1952), it was to combat the ravages of this one species, known only as a museum specimen from 1819 to 1850, that the modern sprayer had its beginning. Prior to this period in the United States, there were no hand or powered sprayers, nor did chemical insecticides, as we know them today, exist. Following the development of Paris green and other arsenicals, many groups of compounds were used to hold this pest in check until after World War II. Since then the situation has changed drastically; the recent history of control efforts involves the introduction and eventual failure of innumerable insecticides because the Colorado potato beetle became resistant.

BIOLOGY AND FEEDING HABITS OF THE COLORADO POTATO BEETLE

Adult Colorado potato beetles are three-eights inch long with ten black lengthwise stripes along the elytra; thus the specific name decemlineata. According to Essig (1952) the beetles hibernate in the central and northern United States and Europe but aestivate during the dry season in the tropics. The overwintering adults dig in the soil to a depth of several inches, seldom more than twenty to twenty-five cm (Metcalf et al., 1962). Upon emergence in the spring, adults feed and mate, and females deposit bright orange, oval-shaped eggs on early planted potatoes. These eggs are normally deposited on undersides of leaves in masses of ten to forty eggs. An adult female can deposit over three hundred eggs in her life span of four to five weeks. The eggs hatch in about a week. The larvae go through four instars in about three weeks, eventually attaining a length of 0.5 to 1.25 centimeters (cm). During larval development, the insects feed almost continuously, stopping only to molt. Mature larvae drop off the leaves and enter the soil for pupation. The pupal stage lasts five to ten days, after which a new generation of adults emerges.

There are one to three generations per year depending on latitude (Peairs and Davison, 1956). Hofmaster (1979) indicated that in Virginia there are three generations per year. Overwintered adults begin feeding and oviposition in late April with subsequent larval feeding during May. Peak adult emergence of the summer generation occurs around June 15 at Painter, Virginia, and these adults cause much defoliation before harvest time of early-maturing varieties. The cycle is then repeated on late-maturing potatoes, although damage by second-generation beetle may be reduced because little oviposition occurs when the maximum temperature reaches 80°F. The second-generation pupae aestivate until late summer and early fall when the second generation adult beetles emerge and deposit eggs, and the cycle is repeated for a third time on fall potatoes. Overwintering is by adults from the third generation. Farther north, the beetle has two generations per season, and in southern Maine there is only one generation.

Although potato is the preferred host, the beetles may feed and survive on other members of the nightshade family, including eggplant, tomato, pepper, tobacco, groundcherry, Jimsonweed, thorn apple, henbane, horsenettle, and belladonna and other plant hosts such as thistle and mullein (Metcalf et al., 1962; Westcott, 1964; Hofmaster, 1979; Linduska, 1978; Schalk and Stoner, 1979).

DISPERSION AND DISTRIBUTION OF THE COLORADO POTATO BEETLE

Dispersion of the Colorado potato beetle prior to the introduction of the potato into the West probably was limited by the spread of its native hosts, *Solanum datura* and *S. rostratum* (Sanderson, 1912; Comstock, 1950). By the mid-1800s, the western frontier of the potato culture had reached the native habitat of the beetle, which began to feed on the potato foliage. The insect thrived on potato, and its distribution began extending to the east and north, following the distribution of its new host. By 1859, the beetle had reached a point one hundred miles west of Omaha, Nebraska, crossing into Illinois by 1864 and reaching the Atlantic coast in 1874 (Sanderson, 1912). It continued to travel about fifty to eighty-five miles per year. By 1912 it was found wherever potatoes were grown in the United States except in parts of Florida, Nevada, and California (Sanderson 1912; Comstock, 1950; Westcott, 1964). A more detailed discussion of the easterly movement is given by Johnson (1969). The Colorado potato beetle also spread rapidly throughout Europe following its introduction into France in 1922.

DEVELOPMENT OF SPRAY EQUIPMENT

Between 1850 and 1860, many early settlers were threatened by starvation as hordes of Colorado potato beetles, which had adapted to their new source of food, devoured field after field of potatoes. At that time the only means of control was to hand-pick adults, larvae, and egg masses. By the time the insect reached the Atlantic seaboard, early work of pioneering entomologists had proved that it was vulnerable to a poison later known as Paris green (Bronson and Anderson, 1952). Because there was no equipment to apply these early arsenical insecticides, they were applied by hand in dry or slurried form. Makeshift application apparatus included whisk brooms, tin cans with holes punched in them, and other simple devices.

Between 1880 and 1894, John Bean of California, D. B. Smith of New York, and Brandt Brothers of Minnesota developed and then improved the first hand sprayers. Before the turn of the century, hand machines included barrel pumps, bucket pumps, and hand-operated knapsack sprayers. Powder guns with hand-cranked geared drives such as the Leggett powder gun came into vogue specifically to kill beetles in large potato fields using clear Paris green. Paddock (1897) and Sanderson (1912) give excellent and thorough accounts of these early sprayer types.

The first wheel-drawn sprayers had steel wheels connected to a pump, which could develop up to 300 pounds per square inch (psi). The sprayers were equipped with a wooden tank having a capacity of fifty to two hundred gallons and were used to apply lead arsenate and Bordeaux mixtures. These early sprayers had small trailing booms equipped with nozzles for row crop spraying. In 1895 a traction duster came into use. Early field trials to control beetles with traction sprayers were conducted at the Virginia Truck Experiment Station by L. C. Corbett (Sanderson, 1912).

The first successful power sprayer using a steam engine appeared in 1894 in Connecticut, followed in 1900 by the first complete gasoline-powered sprayer. Airplanes were first used about 1926-1931, and the first successful airblast sprayer, the prototype of a modern-day speed sprayer, was made around 1937. Since much of this equipment was initially developed or extensively used for Colorado potato beetle control, it is apparent that this beetle was largely responsible for the early development of pesticide application methodology. This pest may be largely responsible for creation of the pesticide industry also.

CONTROL OF THE BEETLE

The First Hundred Years

Between 1865 and 1965 hundreds of compounds, singly or in combination, were utilized to control the Colorado potato beetle. These chemical insecticides encompass several groups, including the first arsenicals, fluorides, botanicals, chlorinated hydrocarbons, carbamates, organophosphates, and other compounds. In the early use of arsenicals, between 1864 and 1875, Paris green, the aceto-arsenite of copper, was used in the United States and preferred over London purple. London purple, manufactured from aniline dyes, contained varied amounts of arsenite of lime. Because of its variable arsenic composition, its high water-soluble arsenic content, and the resulting injury to plant foliage, London purple was not extensively used. Arsenicals were slow acting, and the free arsenic in their formulation made them inherently phytotoxic; thus they had to be mixed with lime to protect the foliage. Early mixtures of Paris green used concentrations of 0.4 percent suspensions with added lime (Brown, 1951).

By the 1900s rate differences were appearing, for Sanderson (1912) reported that one pound Paris green plus one pound freshly slaked quicklime in fifty gallons of water would kill larvae, but often two to three pounds of arsenic was needed to destroy the adult beetles. Many growers at the turn of the century switched to lead arsenate because of the phytotoxicity of Paris green. Lead arsenate used at 0.6 percent suspensions was also a mixture of arsenicals. Two compounds commonly used, acid lead arsenate (dilead-ortho arsenate) and basic lead arsenate (lead hydroxy-arsenate), contain very little water-soluble arsenic pentoxide and so damaged plants relatively little (Carter, 1952). Basic lead arsenate was safer to use on plants than the acid form but generally was not as toxic to insects. Because of inadequate spray equipment and poor coverage, the arsenicals provided erratic control. They were also difficult to mix and apply and had a short residual life on potato foliage. Of the botanical compounds tried such as veratrine alkaloids (Sabadilla, hellebore), ryania, rotenone (derris or cube), and others, only rotenone had sufficient residual properties to be effective (Brown, 1951).

The introduction of the Colorado potato beetle into Europe was followed by the discovery in Switzerland in

1939 that both larvae and adults could be destroyed by DDT. Paul Miller of J. R. Geigy, S.A., of Basel, Switzerland, was awarded the 1948 Nobel Prize in Physiology and Medicine for the discovery of the insecticidal properties of DDT. Hitchner (1952) provided an interesting account of this event. In 1939, the potato crop of Switzerland was seriously threatened by this beetle. J. R. Geigy made available to Swiss entomologists a sample of DDT for testing. The first tests against the Colorado potato beetle confirmed company findings and resulted in the control of this destructive insect.

Other chlorinated hydrocarbon insecticides were soon developed. Brown (1951) summarized the performance of these materials. BHC initially was found to be superior to DDT, but its effect was transitory. Aldrin, dieldrin, and heptachlor in 0.025 percent suspensions were as effective as DDT used in 0.1 percent suspensions. Chlordane was about as effective as DDT against the Colorado potato beetle. Excellent control was obtained with these compounds during the mid-to late 1940s and early 1950s, successfully replacing the arsenicals and botanicals. Other compounds utilized with success during this period were foliar and soil treatments of dieldrin, heptachlor, and methoxychlor.

In 1953, however, Hofmaster (1956a) reported the beetle's resistance to DDT and in 1958, the beetle was discovered to be resistant to dieldrin and other chlorinated hydrocarbon insecticides (Hofmaster and Waterfield, 1965). From 1960 to 1970 the Colorado potato beetle emerged again as a limiting factor in potato production. Resistance problems were reported in many northeastern potato-growing areas, although as late as 1970 Founk and McClanahan (1970) reported no evidence of resistance to DDT in laboratory toxicology studies with Colorado potato beetle larvae in Ontario. In a similar study six years later, Harris and Svec (1976) found resistance of third-stage larvae to aldrin, DDT, endrin, and endosulfan.

1965 to the Present

Since the development of DDT resistance, the Colorado potato beetle has consistently been the prime agricultural pest in potato-producing areas of Virginia, New Jersey, Long Island, Connecticut, and Rhode Island. Thus beetle population pressures in these areas are more severe than in any other sections of the United States. With the demise of DDT and most other chlorinated hydrocarbon insecticides brought about by resistance and banning, new insecticide groups have been employed against the beetle. Although newer carbamate, phosphate, and other synthetic organic insecticides have been used, their residual activity is not as effective as the chlorinated hydrocarbons. Research began in the mid-1950s to find soil treatments for supplementing foliar sprays. Two groups of soil insecticides became available for this purpose: the older nonsystemic chlorinated hydrocarbons such as aldrin, dieldrin, and heptachlor, and new systemic compounds, mostly in the carbamate and phosphate groups. Hofmaster (1956b) found that during spring planting, temperatures were low enough that overwintering adults, seeking protection under clods of soil, came in contact with residual insecticides and were killed. The advent of the systemic insecticides such as disulfoton (DiSyston), aldicarb (Temik), fensulfothion (Dasanit), phorate (Thimet), and carbofuran (Furadan) also made possible the introduction of the chemical into growing plant parts.

Growers soon recognized that systemic compounds would provide Colorado potato beetle control for a period of fifty to sixty days, so that only a few foliar sprays were needed to supplement control to harvest. The systemics, like the arsenicals and DDT, gave the illusion of such good overall kill that for several years the Colorado potato beetle was not considered a problem in many areas; however, because of continual insecticide exposure, once again the beetle populations developed resistance.

At present, aldicarb is the only generally effective soil treatment in the Northeast, and the only generally effective foliar treatments are fenvalerate and permethrin.

HISTORY OF CONTROL IN VIRGINIA

In Virginia, early reports are somewhat sketchy so it is difficult to determine when the Colorado potato beetle appeared. Chittenden (1907) speculated that the potato beetle was present in the District of Columbia and probably in the adjacent areas of Virginia in 1873. The following year the beetle reached the Atlantic seaboard and was definitely established in Virginia. In this locality the potato was abundant; approximately thirty thousand acres of potatoes were produced annually, mostly in the Norfolk-Eastern Shore area. Production of the spring crop on the Eastern Shore peaked to slightly over sixty-two thousand acres between 1922 and 1926 and now is somewhere around twenty-five thousand acres. In addition, a fall crop of varying acreage is produced, primarily for seed.

The potato beetle quickly became a menace to the Virginia potato crop. Alwood (1893) characterized the beetle as a persistent pest of Virginia potatoes, "so well known that almost every child recognizes it." Paris green was widely used during the early years to protect the crop. Alwood, in investigating the use of London purple, lessened the phytotoxic effect of this compound and increased its stability by adding limewater to the spray mixture. He strongly advised the use of sprays over dusts, even though Paris green dusts had been used on potato beetles in Virginia for almost twenty years. London purple proved rather erratic in its performance, however, and Paris green had the definite limitations of phytotoxicity, difficulties in keeping suspension, short residual activity, and expense. As early as 1898 Phillips and Price (1899) tested other arsenicals in Virginia in an attempt to devise a more satisfactory control measure.

Popenoe (1909) called the potato beetle the most important pest of potatoes in Virginia by far and urged a change in insecticidal application. He pointed out the limitations of early practices in which Paris green was mixed with land plaster in the proportion of one "to" ten or one "to" twenty and dusted onto the plants by means of a burlap bag. This method was laborious, costly, and gave poor coverage. Sprays were demonstrated to be three to four times cheaper and much more effective, and they did not impart acidity to the soil such as occurred from the use of land plaster. A by-product of the increased soil acidity from the use of land plaster surfaced when it was observed that a "disease" occurred in the cabbage crops that followed potatoes, signifying rotation problems following the use of certain chemical treatments. Popenoe also observed that Bordeaux mixture acted as a safener and prevented foliage injury when it was added to the arsenicals and that it also repelled potato beetles. He was instrumental in convincing a number of farmers to shift from dust to sprays.

Popenoe (1909) also made a number of observations on the life history of the potato beetle in Virginia, stressing that the overwintered beetles were literally sitting around waiting for the potato plants to emerge and that as many as twenty beetles were present on a single hill, completely "stemming" the plant. He was surprised to find that the beetles did not oviposit on the fall crop and that then damage was caused entirely by adults. This is not true today because many eggs and larvae develop on the fall crop.

By about 1909, Paris green was being replaced in Virginia by lead arsenate, which was less harmful to the plants and adhered to the foliage better. Chittenden (1909) states "The number of sprayings (lead arsenate) to be applied depends upon local and seasonal conditions. Two or three will usually suffice for spray work, especially if applied just after the eggs of the potato beetle have hatched, and the same number should be applied for the second generation of such an insect." Nevertheless the severity of the Colorado potato beetle problem in Virginia continued to increase, and Smith somewhat later (1915a) reported that it was still the most important insect pest of potatoes. Smith also emphasized that the potato beetles emerged prior to the plants and stemmed them, as Popenoe (1909) reported and as they do now. Eventually there was a gradual shift to the calcium arsenate formulations, probably because they were cheaper and because the insect problems changed.

Dusts were quite popular in Tidewater, Virginia, and had their beginning when the Colorado potato beetle first became a menace to the potato crop. Originally Paris green was mixed with some carrier such as hydrated lime, gypsum, or flour and was applied by sifting the material through a burlap sack directly onto the plant. This method was rather erratic but showed enough promise that traction-type dusters eventually came into being. This also led to the development of improved dust formulations. One such mixture contained 23 percent arsenious oxide, derived chiefly from Paris green, and was dusted on the plants with a small traction duster. Zimmerly et al. (1921) reported this mixture to be very effective in controlling the beetles on the parts of the foliage covered, but because of the small amount used and the lack of uniform coverage, considerable injury occurred on the lower and less exposed leaves of the plants. Dusts containing 20 percent calcium arsenate or 20 percent lead arsenate applied 20 pounds per acre gave better results. These dusts, especially calcium arsenate, were widely used until the development of DDT. In fact, a considerable acreage was treated with calcium arsenate in the mid-to late 1940s. Although the Colorado potato beetle problem continued, this pest was controlled with the lead or calcium arsenate treatments.

Attention was diverted from the potato beetle for several years when another insect, the potato tuberworm, *Phthorimoea operculella* (Zeller), appeared in such numbers on the Eastern Shore in 1923 that a quarantine was threatened on Eastern Shore potatoes (Spencer and Strong, 1925). Beginning around 1930, still another insect pest, the potato flea beetle (*Epitrix cucumeris* Harris), became more of a problem than the potato beetle on the Eastern Shore of Virginia and seriously threatened the potato industry. Unlike the Colorado potato beetle, potato flea beetle larvae develop underground where they feed on the tubers, causing scarring and reduction in grade or even loss of sale.

An intensive research program launched by Dr. Lauren D. Anderson in 1932 resulted in improved insecticidal treatments, formulations, and application methods. Anderson and Walker (1936) concluded that a 2-4-6-50 calcium arsenate-Bordeaux spray applied at one hundred gallons per acre was the most efficient treatment available. Such treatments also held potato beetles in check to such an extent that they were of relatively minor importance for the next fifteen to twenty years. The contributions of Lauren Anderson to the potato industry in Virginia cannot be overemphasized.

Following World War II and the adoption of DDT as the universal insecticide, potato beetles almost disappeared from the Eastern Shore. By 1953, however, scattered colonies of potato beetles appeared in DDT-treated fields, and by 1955 extensive injury occurred in many areas. In some cases potato beetles had increased to such an extent that even newly harvested tubers were damaged by feeding of the adults.

Tests conducted in 1955 at Painter, Virginia, by Hofmaster (1956a) showed that DDT was no longer completely killing larvae. Table 1, a compilation of selected results of several of Hofmaster's published insecticide trials, shows that in 1955 DDT gave only 78 percent control of late instar larvae, and over the next several years it became even less effective. Potato growers continued to get partial control from 1957 through 1959, but by 1960 DDT had faded completely in some areas, a situation reflected by the 5 percent mortality in Nandva (Table 1).

Dieldrin had largely supplanted DDT as the favored potato foliage spray by 1955 and was so effective that once again potato beetles became quite scarce. The efficacy of one dieldrin foliage spray, used primarily for flea beetle control, was augmented by soil treatments of aldrin, dieldrin, and heptachlor, which themselves had been demonstrated to provide a significant degree of flea beetle control (Hofmaster, 1956b).

Observations at the Eastern Shore Station at Painter seemed to show that damage by Colorado potato beetle larvae was reduced in areas treated by soil insecticides. Detailed counts confirmed this; the number of hills infested with larvae was reduced 73 to 74 percent below the untreated check by preplant broadcast soil treatments of aldrin and heptachlor (Hofmaster, 1956b).

A number of growers also reported an apparent decrease in Colorado potato beetle damage to potatoes receiving soil applications of aldrin or heptachlor. Upon examination it was estimated that potato beetle damage had been reduced as much as 90 percent as compared to adjacent untreated fields. Because the Colorado potato beetle larvae from the spring brood enter the soil where they pupate and emerge as adults, it was thought that soil treatments further reduce populations. Data obtained from cage tests (Hofmaster, 1956b) show that aldrin and dieldrin at 2.5 lbs active ingredient per acre and dieldrin at 1.5 pounds (lb) reduced adult emergence 97.7 percent or more. In addition, none of the adults that emerged from the insecticide treated plots was alive more than twenty-four hours.

Despite the overall excellent results obtained with dieldrin, new problems began to emerge. In 1957, tests at the Eastern Shore Station showed that dieldrin gave less rapid knockdown of potato beetle larvae than in previous years; the kill rate was 96 percent at seventy-two hours compared to nearly 100 percent control in three hours in previous years. Increasing the rate of dieldrin to 0.5 lb per acre resulted in rapid knockdown in 1956, but in 1959 there was poor knockdown and 98 percent mortality at the 0.5 lb rate. Based on these results, it was predicted that widespread dieldrin resistance might occur by 1960--and it did. Two applications of a 1.5 percent dieldrin dust at 30 to 35 lbs per acre were almost totally ineffective near Nandva,

Table 1
Mortality of Late Instar Colorado Potato Beetle Larvae Seventy-two Hours after Pesticide Applications, Painter, Virginia (in percent)

Material	1955[a]		1956[b]		1957	
	Rate*	Mortality	Rate	Mortality	Rate	Mortality
DDT	(0.5)	78	(0.5)	79	(0.5)	80
Dieldrin	(0.2)	100	(0.2)	100	(0.2)	96
Diazinon	(0.5)	99	-	-	(0.5)	93
Endrin	(0.2)	100	-	-	(0.2)	100
Heptachlor	(0.5)	98	-	-	(0.5)	99
Phosdrin	(0.2)	99	-	-	(0.2)	99
Carbaryl	-	-	-	-	(0.5)	100
Dibrom	-	-	-	-	-	-
Guthion	-	-	-	-	-	-
Trithion	-	-	-	-	-	-
Parathion	-	-	-	-	-	-
Aldrin	(.15)	100	-	-	(0.5)	98
Malathion	-	-	-	-	-	-
Phosphamidon	-	-	-	-	-	-
Thiodan	-	-	-	-	-	-

*Active ingredient/acre.
[a]Ninety-six hour count.
[b]Forty-eight hour count; adults tested.

Virginia, in May 1960. Examination of the potato field on May 17, 1960, showed potato beetle larvae of all sizes present, with an average of approximately twenty-five per plant. Emergency screening measures demonstrated that azinphos-methyl at 0.4 lb per acre, mevinphos at 0.5 lb per acre, and carbaryl and naled at 1 lb per acre all gave 100 percent or nearly 100 percent knockdown at two hours. Naled and carbaryl were utilized in field treatments and gave excellent results.

In order to verify the action of some of the insecticides over a longer period of time, approximately ten thousand third instar potato beetle larvae were collected near Nandva and taken to the Painter Station, where they were placed on potato plants, allowed to become acclimated, and then treated in a controlled test (Table 1). The excellent results obtained with naled, mevinphos, azinphos-methyl, and carbaryl were substantiated. Furthermore, it was shown that DDT and dieldrin applied at 0.5 lb active ingredient per acre, had mortalities of only 5 and 10 percent, respectively and had become practically worthless as potato beetle insecticides. In addition to aldrin, dieldrin, and DDT, by 1961 endosulfan activity had deteriorated to limited effectiveness. Furthermore, the chlorinated hydrocarbon soil treatments of aldrin, dieldrin, and heptachlor had lost most of their effectiveness.

Azinphos-methyl and carbaryl, adopted as foliar

1959		1960[c]		1961	
Rate	Mortality	Rate	Mortality	Rate	Mortality
(0.5)	64	(0.5)	5	(0.5)	42
(0.5)	98	(0.5)	9	(0.5)	5
-	-	(0.5)	10	(0.5)	10
-	-	(0.2)	16	(0.4)	13
-	-	-	-	-	-
-	-	(0.5)	100	(0.5)	100
(0.5)	100	(1.0)	100	(1.0)	100
(0.5)	100	(1.0)	100	(1.0)	100
-	-	(0.4)	100	(0.5)	100
-	-	(1.0)	96	(1.0)	98
-	-	(0.5)	11	(0.5)	7
-	-	(0.5)	9	(0.5)	8
-	-	-	-	(1.0)	8
(0.5)	100	(0.5)	100	(0.5)	100
(0.5)	100	(0.5)	86	(0.5)	68

[c]Third instar larvae collected in Nandva, and tested at Painter.

sprays in 1960, were widely used with excellent results for a number of years. Their activity, along with that of carbofuran, is traced from 1969 through 1979 in Table 2, which shows that a decline in carbaryl activity began about 1970. Azinphos-methyl persisted somewhat longer but started to show a decrease in activity in 1974. Resistance has been variable throughout the state, though azinphos-methyl is still giving good potato beetle control in some sections of the lower Eastern Shore. Carbofuran sprays began to show signs of weakening activity against adults in 1976, but it was not until 1978 that really poor larval control developed.

A new method of chemical treatment became available in 1960. Hofmaster and Dunton (1961) demonstrated that phorate granules applied in the furrow at planting would give a high degree of Colorado potato beetle control. This organophosphate, labeled and recommended for flea beetle control in 1961, proved very effective against flea beetles. It also performed so well against potato beetles that although the beetle was not named on the label, much of the potato acreage was treated with phorate at 2 to 3 lbs per acre with this pest in mind. So effective were the phorate soil treatments that very few foliar sprays were applied despite the fact that the systemic effectiveness was lost about a month before harvest, and there was no protection left against potato beetles, second brood flea beetles, variegated cutworms, or potato tuberworms.

Two new systemics, aldicarb and carbofuran, were first tested against the Colorado potato beetle in 1964, with promising results. In 1969 Hofmaster initiated a series of tests evaluating these and other insecticides. The results (Table 2) show that in 1969 all the systemics gave good early season control, but other results, not included, show that by June 25, thimet and carbofuran were far less effective then aldicarb. Granular carbofuran was not used to any extent commercially in Virginia until 1975, although the flowable formulation was in use before that time. By 1976 there was evidence of resistance to granular carbofuran in experimental plots and in commercial fields as well. Approximately a thousand acres in Accomac County that had been treated with granular carbofuran had high beetle populations in 1977. Again in 1978 many growers used carbofuran instead of aldicarb primarily because of carbofuran's lower cost and its effectiveness against European corn borer. By 1978, widespread failure of the carbofuran applications caused an emergency situation in Virginia, since there were no effective foliar insecticides that were labeled for potato use.

In response a 24-C state label was obtained on June 13 for the use of oxamyl in Virginia, and the Environmental Protection Agency granted an emergency

Table 2
Colorado Potato Beetle Larvae per Ten Hills in June, Painter Virginia (in rate of pounds AI per acre)

Insecticide	Rate	1969	1970	1971	1972
Granular comparisons					
Temik 15G	3.0[a]	0.0[a]	0.0	0.0	--
Furadan 10G	3.0[b]	0.0	0.0	0.0	--
Thimet 10G	3.0	1.8	0.0	7.8	--
Guthion 2E (foliar)	0.5[c]	3.3	0.0	0.3	--
Control		370.8	Def.	318.0	--
Spray formulations					
Furadan 4F	0.5	0	0	0	0
Sevin 80S	1.0[d]	0	18	2	11
Guthion 2E	0.5[c]	0	1	0	0
Control		Def.	Def.	327	323

Note: Primarily Pungo variety of potatoes; usually planted March 15-25, sprayed at seven- to ten-day intervals.
[a]In 1969-1971 Temik 10G was applied at a rate of 2.0 lb AI per acre.
[b]In 1969-1971 Furadan 10G was applied at a rate of 2.0 lb AI per acre.

exemption on July 11 for the use of permethrin. Oxamyl and permethrin were used effectively for the remainder of the 1978 season. Despite demonstrated carbofuran resistance, one of the most effective adult treatments consisted of a combination of 0.5 lb oxamyl plus 0.5 lb carbofuran. Recognizing that a serious potato beetle problem would occur in 1979, attempts were made to secure early labeling of permethrin and fenvalerate. After considerable controversy, permethrin was approved on April 20, 1979, and fenvalerate on May 18, 1979. These synthetic pyrethroids (0.1 to 0.2 lb active ingredient per acre) along with the oxamyl-carbofuran combination, gave excellent control of potato beetles up to July 15. There were some indications of problems with adults in July and early August, but the potatoes were either harvested or senescent by that time.

Because of the problems encountered with carbofuran in 1978, many growers shifted to aldicarb in 1979. Of the ten thousand acres in Accomac County treated with aldicarb, there were no cases of beetles' severely damaging plants at emergence as they did in carbofuran-treated areas. Of the potatoes planted in March with aldicarb, a few fields had to be treated with foliar sprays by May 15-17 and all fields in the area needed foliar sprays, but carbofuran granules were still giving satisfactory control in the lower Eastern Shore.

1973	1974	1975	1976	1977	1978	1979
--	0.0	0.3	0.0	0.0	2.0	0.0
--	0.0	0.3	5.0	28.3	21.3	73.3
--	7.3	16.0	0.0	4.0	32.3	120.7
--	0.0	8.3	1.3	0.5	20.3	28.0
	335.0	300.3	178.0	Def.	238.3	184.7
1	1	6	0	1	43	18
1	32	21	21	9	98	66
1	10	8	4	1	20	28
372	275	241	206	280	324	222

[c]In 1978 Guthion 2E was applied as a foliar spray at 1.0 lb AI per acre.

[d]In 1969-1973 Seven 80W was applied at a rate of 1.0 lb AI per acre.

HISTORY OF CONTROL ON LONG ISLAND

Most of the potato production in New York State is concentrated in two distinct locations: Long Island and northern Steuben County. Potatoes are grown throughout the state, however, and represent the state's largest vegetable crop. New York State currently ranks eighth in total fall production in the United States. From the late 1860s through the early 1900s, New York was the nation's leading potato producer, reaching a peak of 442,254 acres in 1904. From 1945 through 1949, Long Island production peaked at over 11 million hundredweight annually grown on approximately 63,000 acres. Since then acreage and production have subsided substantially to approximately 23,000 acres and 6.5 million hundredweight. Currently Long Island accounts for half of New York's production.

Records show that potatoes were introduced in New Hampshire by 1719, and it is assumed that not long after they were introduced to the Long Island area, although there was little commercial production until the mid-1800s. At the outset, commercial potato production was limited to small potato fields of one to a few acres since mechanical planters and harvesters were not yet developed. As these tools made their appearance, expanded acreage became possible. This increase in potato acreage preceded by a few years the arrival of the Colorado potato beetle on Long Island and offered an ever-increasing concentration of preferred host material. A quotation from one of Long Island's pioneering commercial producers serves to illustrate the state of the art in potato production during 1880: "My first recollection of working is of dropping potatoes on a little farm at Westhampton--potato planters were unheard of. I had to stick them into the side of a furrow. They were plowed in. The potato beetle had just become a pest on Long Island, and we didn't know how to combat it. Father paid my sister and me one cent a hundred for the old shell-back bugs. And a nest of eggs counted the same as a bug." The first appearance of the potato beetle on Long Island was noted in the Coney Island area in 1874 and within a few years had become an economic problem. Since the first efforts to control the beetles by hand picking, Long Island growers have used a diverse array of chemicals to achieve satisfactory control.

During the earliest periods of beetle proliferation, Long Island growers adopted the insecticide and spray technology utilized by western growers. Sprays and dusts of Paris green were the most effective means of controlling the beetle; however, even the inclusion of lime in the spray mixture did not completely mitigate the phytotoxic effect of this arsenical on the potato plant. Dilute formulations of sodium arsenate, with Bordeaux as a safener, were also used to some extent, but this insecticide had some phytotoxic side effects. Lead and calcium arsenates were substituted and found to be much safer to plant foliage. Somewhat higher concentrations of calcium arsenate and sodium arsenate were used to achieve the same control. Just prior to the advent of DDT, concentrations of calcium arsenate were radically increased to as high as 11 lbs in 100 gallons (gal) of water to overcome what appeared to be resistance of the potato beetle. During this period a number of growers used Cryolite at 6 lbs per 100 gal of water as a substitute for calcium arsenate with limited success.

DDT became available for use in 1945 on a limited basis and was in full supply the following year. By 1951, the potato beetle was virtually nonexistent as an economic pest. But early in 1952 potato beetle resurgence was reported in western Long Island, and additional accounts of DDT failures were soon noted on scattered farms throughout the potato--growing area. By 1954, DDT resistance was widespread, and many insecticide mixtures were tried, including a commonly used mixture of DDT and rotenone. Growers who reverted to the use of calcium arsenate met with little success. Methoxychlor was found to provide effective control at 1 lb active ingredient per acre, but the cost was relatively high compared to DDT, and growers were reluctant to go to a full schedule of applications.

In 1954, dieldrin and heptachlor were labeled for use and provided a needed respite to the beetle problem, but by 1957 dieldrin resistance was evident, and heptachlor became more widely used. At that time endrin was also found to be effective; however, its value was to be short-lived. In 1958, endrin resistance developed throughout Long Island, and many growers again used methoxychlor. In 1959 when azinphos-methyl and carbaryl were registered for use, they immediately became the preferred controls at 4 ounces (oz) active ingredient per acre. Carbophenothion appeared in 1960 and was recommended for a few years but could not compete as economically or effectively as azinphos-methyl or carbaryl. In 1961, phosphamidon was made available but did not gain wide acceptance.

Potato beetle resistance to carbaryl was apparent in a few locations in 1963, and the dosage was increased to 8 and then to 12 and 16 oz per acre. The higher concentrations were more effective, but damaging numbers of larvae and adults still survived. The following year, a limited number of complaints were recorded concerning the inadequacy of azinphos-methyl at the 4 oz rate. The dosage was increased to 8 oz per acre and toward season's end 4 oz of parathion was added to boost control.

During these years the number of growers deserting carbaryl usage rapidly grew; by 1971 none remained, even after trying to obtain control at 24 oz active ingredient per acre. By 1970-1971 azinphos-methyl had failed to provide good larval control at 8 oz, and the dosage was increased to 12 oz active ingredient. Control was generally effective for the next two years, although there were a number of failures, even at the increased rate. The addition of parathion at dosages as high as 16 oz active ingredient per acre was only a temporary palliative.

The virtual inability of azinphos-methyl sprays to control the potato beetle led to disastrous results in 1973. Spray material costs alone in 1973 averaged approximately $52 per acre compared to average costs of about $9 during the preceding few seasons, and control was generally poor in spite of these high expenditures. In that year phosmet and monocrotophos (Azodrin) were introduced into the commercial spray program, but they were completely inadequate. The failure of both compounds despite promising test results were attributed to the development of cross-resistance.

Late in the 1973 season, a temporary permit was issued allowing the use of a single application of carbofuran, but unfavorable climatic conditions and extremely severe insect population pressure caused mixed results. Full use of carbofuran provided economic control in 1974 and in 1975 as well, but by 1976 signs of resistance were evident. The addition of endosulfan to sprays of carbofuran apparently had a synergistic effect and enabled the use of carbofuran during the next growing season. Once again, in 1978 there were many grower complaints about the ineffectiveness of chemical control, and the New York State Department of Environmental Conservation issued a special local-need registration for the use of oxamyl as a foliar spray for potatoes. Although this chemical had been tested with considerable success, commercial successes were few, and test results the following year confirmed that the potato beetle was resistant. It was also determined that distinctly different strains existed on the north and south forks of eastern Long Island, with the southern beetles being more susceptible to oxamyl. Apparently because fewer insecticide sprays are generally applied on the south fork, the selection of insecticide-resistant potato beetles is markedly slowed.

Granular systemic insecticides were introduced in Long Island in 1973 with the first small-scale use of phorate and disulfoton by commercial growers. Disulfoton gave unsatisfactory results in 1974, and phorate provided control only for about two weeks after plant emergence. Most commercial growers did not use granular insecticides until 1975. The first widespread use was of carbofuran and, shortly thereafter, aldicarb. The effectiveness of granular carbofuran declined dramatically during the following two years, and aldicarb has been substituted by virtually all potato growers because of its continued effectiveness. Complete failure of carbofuran to control colonizing beetle was noted in 1978, the same year that the foliar formulation failed throughout the area.

In 1979 a new group of compounds, the synthetic

pyrethroids, was introduced through an emergency exemption granted by the EPA. Both permethrin and fenvalerate were generally effective foliar sprays used to provide mid to late-season control in conjunction with the early-season protection afforded by aldicarb. Table 3 gives a historical record of materials used on Long Island and time of failures.

Table 3
Insecticides Used Against Colorado Potato Beetle on Long Island

Material	Year Introduced	First Failure	General Failure
Arsenicals	1880	1940's	
DDT	1945	1952	1953
Methoxychlor	1953		
Dieldrin	1954	1957	
Heptachlor	1954		
Endrin	1957	1958	1958
Guthion	1959	1964	1973
Carbaryl	1959	1963	1970
Phosphamidon	1961		
Parathion	1964		
Azodrin	1973	1973	1973[a]
Phosmet	1973	1973	1973[a]
Carbofuran	1974	1976	1978
Permethrin	1978		
Fenvalerate	1978		
Oxamyl	1978	1978	1979
Phorate	1973	1974	1974
Disulfoton	1973	1974	1974
Aldicarb	1976		

[a]Cross-resistance.

CURRENT SITUATION

Over a century has passed since the insecticides were first used against the Colorado potato beetle. During this period hundreds of inorganic, botanical, and synthetic organic chemicals have been evaluated against this pest. Over the years, dozens of effective insecticides have been used, including arsenicals, chlorinated hydrocarbons, organophosphates, and carbamates. The response of the beetle has been remarkably uniform: it has developed resistance to all classes of materials used against it. The similar records in Virginia and Long Island indicate that resistance should now be expected within a few seasons of the introduction of a new insecticide. In these areas and for much of the Northeast, the hopes for early season potato beetle control rest on a single registered insecticide, aldicarb, and on the probability that the EPA will once again grant permission to use foliar applications of new materials before full labels are gained as was the example with the synthetic pyrethoid permethrin. Fenvalerate has not obtained full federal labeling as a foliar insecticide for use on potatoes for CPB control. On Long Island resistance is not the only problem facing growers. Trace amounts of aldicarb and carbofuran were found in drinking water in 1979, a situation jeopardizing their use. The usage label was withdrawn in 1980.

Factors other than insecticides must be evaluated for their effect upon the Colorado potato beetle. Work is underway on control with sprays of entomophagous fungi and selection of resistant potato cultivars, but these investigations are in the earliest stages. Other research on cultural control, insect growth regulators, and biological control must be pursued.

Characteristics of current potato varieties should be considered as part of any control program. For example, early-maturing potatoes such as Norland and Superior may require half the foliar sprays needed to protect late-maturing varieties such as Sebago and Green Mountain. These late varieties also allow development of an additional generation of potato beetles in northern areas. The use of irrigation has also effectively extended the growing period of the potato, providing suitable host material for the Colorado potato beetle well into the fall. Irrigation has also moderated the adverse effects of hot, dry summer weather upon the beetle.

Another cultural practice that affects beetles is vine killing. Mechanical destruction of vines generally has given way to chemical vine killing, and in many cases vines are not killed at all. Early vine-killing chemicals included sodium arsenite, dinoseb (DNBP), copper sulphate, and tank mixes of caustic soda and white arsenic. Sodium arsenite was an effective insecticide, and other materials adversely affected potato beetles as well. In recent years, these materials have been replaced with other vine-killing chemicals that are only weakly insecticidal. In those fields where vines are not killed, Colorado potato beetles have additional time to complete feeding and development before overwintering.

Other chemicals used in potato production are important. For a long time Bordeaux misture was the sole control for late blight and other potato diseases. This mixture was known to have a repellent effect on the potato beetle. But this copper-based fungicide has been replaced with organic fungicides such as mancozeb, polyram, captafol (Difolitan), and chlorothalonil, none of which exhibits this repellency. On the other hand, these fungicides have an effect on the naturally occurring entomophagous fungi such as _Beauveria bassiana_ and _Metarrhizium anisopliae_, which attack the potato beetle. The possibility also exists that carbamate fungicide use may confer a measure of carbamate insecticide resistance.

Although the science of pest management has made

great progress in the past two decades, the problems caused by insects and their control seem greater than ever. The answer to the Colorado potato beetle problem as well as other pest control challenges we face in the 1980s is clear. The state of the art of scientific investigations into crop protection must be evaluated in light of the most recent advances in technology. There must be prudent manipulation of cultural, biological, and other techniques used in potato production to blend harmoniously with efficient chemical controls and reduce the devastatingly high populations of the Colorado potato beetle, lessen the adverse effects on the environment, and provide producers the opportunity to grow a crop of potatoes profitably.

REFERENCES

Alwood, W. B., 1893, Injurious Insects and Diseases of Plants with Remedial Measures for the Same, Va. Agr. & Mechanical College Agr. Expt. Sta., Bull. 24:10, 15-19.

Anderson, L. D., and H. G. Walker, 1936, Control of the Potato Flea Beetle (*Epitrix cucumeris* Harris), Va. Truck Expt. Sta. Bull. No. 92.

Bronson, T. E., and E. D. Anderson, 1952, Choosing and Using Hand Equipment, in Insects, The Yearbook of Agriculture 1952.

Brown, A. W. A. 1951. Chemical Control of Insects Feeding on Plants. In Insect Control by Chemicals. John Wiley New York pp. 574-667.

Carter, R. H., 1952, The Inorganic Insecticides, in Insects, The Yearbook of Agriculture 1952, pp. 218-222.

Chittenden, F. H., 1907, The Colorado Potato Beetle, USDA Circular 87:1-15.

Chittenden, F. H., 1909, Some Insects Injurious to Cabbage, Cucumbers and Related Crops, Va. Truck Expt. Sta. Bull. No 2, Sept. 29.

Comstock, J. H. 1950, Coleoptera: An Introduction to Entomology, Comstock, Ithaca, N.Y., pp. 464-546.

Essig, E. O., 1952, How Insects Live, in Insects, The Yearbook of Agriculture 1952, USDA Publ. 413.

Founk, J., and R. J. McClanahan, 1970, Laboratory Studies on the Toxicity of Insecticides to Larvae of the Colorado Potato Beetle, J. Econ. Entomol. 63:2006-2007.

Harris, C. R., and H. J. Svec, 1976, Susceptibility of the Colorado Potato Beetle in Ontario to Insecticides, J. Econ. Entomol. 69:625-629.

Hitchner, L. S., 1952, The Insecticide Industry, in Insects, the Yearbook of Agriculture 1952, House Doc. 413:450-457.

Hofmaster, R. N., 1956a, Resistance of the Colorado Potato Beetle to DDT, Veg. Growers News 10:3.

Hofmaster, R. N., 1956b, Soil Insecticides i- Vegetable Crop Production, Bull. State Board of Agriculture, Dover, Delaware 46:13-20.

Hofmaster, R. N., 1957, Control of the Colorado Potato Beetle, Veg. Growers News 11:2-3.

Hofmaster, R. N., 1961, Control of Resistant Colorado Potato Beetles, Station to Station Research News, Union Carbide Chemicals Co. 9(3).

Hofmaster, R. N., 1979, The Colorado Potato Beetle Problem, Veg. Growers News 33:3-4.

Hofmaster, R. N., and E. M. Dunton, 1961, Soil Application of Insecticides for Control of Foliage Nests of Irish Potato, Am. Potato J. 38:341-345.

Hofmaster, R. N., and R. L. Waterfield, 1965, The Colorado Potato Beetle Problem in Virginia and Present Status of Control Measures, Trans. Peninsula Hort. Soc. 1965:20-26.

Johnson, C. G., 1969, Migration and Dispersal of Insects by Flight Methuen, London, 763p.

Linduska, J. J., 1978, Evalution of Soil Systemics for Control of Colorado Potato Beetle on Tomatoes in Maryland, J. Econ. Entomol. 71:647-649.

Metcalf, C. L., W. P. Flint, and R. L. Metcalf, 1962, Insects Injurious to Vegetable Gardens and Truck Crops, in Destructive and Useful Insects, 4th ed., McGraw-Hill Book Co., New York, pp. 598-683.

Paddock, Wendell, 1897, Spray Pumps and Spraying, New York Agr. Exp. Sta. Bull. No. 21, 15p.

Peairs, L. M., and R. H. Davison, 1956, Insects Injurious to Solonaceous Crops, in Insect Pests of Farm, Garden, and Orchard, 5th ed., John Wiley and Sons, New York, pp. 111-113.

Phillips, J. L., and H. L. Price, 1899, The Nature and Use of Certain Insecticides, Virginia Agr. Expt. Sta. Bull No. 97.

Popenoe, C. H., 1909, The Colorado Potato Beetle in Virginia in 1908, USDA Bur. Ent. Bull. 82:1-8.

Sanderson, E. D., 1912, Insects Injurious to Potatoes and Tomatoes, in Insect Pests of Farm, Garden and Orchard, John Wiley and Sons, New York, pp. 291-294.

Schalk, J. M., and A. K. Stoner, 1979, Tomato Production in Maryland: Effect of Different Densities of Larvae and Adults of the Colorado Potato Beetle, J. Econ. Entomol. 72:826-829.

Smith, Loren, B., 1915a, Control of the Colorado Potato Beetle (Leptinotarsa decemlineata Say), Va. Truck Expt. Sta. Bull. No. 14. 2p.

Smith, Loren B., 1915b, Second Report Control of the Colorado Potato Beetle (Leptinotarsa decemlineata Say), Va. Truck Expt. Sta. Bull. No. 17.

Smith, Loren B., 1917, Control of Some Garden Insects, Va. Truck Expt. Sta. Bull. No. 23.

Spencer, H. and W. O. Strong, 1925, The Potato Tuberworm, Va. Truck Expt. Sta. Bull. No. 53. 3p.

Westcott, C., 1964, The Gardeners Bug Book, 3d ed. Doubleday and Co., Garden City, New York, 625p.

Zimmerly, H. H., Fred W. Geise, and C. R. Willey, 1921, Dusting Vegetable Crops, Va. Truck Expt. Sta. Bull. No. 34. 3p.

Insecticide Resistance of the Colorado Potato Beetle, *Leptinotarsa decemlineata* (Say)

Andrew J. Forgash

The science of insect pest management is evolving rapidly, and many of the developing strategies undoubtedly will become standard procedures in agricultural production and wherever else arthropods constitute a threat to humans. Nevertheless, elimination or alleviation of most pest problems will require some use of pesticides, and in many cases control will depend predominately on the use of insecticides or acaricides (Georghiou, 1972). A 1975 USDA report (Blair, 1975) on twenty-six pest management projects stated that pesticides were necessary to every such program and that there were no producer-supported pest management programs in the United States that could be classified as successful where pesticides play less than an important role. Furthermore, over the past forty years there has been a sustained annual increase in the use of pesticides in the United States and other developed nations and in many of the developing countries as well. This upward trend will continue for at least the next few decades, despite the implementation of alternative control methods, including integrated management procedures (Glass, 1977). Therefore it appears certain that arthropod resistance to pesticides will remain a serious consideration in agricultural production in the foreseeable future.

Thirty-three years have passed since the first report of insect resistance to DDT (*Musca domestica* (L.); Sacca, 1947), at which time there were only 12 arthropod species with resistant strains. By 1975, however, the number had risen to at least 305, plus 59 unconfirmed

---

This work was performed as part of NJAES Project No. 08149, supported by the New Jersey Agricultural Experiment Station and Hatch Act Funds.

I gratefully acknowledge the many hours of technical assistance provided by Claudia Golenda, George Kresnosky, Kenneth Medeiros, Jeffrey O'Neill, and Sonny Ramaswamy.

field observations, with no indication of a decrease in the rate of appearance of new resistant strains (Figure 1). The reported species represent 14 orders and 84 families and include compounds from all the major categories of proprietary insecticides and acaricides (Georghiou and Taylor, 1976). The 120 additional species of resistant agricultural pests reported between 1967 and 1975 represent a 114 percent increase for the eight-year period and consist of increases of 57 percent for hemipterans, 79 percent for arachnids, 70 percent for lepidopterans, and 143 percent for coleopterans. Of the 56 resistant species of beetles, 19 are curculionids and 12 are chrysomelids (Table 1).

Resistance of the Colorado potato beetle (CPB) to synthetic organic insecticides was first noted with DDT in New York State in 1952 (Quinton, 1955). In the early 1950s additional CPB control problems with DDT were reported from North Dakota, Virginia, and Minnesota (Post, 1954; Hofmaster et al., 1967; Cutkomp et al.,

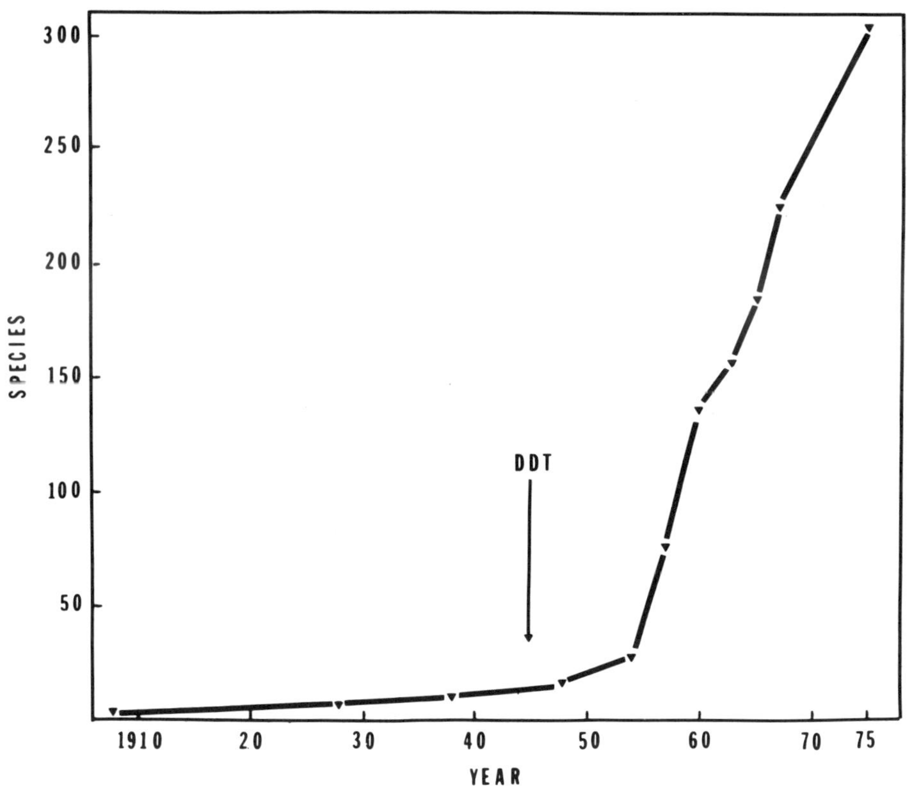

Figure 1
Increase in number of resistant Arthropod species over time (data from Georghiou and Taylor, 1976, Table 1).

Table 1
Insecticide Resistance of Agricultural Pest Arthropods

Arthropod Type	Number of Resistant Species, 1975	1967-1975 Percentage Increase in Eight Years
All species	225	114
Coleoptera	56	143
Lepidoptera	52	79
Acarina	33	79
Hemiptera	51	57
Chrysomelidae	12	-
Curculionidae	19	-

Source: From Georghiou and Taylor, 1976.

1958). Resistance to dieldrin was recognized in 1958 in Virginia, followed by resistance to other chlorinated hydrocarbons (Hofmaster et al., 1967). In recent years, severe control problems with many organophosphorus (OPs) and carbamate compounds have been encountered in several northeastern states.

In Europe, resistance to DDT was first reported from Spain in 1955, and this was followed by similar accounts from many other European countries (Brown, 1971). Since then, some European CPB populations have become resistant to lindane, cyclodienes, and various OPs and carbamates (Lakocy, 1976).

In Canada, DDT was discontinued for CPB control in 1962 due to widespread resistance of the pest. Now some populations in Ontario are resistant not only to DDT and dieldrin-type compounds but also to endosulfan (Harris and Svec, 1976); however, unlike the situation in certain U.S. and European CPB populations, Canadian populations of chlorinated-hydrocarbon resistant beetles appear to remain susceptible to many OP and carbamate insecticides.

LABORATORY STUDIES OF COLORADO POTATO BEETLE RESISTANCE

It was apparent from the numerous reports of control failures that many New Jersey populations of CPB have developed resistance to a variety of insecticides, but information on the actual levels of resistance to these compounds was absent or unavailable. In the fall of 1978, various types of insecticides were tested by topical application in the laboratory for toxicity to beetles collected from Bridgeton, Centerton, and Vineland, New Jersey. These tests indicated the following LD50s: fenvalerate 1 μg/beetle, permethrin 2, chlorfenvinphos 5, lindane 30, azinphosmethyl more than 80, stirofos more than 100, carbofuran more than 100, carbaryl more than 400, and DDT more than 400. With the exception of DDT, we could not estimate the degree of tolerance or resistance of these collections to the various compounds because of the unavailability of data for susceptible populations. In the case of DDT, however, it appears that the New Jersey beetles are at least forty-fold resistant compared to susceptible (LD50 about 10 µg) European beetles (Food and Agricultural Organization, 1974).

In 1979, through the generous cooperation of Richard Storch of the University of Maine, we received several shipments of field-collected Maine CPB, which do not present control problems in the state. The Maine beetles provided a means of measuring the degree of resistance of New Jersey beetles to specific compounds.

Starting with the overwintering generation in the spring of 1979, we collected beetles throughout the season from potatoes and eggplants at various locations in New Jersey and returned them to the laboratory for testing. The tolerance of New Jersey beetles and northern Maine specimens to topical doses of various insecticides are shown in Table 2. The New Jersey beetles show some degree of resistance to all compounds, even fenvalerate, but especially to azinphosmethyl (about one-hundred-fold) and carbofuran (more than five-hundred-fold).

There also appear to be tolerance differences between northern Maine and central Maine beetles, indicating the possibility of incipient resistance in populations from central Maine where it has been necessary to apply insecticides for CPB control (Table 3).

Because of the difference in fenvalerate tolerance between New Jersey and Maine beetles, we should be alert

Table 2
Topical Toxicity of Insecticides to Adult Colorado Potato Beetles from New Jersey and Northern Maine, Spring 1979

Insecticide	LD50 (µg/beetle)[a]	
	New Jersey	Northern Maine
Fenvalerate	1	0.12
Permethrin	2	-
Chlorfenvinphos	5	0.50
Aldicarb	12	2.0
Lindane	25	1.2
Azinphosmethyl	30	0.30
Chlorpyrifos	50	-
Carbofuran	>100	0.22
Mobam™	> 40	-
Oxamyl	> 80	40.0
Endosulfan	> 80	-
Phosmet	> 80	-
Sulprofos	> 80	-
Terbufos	> 80	-
Stirofos	>100	80.0
Carbaryl	>400	-[b]
DDT	>400	80.0

[a] Samples of mixed sexes were tested.
[b] Twenty-five percent mortality.

Table 3
Topical Toxicity of Insecticides to Colorado Potato Beetles from Northern Maine, Central Maine, and New Jersey

Insecticide	LD50 (μg/beetle)[a]		
	Northern Maine	Central Maine	New Jersey
Fenvalerate	0.12	0.30	1.0
Azinphosmethyl	0.30	1.5	30.0
Aldicarb	2.0	-	12.0
Carbofuran	0.22	1.5	>100.0
Oxamyl	40.0	-	> 80.0
Chorfenvinphos	0.5	1.7	5.0
Lindane	1.2	2.0	25.0
DDT	-	160.0[c]	>400.0
Stirofos	80.0[b]		>100.0

[a] Samples of mixed sexes were tested.
[b] Twenty-three percent mortality.
[c] Twenty-five percent mortality.

to the possibility of early resistance to the pyrethroids, which are the principal materials for CPB control in New Jersey.

Piperonyl butoxide, an inhibitor of mixed-function oxidases, will synergize certain types of insecticides in various other insect species, being particularly effective in strains that are resistant to the insecticides. Our preliminary work with CPB shows that fenvalerate can be synergized in New Jersey beetles by piperonyl butoxide, indicating the presence of mixed-function oxidases capable of detoxifying the pyrethroids. However, unlike many other species, to produce the effect in CPB it is necessary to apply the synergist in advance of the insecticide, with a two-hour pretreatment superior to one hour; the necessity for pretreatment probably is related to the penetration rate of the synergist. With other insecticides, however, we detected only slight synergism for carbofuran and azinphosmethyl and none for chlorfenvinphos, aldicarb, lindane, phosmet, or carbaryl (data not shown); possibly a synergist pretreatment time longer than one hour may be necessary to obtain synergism of some of these compounds, particularly the carbamates, which are readily synergized by piperonyl butoxide in resistant strains of other insect species (Table 4).

There have been reports of field control successes with combinations of phosmet and fenvalerate at dosages considerably below those required when either compound is used alone (Hofmaster, unpubl.), suggesting the possibility of interaction (for example, synergism or potentiation) of fenvalerate with other compounds. We attempted to demonstrate this in the laboratory using topical applications of fenvalerate in combination with phosmet and also with azinphosmethyl or sulprofos; however, there was no evidence of interaction. The toxicities of the combinations were no greater than the sum of the activities of the two components of the

Table 4
Synergism of Fenvalerate by Piperonyl Butoxide in New Jersey Colorado Potato Beetles

Treatment	% Mortality[a] 1 Day	7 Days
0.6 µg/beetle Fenvalerate alone	43	10
0.2 µg Fenvalerate plus synergist[b]	57	33
0.3 µg Fenvalerate plus synergist[b]	63	37
0.6 µg Fenvalerate plus synergist[b]	93	67
0.6 µg Fenvalerate plus synergist[c]	90	83

[a] Moribund and dead beetles.
[b] Treated topically with piperonyl butoxide one hour prior to fenvalerate application.
[c] Treated topically with piperonyl butoxide two hours prior to fenvalerate application.

Table 5
Response of Colorado Potato Beetles to Combinations of Fenvalerate plus Phosmet, Azinphosmethyl, or Sulprofos

Treatment	% Mortality[a] 5 Days	7 Days	10 Days	Expected Additive Toxicity at 10 Days
Fenvalerate 0.5 µg/beetle	80	44	32	-
Phosmet 40 µg	36	32	52	-
Fenvalerate 0.5 µg plus phosmet 40 µg	52	44	40	84
Sulprofos 80 µg	24	40	36	-
Fenvalerate 0.5 µg plus sulprofos 80 µg	80	72	76	68
Azinphosmethyl 40 µg	-	20	23	-
Fenvalerate 0.5 µg plus azinphosmethyl 40 µg	-	23	23	55

[a] Moribund and dead beetles.

mixture when applied alone (Table 5). Perhaps a different test procedure, such as pretreatment with one or the other compound, or feeding experiments would provide evidence of interaction.

Although problems in the control of the CPB are prevalent throughout the commercial growing areas of New Jersey, we had no information on whether the levels of resistance are comparable in different populations. Thus in 1979, we began to test samples of adult beetles from different New Jersey locations for sensitivity to topical applications of various insecticides. The data in Table 6 indicate that tolerances to representative compounds from various classes of insecticides are similar among the populations from the five areas sampled, with the possible exception of the Cohansey beetles, which appear

Table 6

Toxicity of Insecticides to Different New Jersey Populations of Colorado Potato Beetle

Insecticide	LD50 (µg/beetle)				
	Cohansey	Salem	Centerton	Vineland	Manahawkin
Fenvalerate	2.5	1.2	1.0	1.0	-
Chlorfenvinphos	16	15	25	17	-
Azinphosmethyl	20(55%)	30(53%)	15	30	16(20%)
Carbofuran	50(5%)	100(33%)	100(18%)	100(27%)	100(10%)
Aldicarb	15	-	17	13	-
Lindane	30(60%)	30(87%)	20	25	-

to be less susceptible to fenvalerate than are beetles from other areas, although the measured differences may not be significant. Monitoring of this nature, particularly for changes in pyrethroid tolerance, will be intensified in the future.

We were also interested in determining whether there is a change in the insecticide resistance of a population during the fall-spring interval. Data in Table 7 indicate that overwintering beetles were somewhat more tolerant prior to diapause in the fall of 1978 than shortly after emergence the following spring, particularly to fenvalerate, chlorfenvinphos, and lindane.

We monitored changes in tolerance of a population in succeeding generations over the season. The data in Table 8 suggest that fenvalerate susceptibility had not changed from the level present in the population prior to overwintering the previous year. Resistance to carbofuran, lindane, carbaryl, and DDT remained consistently high, although there appears to have been a slight increase in resistance to azinphosmethyl and a five-fold increase in chlorfenvinphos tolerance.

Soil treatments with aldicarb generally continue to give satisfactory protection of potato plantings against CPB in New Jersey during the early part of the season,

Table 7
Toxicity of Insecticides to New Jersey Colorado Potato Beetles prior to Fall Diapause and after Spring Emergence, Centerton

Insecticide	Topical Dose (μg/beetle)	% Mortality	
		1978 Fall Beetles	1979 Spring Beetles
Fenvalerate	2	75	100
Chlorfenvinphos	5	50	100
Azinphosmethyl	40	37	27
Carbofuran	40	40	43
Lindane	40	30	87
Carbaryl	400	10	7
DDT	400	20	47

Table 8
Susceptibility of a New Jersey Colorado Potato Beetle Population to Insecticides over One Season, Centerton

Insecticide	LD50 (μg/beetle)		
	Overwintered Adults	First Generation	Second Generation
Fenvalerate	2(100%)	1.0	1.3
Chlorfenvinphos	5(93%)	25	25
Azinphosmethyl	20	15	45
Carbofuran	40(37%)	100(18%)	100(20%)
Lindane	40(83%)	20	20
Carbaryl	200(3%)	200(10%)	200(20%)
DDT	400(50%)	400(0%)	-

despite their being considerably more tolerant than Maine specimens. Because failure of aldicarb control would greatly exacerbate the overall resistance problem, we are particularly concerned about changes in CPB tolerance to this compound. Data in Table 9 indicate that beetles from the three locations sampled are comparable in sensitivity to aldicarb at the LD50 level; however, the differences at the LD95 level suggest that tolerance may be increasing in some populations.

During the 1979 season, we were provided beetles from farms in Salem County, New Jersey where pyrethroid applications were not satisfactorily controlling the beetle. Topical tests with fenvalerate showed that Salem beetles were no less sensitive than beetles from other areas of the state where control was satisfactory. This result led us to be concerned about the role that inadequate coverage might play, particularly when the insecticide is being applied by aircraft at the rate of five gallons of finished formulation per acre, so we calculated on the basis of a flat acre of leaves the quantity of fenvalerate that would deposit per unit area of foliage if all of the applied material landed on target. Then we treated measured sections of eggplant leaves with this maximum or with one-half, one-quarter, or one-tenth that amount and placed adult beetles on the treated foliage for observation. The importance of thorough and efficient application can be seen in the results presented in Table 10, where it is evident that the poorer the foliar coverage, the greater the beetle recovery and survival.

DISCUSSION

The development of resistance to a pesticide is an evolutionary process--a kind of unnatural selection--in which repeated exposure of a population to the pesticide (the selecting agent) gradually eliminates those members incapable of tolerating the usual doses of the chemical, until finally the population consists largely of individuals that through one or more means are practically immune to the action of the toxicant. This altered population may show cross-resistance or multiple resistance, the former pertaining to a population in which

Table 9
Aldicarb Tolerance in Different Populations of Colorado Potato Beetles

Population Source	Approx. LD50 (μg/beetle)	Approx. LD95 (μg/beetle)
Northern Maine	2	8
Vineland, N.J.	13	60
Cohansey, N.J.	15	80
Centerton, N.J.	17	100

Table 10
Spray Coverage

Volume (μl) of Fenvalerate Spray per 42 sq. cm. Eggplant Foliage[b]	% Mortality[a]			
	Test 1		Test 2	
	1 Day	7 Days	1 Day	7 Days
2	100	10	67	17
5	100	60	83	23
10	100	60	93	40
20[c]	100	90	93	47

[a]Moribund and dead beetles.
[b]The finished formulation was equivalent to 0.2 lb fenvalerate (AI) in 5 gal.
[c]The maximum quantity of spray that could deposit if all of a 0.2 lb per 5 gal application landed on an acre of foliage.

a single defense mechanism confers resistance against various pesticides and the latter to one in which resistance to various compounds is conferred by different, coexisting mechanisms (Georghiou, 1972).

From resistance studies with species other than the CPB (primarily the house fly and mosquitoes) the known resistance mechanisms are enhancement of biochemical and physiological barriers to intoxication, decrease in sensitivity of the target site, and modification of behavior leading to reduced exposure to the toxicant (Georghiou, 1972). Of these, metabolic detoxication is the preeminent means whereby insects resist insecticides.

With the possible exception of the cyclodienes, resistance to organochloride, organophosphorus (OP), and carbamate insecticides appears to be based on the same, or similar, backgrounds of ancillary genes. Consequently substitution of one chemical for another within these groups tends to preserve or even enhance the existing resistance. The resultant multiple resistance thus reflects the past history of insecticide selection within a population. Cross-resistance limits the choice of available insecticides, and multiple resistance precludes return to insecticides used previously (Georghiou, 1972).

Generally DDT resistance is accompanied by resistance to only those compounds that are structurally similar to DDT, although in several species there has been cross-resistance to pyrethroids (usually of a small degree), apparently conferred through a gene for knockdown resistance (Elliott et al., 1978). Resistance to dieldrin or analogs usually is accompanied only by resistance to other cyclodiene compounds and to lindane (Brown, 1971).

Resistance to one OP compound usually confers resistance to many other OPs and also to certain carbamates. Strains that become resistant as the result of selection with one carbamate generally are also resistant to many other carbamates and to many OPs as well. Resistance that develops as the result of selection with OP compounds does not impart resistance to the pyrethroids, but resistance to one pyrethroid will carry resistance to others of its kind.

Once a population becomes resistant to one insecticide, usually the development of resistance to alternative materials is accelerated. In the case of the CPB, reports from Poland (Lakocy, 1976) suggest that DDT resistance facilitates the development of OP resistance, while lindane resistance expedites the buildup of resistance to carbamates. It seems that resistance develops more rapidly for the carbamates than for the OPs.

From knowledge of resistance phenomena in general and the history of control treatments, I deduce that DDT resistance of New Jersey CPB populations is the result of DDT treatments of years ago, whereas lindane resistance is the consequence of past selection by cyclodiene-type compounds, and endosulfan resistance is the product of earlier cyclodiene treatments and subsequent use of endosulfan itself. As has been shown for other resistant species, the alleles for the resistance mechanisms to these compounds apparently have been retained in the population after substitution of alternative insecticides, such as OPs and carbamates. Resistance to OPs probably resulted from repeated exposure to azinphosmethyl, while carbamate resistance derived from carbaryl and carbofuran selection.

In view of the CPB's extraordinary resistance to carbofuran (about five-hundred-fold) and high resistance to other carbamates, it is surprising that the increased tolerance to aldicarb is so low (six fold). Obviously there was minimal cross-resistance to this carbamate, but more importantly, repeated exposure to aldicarb apparently has not increased the level of resistance. The lack of increase is understandable, however, because aldicarb is not used for control on eggplants, and the systemic effect of this chemical in potato fields is lost early in the growing season; thus the overall selection pressure against CPB generations each year is relatively low. Nevertheless this does not preclude the possibility of eventual resistance, and there is some indication that aldicarb tolerance may be increasing in New Jersey populations.

Resistance to DDT and other chlorinated hydrocarbons persists in CPB populations, as in other species, for many years after switching to OPs and carbamates. What will be the eventual effect of pyrethroid usage on the levels of resistance to the compounds that the pyrethroids are replacing? Once again, we can surmise from experiences with other resistant species; for example, reports from Denmark show that house fly resistance to OP compounds persisted for at least three years after the cessation of usage and the institution of pyrethroid treatments (Keiding, 1976, 1977). Since house flies in a dairy barn environment may have as many as sixteen generations per year (as in New Jersey), this means that OP resistance in house flies persisted for at least forty-eight generations after the substitution of pyrethroids. We can only guess what may happen in New Jersey CPB populations where there are but two full generations a year.

The pyrethroids appear to be giving satisfactory control to CPB in New Jersey, although some populations may be less susceptible than others. Nevertheless there is real cause for concern about the future. First, field experiences with house flies show that frequent treatment with pyrethroids leads to resistance in only a few seasons. Second, New Jersey beetles are about eight and three times as tolerant to fenvalerate as those from northern Maine and central Maine, respectively; this correlates with the extent of chemical applications for beetle control in the three areas (Table 3). Third, our own work with house flies shows that selection with pyrethroids (resmethrin) confers a high degree of resistance to many carbamates (Forgash, unpubl.). These facts indicate the possibility that resistance may develop not only to the pyrethroids but to aldicarb as well. It is hoped that judicious use of pyrethroid applications, together with other management procedures, may prevent or at least forestall resistance development to these compounds.

REFERENCES

Blair, B. D., 1975, A Study of Selected Pest Management Projects in the States, Unpublished report, Cooperative Extension Service, Washington, D.C.

Brown, A. W. A., 1971, Pest Resistance to Pesticides, in <u>Pesticides in the Environment</u>, ed. R. White-Stevens, vol. I (part II), Marcel Dekker, New York.

Cutkomp, L. I., A. G. Peterson, and P. E. Hunter, 1958, DDT-Resistance of the Colorado Potato Beetle, <u>J. Econ. Entomol.</u> 51:828-831.

Elliot, M., N. F. Janes, and C. Potter, 1978, The Future of Pyrethroids in Insect Control, <u>Ann. Rev. Entom.</u> 23:443-469.

Food and Agriculture Organization, 1974, Recommended Methods for the Detection and Measurement of Resistance of Agricultural Pests to Pesticides: Tentative Method for Adults of the Colorado Potato Beetle, <u>Leptinotarsa decemlineata</u> Say--FAO Method No. 12. <u>FAO Plant Protect. Bull.</u> 22:112-116.

Georghiou, G. P., 1972, The Evolution of Resistance to Pesticides, <u>Ann. Rev. Ecol. System</u> 3:133-168.

Georghiou, G. P., and C. E. Taylor, 1976, Pesticide Resistance as an Evolutionary Phenomenon, <u>Proc. XV Intern. Cong. Entomol.</u>, pp. 759-785.

Glass, E. H. 1977, Safety as a Factor in the Development of Pest Control Programs, in <u>Pesticide Management and Insecticide Resistance</u>, ed. D. L. Watson and A. W. A. Brown, pp. 183-191, Academic Press, New York.

Harris, C. R., and H. J. Svec, 1976, Susceptibility of the Colorado Potato Beetle in Ontario to Insecticides, J. Econ. Entomol. 69:625-629.

Hofmaster, R. N., R. L. Waterfield, and J. C. Boyd, 1967, Insecticides Applied to the Soil for Control of Eight Species of Insects on Irish Potatoes in Virginia, J. Econ. Entomol. 60:1311-1318.

Keiding, J., 1976, Development of Resistance to Pyrethroids in Field Populations of Danish Houseflies, Pestic. Sci. 7:283-291.

Keiding, J., 1977, Resistance in the Housefly in Denmark and Elsewhere, in Pesticide Management and Insecticide Resistance, ed. D. L. Watson and A. W. A. Brown, pp. 261-302. Academic Press, New York.

Lakocy, A., 1976, Studies on the Development of Acquired and Cross Resistance of Hibernating Adults of the Colorado Potato Beetle (Leptinotarsa decemlineata Say) to Insecticides in the Voivodships of Posnan, Warsaw and Wroclaw during the years 1972-1975, Prace Naukowe Instytutu Ochrony Roslin. 17:97-141 (Abstr.).

Post, R. L., 1954, DDT-Resistant Colorado Potato Beetles? Proc. North Central Branch. Entomol. Soc. Amer. 9:94-95.

Quinton, R. J., 1955, A Study of DDT Resistance in the Colorado Potato Beetle Leptinotarsa decemlineata Say, Ph.D. thesis, Cornell Univ., Ithaca, New York.

Sacca, G., 1947, Sull' Esistenza di Mosche Domestiche Resistential DDT. Riv. Parasitol. 8:127-128.

Geographic Diversity of the Colorado Potato Beetle and Its Infestation in Eurasia

J. de Wilde and *T. Hsiao*

PHYSIOLOGICAL BACKGROUND

The physiological basis of seasonal development of the Colorado potato beetle has been studied by de Wilde and numerous collaborators for about thirty years, and for a similar period we have been interested in the relationship between the beetle and solanaceous plants. We have recognized reproduction and diapause as physiologic states, the regulation of which belongs to the field of environmental endocrinology (de Wilde, 1978), and we have advanced a great deal in analyzing the chain from environmental signals to subcellular response (de Wilde, 1969, 1971; de Kort, 1969; Schooneveld, 1974; Kramer, 1978; Dortland, 1979). Our work on host plant relationships resulted in a scheme of host plant finding (Chin, 1950; de Wilde, 1958) and led us to clarify the role of olfaction and airflow (de Wilde, 1976), culminating in the elucidation of the olfactory signals the insect receives from the plant (Visser, 1979) and the specificity of the chemoreceptors involved (Ma and Visser, 1978).

We wish to thank visiting scientists R. Harmsen, Kingston, Ontario, Canada, and Marek Kozlowsky, Poland; students Erica Boersma, Ans Brouwer, Marlies Hasper, Anita Schoenaker, Annie van der Wielen, Ruud van Gent, Anton Loomans, Paul van Roermund, Jan Veenstra, and Fred Vooren; and technical assistant Helga Privse for their valuable contributions to our research. We also appreciate the assistance of members of the European Plant Protection Organization for providing live specimens of the Colorado potato beetle from various geographic localities. T. H. Hsiao's participation was made possible by research fellowships from the Agricultural University, Wageningen, the Netherlands, and the Netherlands Organization for Pure Scientific Research.

For many years Hsiao has been interested in the substances determining food specificity in the Colorado potato beetle (Hsiao and Fraenkel, 1968a, 1968b; Hsiao, 1969, 1974, 1976, 1978). For five years, de Wilde and Hsiao had worked to establish an association of programs between their laboratories, a situation that enabled considerable progress in work we started in 1976 aimed at obtaining an understanding of the field situation of the Colorado potato beetle on the basis of our previous physiological work. We were especially interested in knowing how much geographic diversity had developed among the Eurasian populations since the beetle was introduced to France in 1921 (Figure 1). For this purpose, in 1976 we obtained samples of a large number of populations throughout the Eurasian area of distribution of the beetle. Twelve of these populations collected between 37° and 52° north latitude were investigated, and six (Wageningen, Netherlands; St. Clair, France; Granada, Spain; Poznan, Poland; Budapest, Hungary; and Athens, Greece) were studied in more detail. Critical photoperiod, latent periods in diapause and oviposition, fecundity, and survival on different host plants were determined.

In addition to laboratory observations, fieldwork was performed by students of the Agricultural University at Wageningen to record seasonality in the field. In addition, observations were made on locomotory activity and the buildup of populations.

ENVIRONMENTAL CONTROL OF REPRODUCTION AND DIAPAUSE

Photoperiodic Response

The Colorado potato beetle populations in the Netherlands (de Wilde, 1955; de Wilde et al., 1959), as well as in Hungary (Jermy and Sáringer, 1955), reared at 25°C ab ovo on young potato foliage, have a photoperiodic response curve with a critical photoperiod of fifteen hours. Above this daylength, 20 to 30 percent of the beetles go into diapause, and 70 to 80 percent feed and reproduce.

In 1978, R. Harmsen, who participated in our research, studied individually the adults that emerged from a culture of seventy-five larvae, reared at 25°C and an eighteen-hour photoperiod. The beetles, obtained from a field population in the Netherlands, varied greatly in their response:

1. No oviposition, entered diapause: 15.
2. Fewer than fifteen eggs laid, entered diapause: 2.
3. More than fifteen eggs laid, entered diapause: 13.
4. Normal oviposition, no diapause: 35.
5. Fewer than fifteen eggs laid, no diapause: 3.
6. No eggs, no diapause: 5.
7. Females died within fifteen days of emergence: 2.

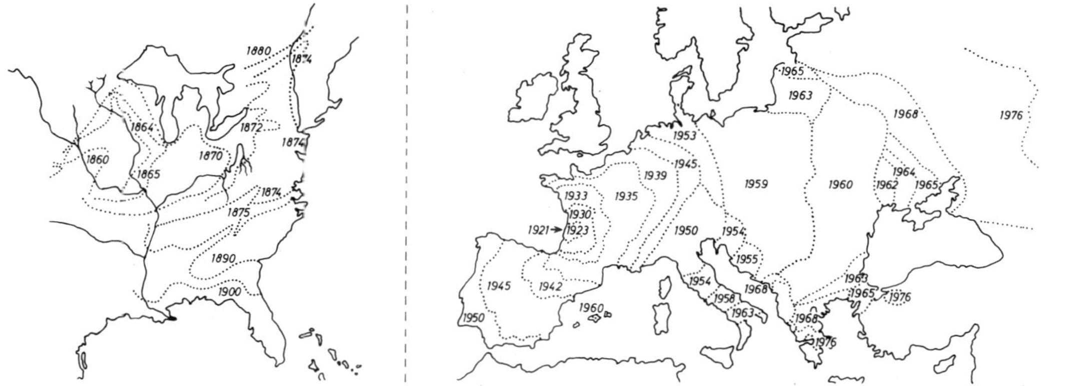

Figure 1
Changes in distribution of Colorado potato beetle in the United States, 1854-1876, and in Europe, 1921-1976 (from Tower, 1906 and Advisory Leaflet 71, Min. of Agriculture, U.K.).

The considerable degree of variation in the long-day response could well be one of the factors ensuring immediate survival of the beetle in newly acquired territories. In contrast to this variability is the short-day response to photoperiods below fourteen hours, which is 100 percent diapause. After hibernation, the beetles that have broken diapause are no longer sensitive to photoperiod, and they reproduce when the temperature exceeds 17°C.

Effect of Temperature

The short-day effect is abolished only by temperatures above 31°C, which shorten the critical period by three to five hours. For laboratory rearing, these are high temperatures, but in the ecoclimate at the canopy of the potato crop, even in the Netherlands, such temperatures periodically occur at midsummer. When the beetles are kept under high humidity at 31°C for fourteen days, diapause can also be broken, providing that at the end of this treatment potato foliage is supplied.

Effect of Host Plant Age

Physiologically aged potato leaves (plants after flowering, leaves with a leathery appearance) under all conditions induce the state of diapause and constitute a veto factor, synchronizing the end of seasonal activity of the beetle with the end of the growing season of the host plant.

While photoperiod synchronizes the activity of the beetle with the astronomic season, the effect of temperature synchronizes this activity with the favorable climatic season, and the effect of host plant age does the same in relation to the host plant season.

HOST PLANT SELECTION

While attraction to the potato plant at a distance is brought about by a combination of green-leaf volatiles, of which hexenol and hexenal are most regular (Visser, 1979), and the same substances in combination with glucose, beta-alanine, and cholesterol probably have a phagostimulatory effect (Ritter, 1968), feeding deterrency can be brought about by alkaloids such as demissine or tomatine (Schreiber, 1958). Although feeding is restricted to the Solanaceae, many species of this family are resistant. Among the wild Solanaceae of Europe, bittersweet (*Solanum dulcamara*) and, to some degree, *Lycium halimifolium*, *Hyoscyamus niger*, and *Atropa belladonna* permit larval growth and adult reproduction.

Among the European cultivated plants that permit the development and reproduction of the beetle is *Solanum* melongena. Remarkably an ornamental plant, Petunia hybrida, is very palatable to the larvae; at the same time it is highly toxic. Solanum nigrum permits only prediapause feeding of adults but is not a larval host plant. Important colonizations of the Colorado potato beetle in Europe, apart from those on potato, can be observed only on bittersweet, but in thirty years of study, this has been observed only once by de Wilde. This is in contrast to the situation in the United States, where Hsiao has observed Colorado potato beetle strains adapted to several wild host plants.

THE WAGENINGEN-UTAH COOPERATIVE PROGRAM

During discussions that started in 1972, the fact emerged that some physiological differences between the Colorado potato beetle populations in Utah and in the Netherlands were apparent. This led to cooperative studies, with reciprocal visits to Wageningen and Utah, for the purpose of elucidating geographic differentiation between populations of the beetle in Europe. Such differentiation would indicate the rate of adaptive changes since the beetle was introduced into Europe in 1921. It was decided to study first the photoperiodic response because clear-cut geographic latitudinal differentiation can be distinguished on this basis. It also is a direct clue to seasonal adaptation. Similar studies subsequently were made in the United States by Hsiao.

A series of wild host plants of the beetle, native to the United States, were grown in a greenhouse in Wageningen, on which beetles from six European populations were tested to determine whether there were differences in acceptance, development, and mortality on these various plants. Work by Hsiao (1978) with beetle populations from the United States had already revealed important differences in this respect.

DIVERSITY IN PHOTOPERIODIC RESPONSE

We have investigated twelve populations from Europe; of these, six populations--three each from Western and Eastern Europe--were especially emphasized in the study. They were from Wageningen, Netherlands; St. Clair, France; and Granada, Spain, in the West; and Poznan, Poland; Budapest, Hungary; and Athens, Greece, in the East. In addition, for comparison we included a laboratory culture from Holland that has been maintained for about twenty-four years in the laboratory. The rearing conditions were 25°C and photophases of ten, twelve, fourteen, and sixteen hours per day. Upon hatching, the larvae were reared in groups of forty and sixty. The potato variety Marijke was used. Resulting adults were maintained by pairs. The adults that entered soil within fifteen days after emergence were considered to be in diapause. Daily egg production of reproducing females was recorded for a seven-day period. From twenty to fifty pairs of adults were used for each photophase.

The results are given in Figures 2 and 3. All of the six populations of the Colorado potato beetle from Europe exhibited photoperiodic response. The southern populations from Granada and Athens showed a lower percentage of individuals entering diapause at the lower photophases of twelve and fourteen hours when compared with those from the more northerly latitudes (Wageningen and St. Clair). The Poznan and Budapest populations were

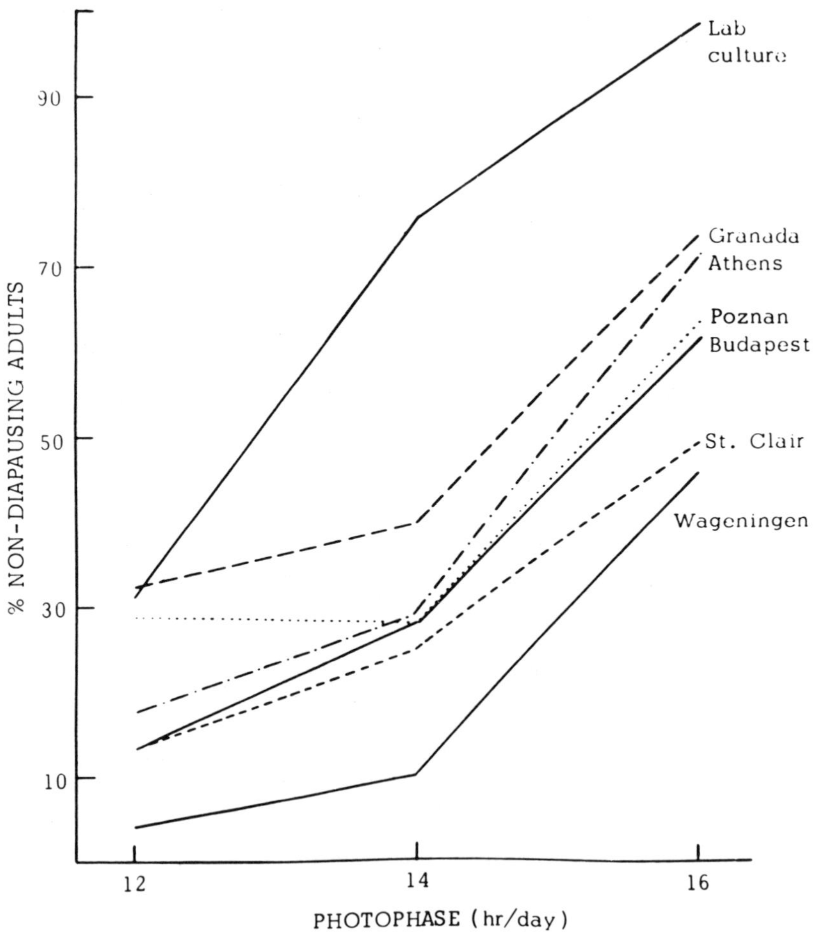

Figure 2
Photoperiodic response of the six populations of the Colorado beetle from Europe, with respect to diapause. For comparison, the response of the Wageningen laboratory population is also given.

intermediate (Figure 2). Even at a sixteen-hour photophase, all six field populations had a rather low percentage of nondiapausing adults, which is somewhat similar to the data obtained with the Utah beetles from North America. The laboratory culture showed remarkably different photoperiodic response compared to the field populations. Most adults did not enter diapause at the fourteen-hour photophase. The critical photophase for this culture is less than thirteen hours. When these data are compared with those collected by de Wilde in the 1950s, it is clear that the critical photoperiod has decreased more than two hours in twenty-four years.

Figure 3 shows the effect of photoperiod on reproduction. All of the northern populations did not reproduce at photoperiods of less than sixteen hours. Only the two southern populations (from Granada and Athens) had some percentage of females that reproduced at twelve and fourteen hours. As a whole, even at a

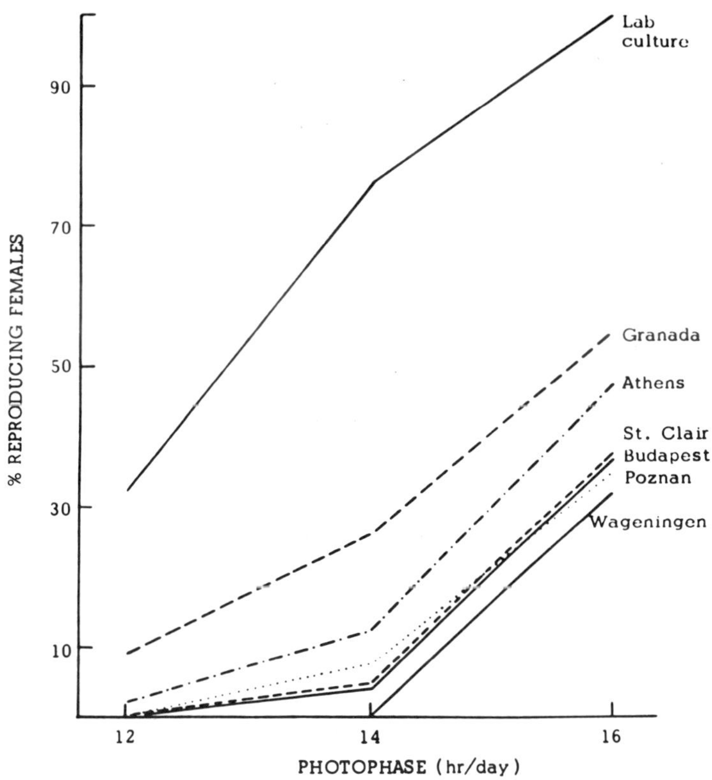

Figure 3
Photoperiodic response of the six populations of the Colorado beetle from Europe, with respect to reproduction. The response of the Wageningen laboratory population is also given.

sixteen-hour photophase, less than 50 percent of the females of the field populations were reproducing. The laboratory culture showed a very high percentage of reproducing females; their fecundity was about twice that of the field populations.

On the whole, a decreasing trend in length of critical photophase seems to be developing among the southern populations of the Colorado potato beetle in Europe, although the differences are much less than those observed in North America, where the Colorado potato beetles have been known since the time of the conquistadores. Clearly an evolutionary time of fewer than fifty years has been insufficient to induce a considerable degree of differentiation in a population of reduced genetic composition, as must have been the offspring of a rather few beetles that had been accidentally introduced to France around 1921.

DIVERSITY IN HOST PLANT SPECIFICITY

The six geographic populations of the Colorado potato beetle from Europe and the two other populations--one from Logan, Utah, and one from the Wageningen laboratory--were included in the study of the host plant relationship. Ten species of solanaceous plants were included in the study. Two species, *Solanum dulcamara* and *S. nigrum*, were collected outdoors in Wageningen. The other eight species--*S. chacoense*, *S. melongena* (long fruit variety), *S. rostratum*, *S. elaeagnifolium*, *S. luteum*, *S. tuberosum*, *Lycopersicon esculentum* (Moneymaker variety), and *Hyoscyamus niger* (two varieties, European and U.S.)--were grown in greenhouses.

All insect rearings were carried out with an eighteen-hour photophase at a constant temperature of 25°C. Newly hatched larvae in groups of thirty to fifty individuals of each population were reared on foliage of each plant species. The foliage was changed once or twice daily as needed. Mortality and growth rate were recorded daily. The growth performance, as expressed in mortality rate and development duration of different stages, and the pupal weights of the two sexes were used as criteria to assess the adaptability of each population to different plants. Several rearings were conducted on each host to provide an accurate measurement of growth performance of each beetle population. With the exception of the series of rearings on *S. elaeagnifolium*, which was conducted at Logan, Utah, because of the availability of this plant there, all experiments were carried out at the Entomology Laboratory in Wageningen.

Of the ten solanaceous plants investigated, *S. nigrum* and *S. luteum* were not suitable as hosts. No individuals of the eight populations were able to develop beyond the fourth larval instar, and the majority of them died during the first and second larval instars. The data on percentage of adult emergence of each population on these plants are graphed in Figure 4 for direct comparison. As a whole, *S. tuberosum* and *S. dulcamara* were among the most suitable hosts for all eight populations, followed by *S. rostratum* and *H. niger* (U.S. variety), which were moderately suitable, while *S. melongena* was marginally suitable. The other four plants that were the least suitable were *S. elaeagnifolium*, *S. chacoense*, *L. esculentum*, and *H. niger* (European variety). With a suitable host, the developmental duration from larval hatching to adult emergence was the shortest, the percentage of individuals developing to the adult stage was the highest, and the weight of individuals of both sexes was greatest. Figure 5 shows the suitability of *S. rostratum* as a host for the eight beetle populations when the three parameters are compared. Plants that were less suitable produced a reverse trend. Accompanied by high mortality, the developmental duration was longer and the pupae were lighter in weight. The tomato was a typical example of a less suitable host, which caused a considerable prolongation of developmental duration for all eight populations.

A marked difference exists between the two varieties of *H. niger*. The U.S. variety, which was much more suitable for all populations, is a biennial, and the European variety is an annual plant. There are also morphological differences between the two varieties, with the European variety having more glandular hairs on the leaf surface.

All field populations showed a considerable variability in their growth performance on the different solanaceous plants tested. It is not immediately apparent that a particular population is more adaptable to several hosts than others; however, the overall better growth performance of the laboratory culture than all field populations is especially significant. This is true with *S. chacoense*, *S. elaeagnifolium*, and *H. niger* (U.S. variety) (Figure 7), indicating a greater degree of adaptiveness of the laboratory culture to several less suitable hosts. It is also apparent that the laboratory culture produces consistently larger individuals than do any of the field populations. The fact that the laboratory culture did not show better growth performance on *S. melongena*, *L. esculentum*, and *H. niger* (European variety) indicates a rather selective adaptability of this population. The laboratory culture has been maintained for the past twenty-four years solely on potato foliage. The greater adaptability of the laboratory culture to a variety of solanaceous plants is a consequence of selection pressure experienced during laboratory rearings.

On the basis of these results, it would appear that host-adapted populations have not yet developed among the geographic populations of the Colorado potato beetle in Europe.

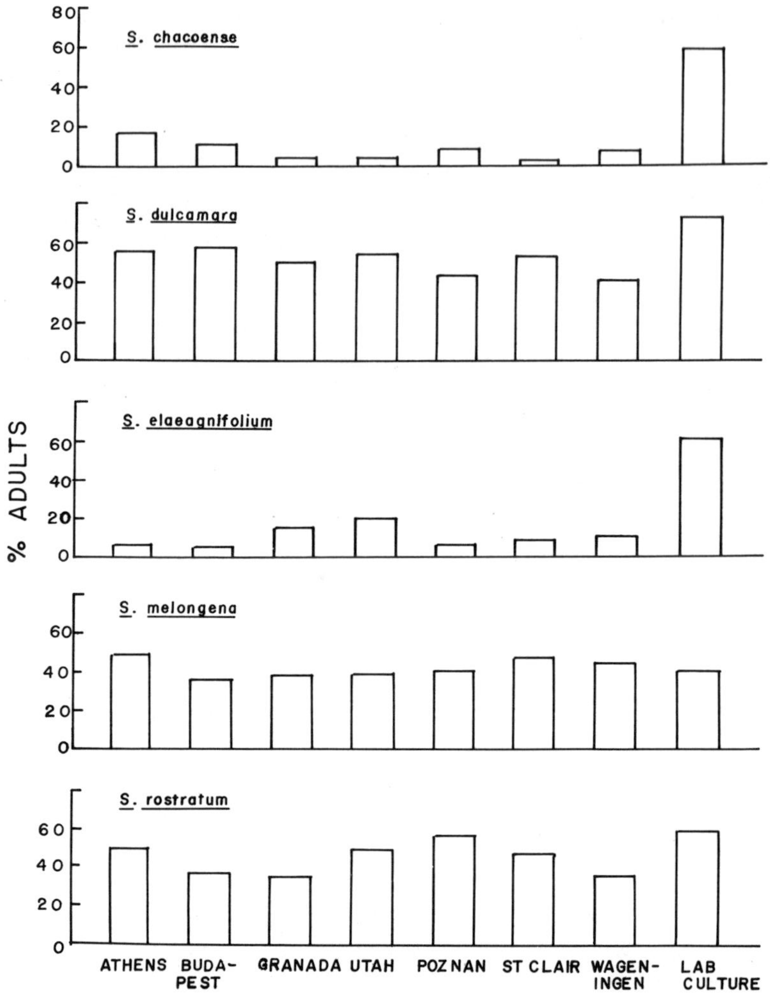

Figure 4
Comparison of rearing success of the six European field populations, the Utah population, and the Wageningen laboratory population on different food plants, expressed as percentage of adult emergence.

FIELDWORK IN EUROPE

In 1978 and 1979, field observations were carried out in different parts of Europe. The following sites were chosen to cover rather extreme climatic conditions between 37° and 52° north latitude: Wageningen, Netherlands (1978, 1979); Surtainville, France (1978,

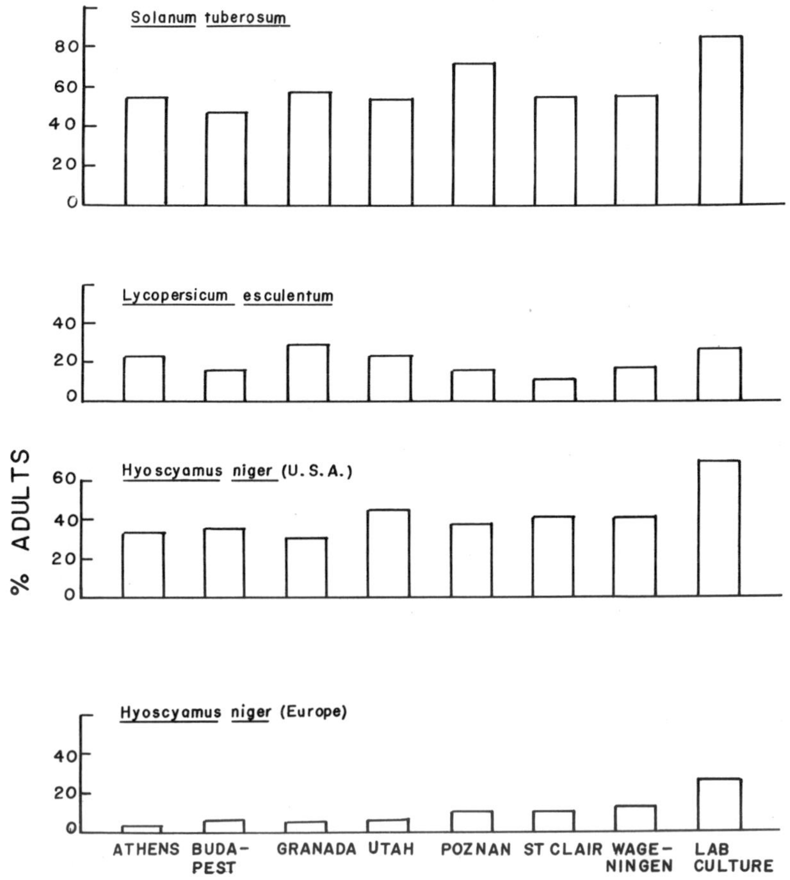

Figure 4 *Continued*

1979); Granada, Spain (1978, 1979); Budapest, Hungary (1978); and Athens, Greece (1978, 1979).

In experimental plots of 10 x 10 meters (m), early and late potato varieties were planted in subplots 5 x 5 m according to the local planting practice. In 1978, one subplot was used for adult and egg counting and one for larval counts. In 1979, the fields were arranged and screened in such a way that one field could be used to record beetles that could have arrived only by flight, and other fields were surrounded by soil traps to catch beetles arriving at or leaving the fields by marching. Beetles were counted every one to three days in 1978 and once weekly in 1979. The results are given in Figures 6 through 10.

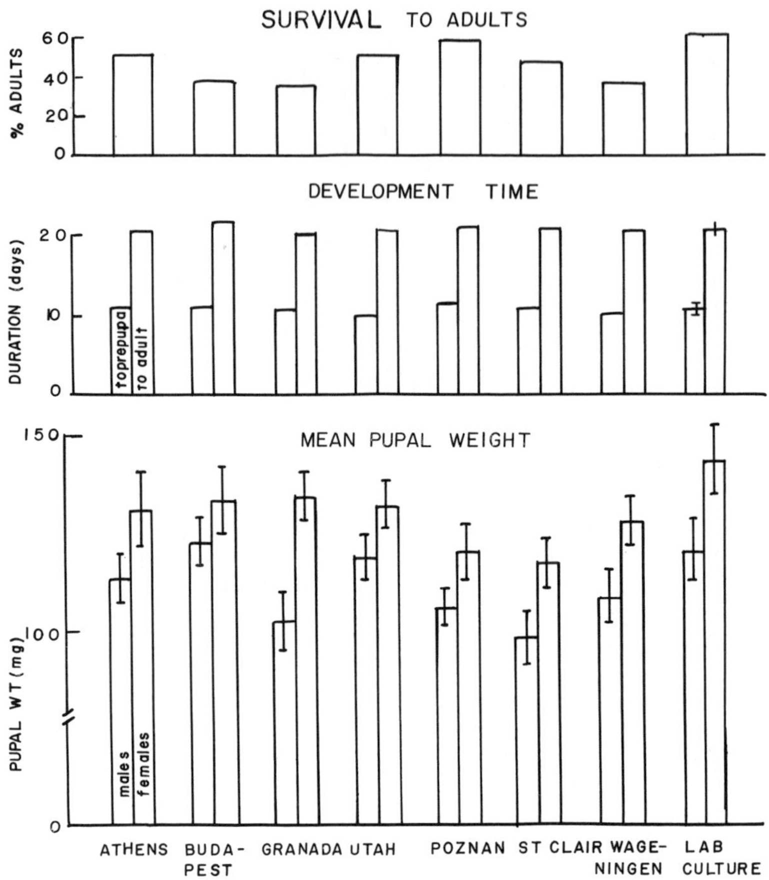

Figure 5
Comparison of suitability of Solanum rostratum as a host for the eight populations in Figure 4.

Adults and Egg Batches

In Figures 6a, 7a, 8a, and 9a, the numbers of adult beetles and egg batches per 100 plants are given. The posthibernation generation can be recognized by its high rate of oviposition. It is clear that in the Netherlands the summer of 1978 was much more tavorable than 1979, when first-generation adults were much less numerous. In both years, oviposition ended by the middle of August. The same is true for Surtainville, France. In Granada, Spain, in 1978, there were three generations, and oviposition did not end before the second week of September. In 1979, the oviposition of the first generation was ended by the harvest of the crop, but in other places (not shown on the graph), a third generation

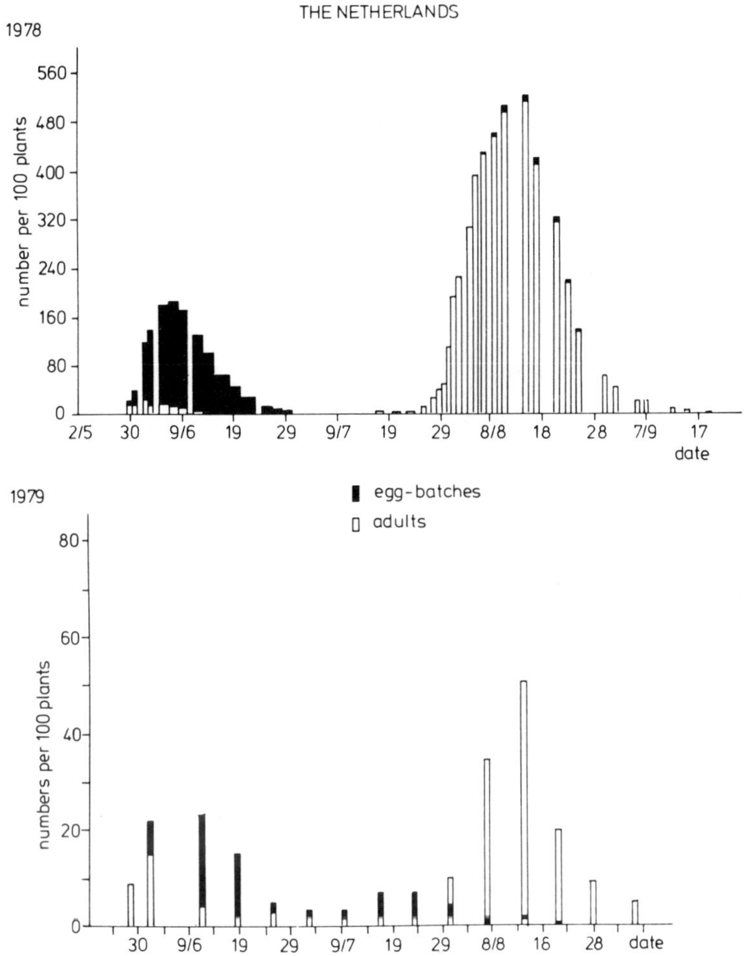

Figure 6a
Number of adults and egg batches counted on one hundred potato plants in Wageningen, Netherlands, 1978, 1979.

could also have developed in newly planted fields. The 1978 observations are consistent with a critical photoperiod below fourteen hours.

Near Athens, Greece, there were three generations. In 1979, the observations were extended from September to December because potato cropping there can continue until Christmas. No eggs were observed after September 15, which is consistent with a critical photoperiod below fourteen hours. Young adult beetles continued to emerge until November, but they did not reproduce.

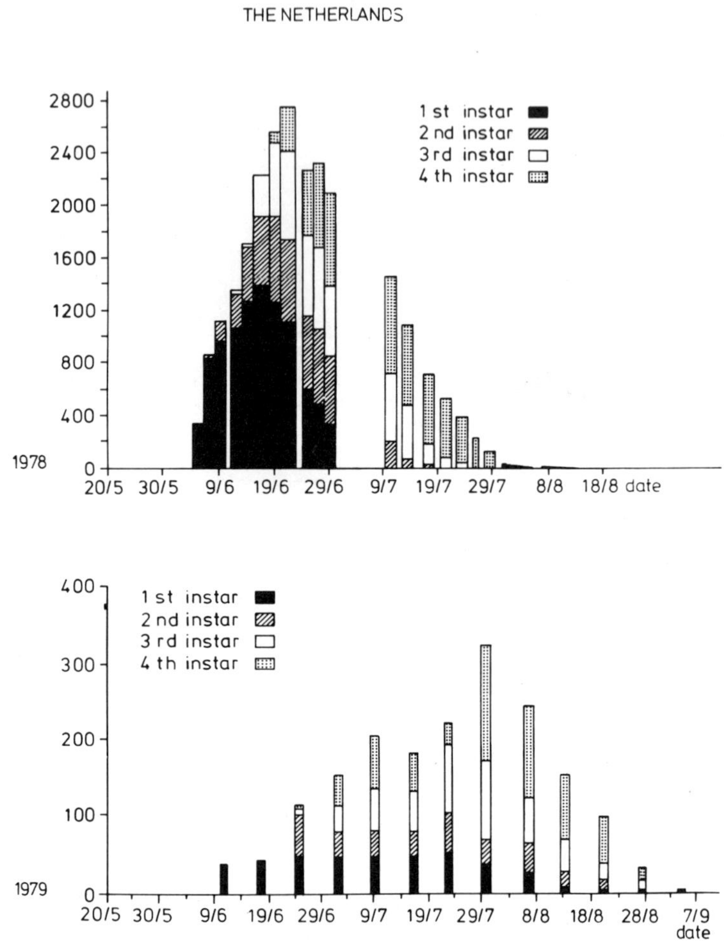

Figure 6b
Number of larvae of several instars per one hundred plants in Wageningen, 1978, 1979.

Larval Instars

The larval populations are shown in Figures 6b, 7b, 8b, and 9b. In the Netherlands in both years, there was only one generation. In the cool summer of 1979 the larval period was very much extended, and the population was also very much less numerous. In France, there was also only one generation. The beetle population in 1979 was more than ten times as numerous as in 1978. This was due to the favorable conditions offered by the experimental plot, as compared with the surroundings, which caused many beetles to be attracted to these plants

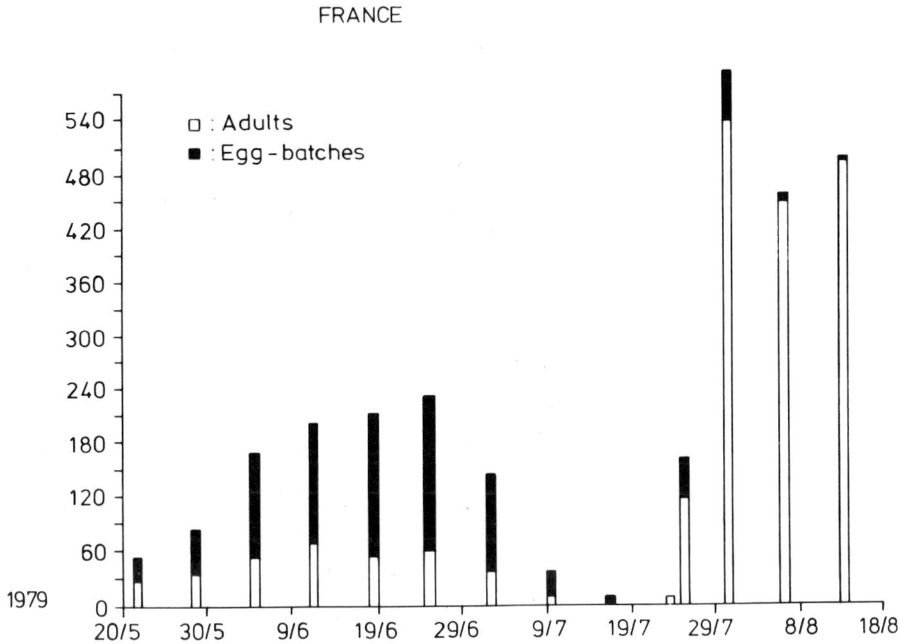

Figure 7a
Number of adults and egg batches counted on one hundred potato plants in Surtainville, Normandy, France, 1979.

in 1979. In Spain, the numbers for 1978 and 1979 were not comparable because in 1978 the number of egg batches was considerably reduced and the observations were discontinued at a later date. In Greece, the second-generation larvae were much more numerous in 1978 than in 1979, probably due to the condition of the plants.

Adult Locomotion

In the Netherlands, beetles are rarely seen in flight, but they have frequently been seen in the Channel zone of France. From Figure 10, giving the results of counts of flights and walking locomotion, it is obvious that walking is a very common means of dispersal. This was also observed in the other field plots, including the Netherlands.

Clear-cut observations from Granada, Budapest, and Athens indicate that hunger flights constitute a considerable part of the flying activity during summer. This was observed in the field as well as in cages. After destruction of the crop or when a harvest

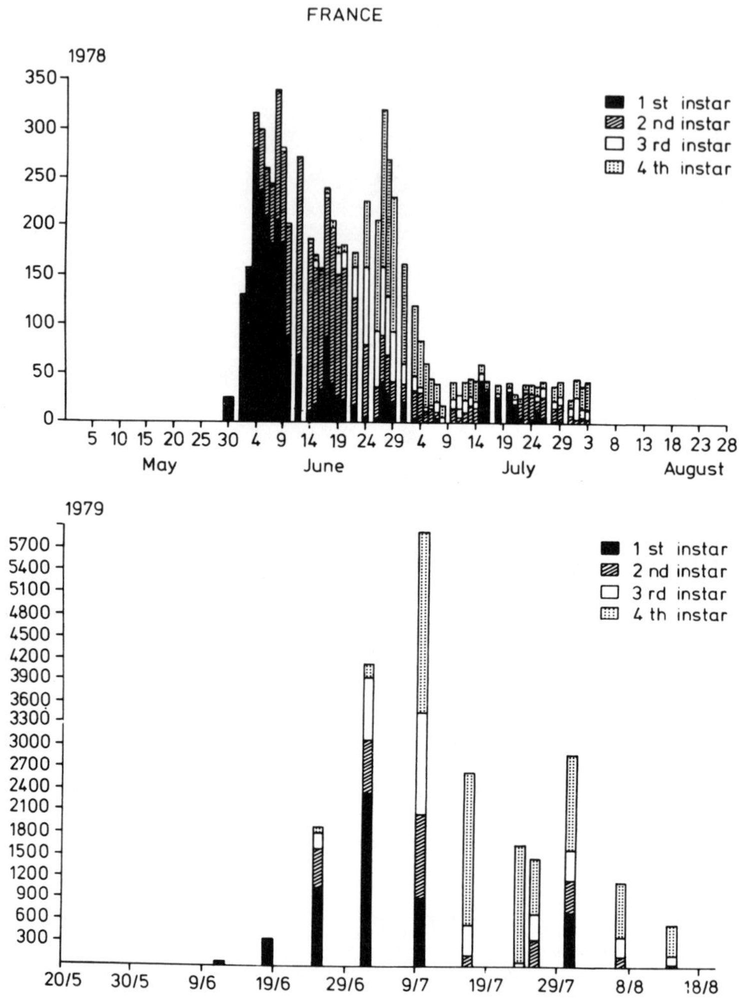

Figure 7b
Number of larvae of several instars per one hundred plants in Surtainville, in 1978, 1979.

interrupts adult feeding, frantic flights are observed, sometimes resulting in mass aggregation of beetles against walls of buildings.

SUMMARY

The Colorado potato beetle in Europe has developed a modest degree of photoperiodic differentiation, with critical periods somewhat lower in southern than in northern populations. The number of generations varies

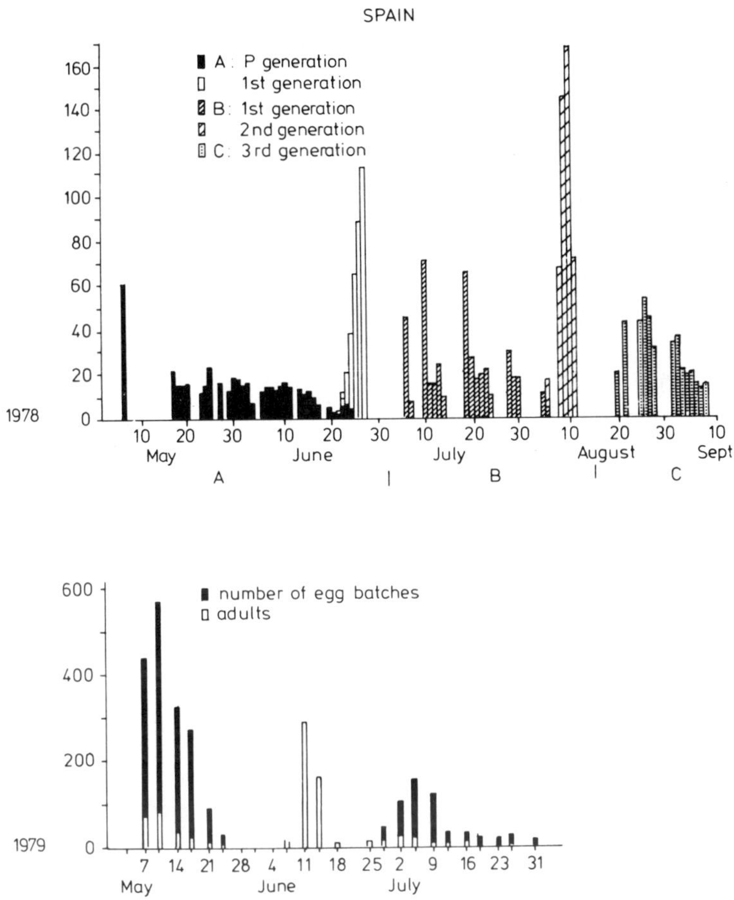

Figure 8a
Number of adults (1978) and adults and egg batches (1979) counted on one hundred potato plants in Granada, Spain.

from one in the north to three in the south, but the number is greatly dependent on the periods of potato growing, which, in turn, vary in southern countries in relation to market conditions. In accord with photoperiodicity, oviposition is generally absent in autumn and winter.

REFERENCES

Chin, C. T., 1950, Studies on the Physiological Relations between the Larvae of Leptinotarsa decemlineata (Say) and Some Solanaceous Plants, Tijdschr. Pl. Ziekt. 56:1-88.

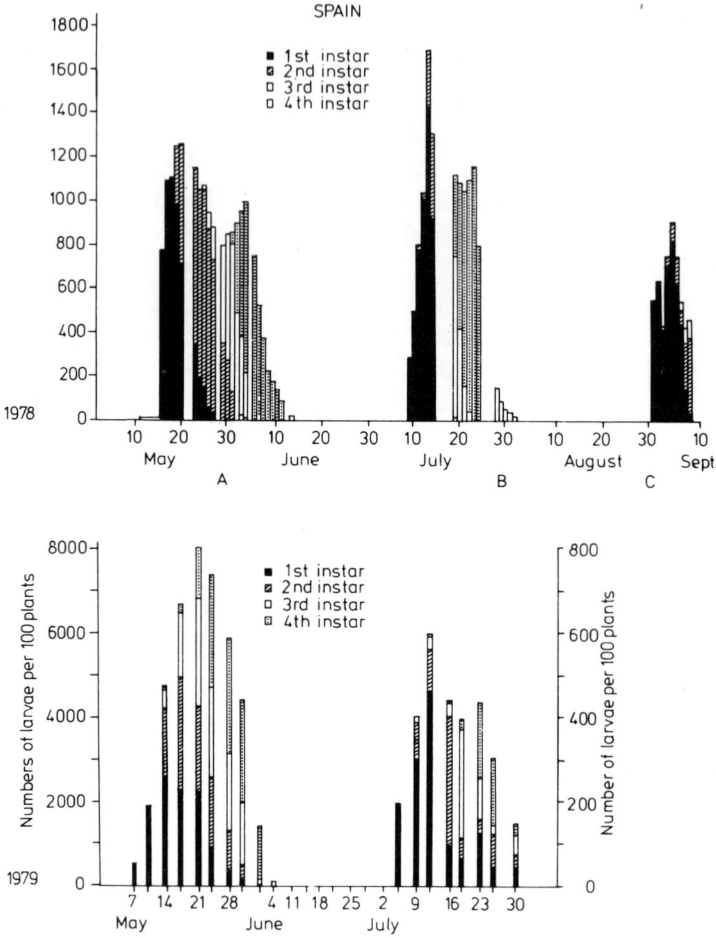

Figure 8b
Number of larvae of several instars per one hundred plants in Granada, 1978, 1979.

de Kort, C. A. D., 1969, Hormones and the Structural and Biochemical Properties of the Flight Muscles in the Colorado Beetle, Med. Landb. Hogesch, Wageningen. 69:1-63.

de Wilde, J., 1955, The Significance of the Photoperiod for the Occurrence of Diapause in the Adult *Leptinotarsa decemlineata* Say, Proc. 1st Int. Photobiol. Congr. 96-101.

de Wilde, J., 1958, Host Plant Selection in the Colorado Beetle Larva (*Leptinotarsa decemlineata* Say), Ent. Exp. Appl. 1:14-22.

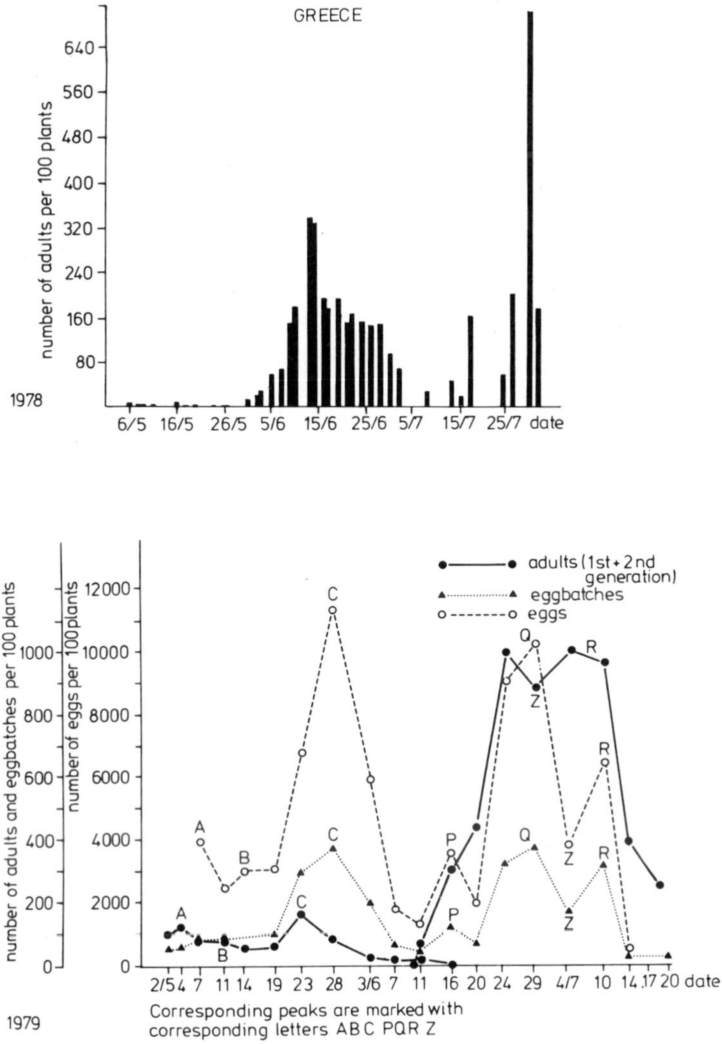

Figure 9a
Number of adults and egg batches counted on one hundred potato plants near Athens, Greece, 1978, 1979.

de Wilde, J., 1969, Diapause and Seasonal Synchronisation in the Adult Colorado Beetle, Leptinotarsa decemlineata Say, Symp. Soc. Exp. Biol. 23:263-284.

de Wilde, J., 1971, Hormones and Insect Diapause in Hormones and the Environment, Mem. Soc. Endocrinol. 18:487-514.

de Wilde, J., 1976, The Olfactory Component in Host-Plant Selection in the Adult Colorado Beetle (Leptinotarsa decemlineata Say), Symp. Biol. Hung. 16:291-300.

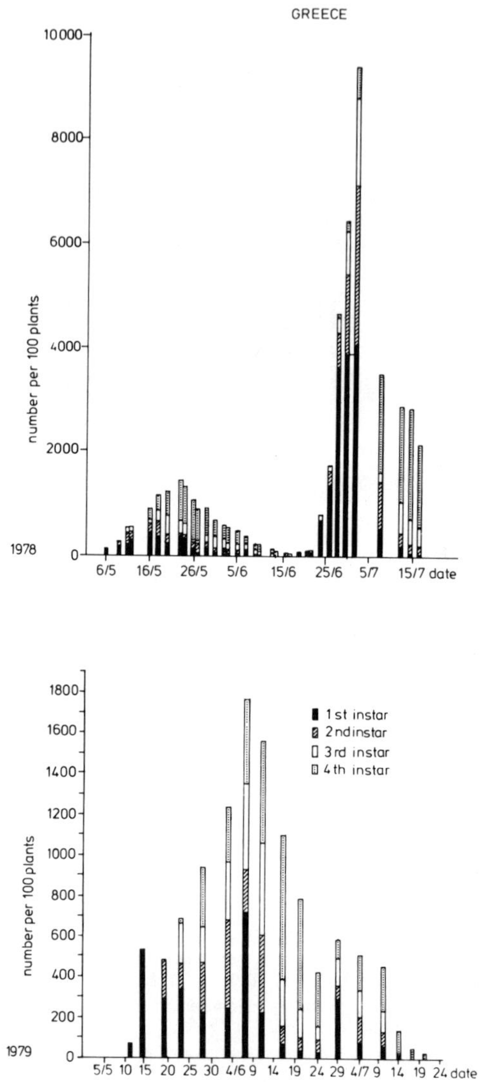

Figure 9b
Number of larvae of several instars per one hundred plants, near Athens, 1978, 1979.

de Wilde, J., 1978, Seasonal States and Endocrine Levels in Insects, in Environmental Endocrinology, pp. 10-19, Springer-Verlag, New York.

de Wilde, J., C. S. Duintjer, and L. Mook, 1959, Physiology of Diapause in the Adult Colorado Beetle (*Leptinotarsa decemlineata* Say). I. The Photoperiod as a Controlling Factor, J. Insect Physiol. 3:75-85.

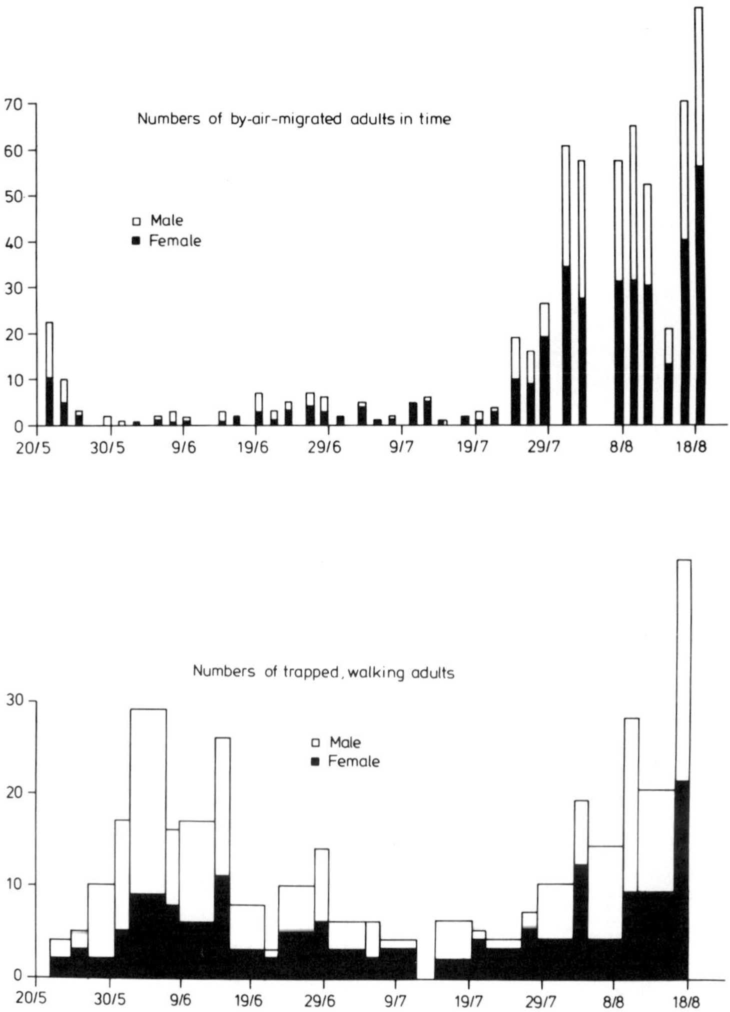

Figure 10
Recordings of flight and marching locomotion on experimental field in France, 1979. Marching on the ground very regularly occurs with a similar periodicity as flight.

Dortland, F. J., 1979, Hormonal Control of Protein Synthesis in the Fat Body of Leptinotarsa decemlineata Say, thesis. Landb. Hogesch., Wageningen. 87p.

Hsiao, T. H., 1969, Chemical Basis of Host Secection and Plant Resistance in Oligophagous Insects, Ent. Exp. Appl. 12:777-788.

Hsiao, T. H., 1974, Chemical Influence on Feeding Behavior of Leptinotarsa Beetles, in Experimental

Analysis of Insect Behaviour, ed. L. B. Brown, pp. 237-248, Springer-Verlag, New York.

Hsiao, T. H., 1976, Chemical and Behavioral Factors Influencing Food Selection of Leptinotarsa Beetle, Symp. Biol. Hung. 16:95-99.

Hsiao, T. H., 1978, Host Plant Adaptations among Geographic Populations of the Colorado Potato Beetle, Ent. Exp. Appl. 24:437-447.

Hsiao, T. H., and G. Fraenkel, 1968a, The Role of Secondary Plant Substances in the Food Specificity of the Colorado Potato Beetle, Leptinotarsa decemlineata (Say), Ann. Entomol. Soc. Am. 61:485-493.

Hsiao, T. H., and G. Fraenkel, 1968b, Selection and Specificity of the Colorado Potato Beetle for Solanaceous and Nonsolanaceous Plants, Ann. Entomol. Soc. Am. 61:493-503.

Jermy, T., and G. Sáringer, 1955, Die Rolle der Photoperiode in der Auslösung der Diapause des Kartoffelkäfer (Leptinotarsa decemlineata Say) und der amerikanischen weissen Bärenspinners (Hyphantria curea Drury), Acta Agron. Acad. Sci. Hung. 5:419-440.

Kramer, S. J., 1978, Regulation of the Juvenile Hormone Titre in the Colorado Potato Beetle, thesis, Landb. Hogesch., Wageningen, 71p.

Ma, W. C., and J. H. Visser, 1978, Single Unit Analysis of Odour Quality Coding by the Olfactory Antennal Receptor System of the Colorado Beetle, Ent. Exp. Appl. 24:520-533.

Ritter, F. J., 1968, Feeding Stimulants for the Colorado Beetle, Meded. Rijksfacult, Landb. Wet. Gent 32:291-305.

Schooneveld, H., 1974, Ultrastructure of the Neurosecretory System of the Colorado Potato Beetle, Leptinotarsa decemlineata (Say).--I. Characterization of the Protocerebral Neurosecretory Cells, Cell Tissue Res. 154:275-288.

Schreiber, K., 1958, Über Einige Inhaltsstoffe der Solanaceen und Ihre Bedeutung für die Kartoffelkäferresistenz, Ent. Exp. Appl. 1:28-37.

Tower, W. L., 1906, An Investigation of Evolution in Chrysomelid Beetles of the Genus Leptinotarsa Carnegie Institution, Washington, D.C.

Visser, J. H., 1979, Olfaction in the Colorado Beetle at the Onset of Host Plant Selection, thesis, Landb. Hogesch., Wageningen.

Ecophysiological Adaptations Among Geographic Populations of the Colorado Potato Beetle in North America

T. Hsiao

Increasingly over the past few years the importance of geographic variations among pest species has been recognized. Differences in ecological, physiological, and behavioral characteristics among geographic populations of a pest are now known to be crucial to the timing and effectiveness of applications of various control strategies. Investigations on the biotype or race of several pest species, such as the European corn borer, the fall webworm, and the screwworm fly are outstanding examples. Such information is indispensable for the successful implementation of any pest management program.

Cosmopolitan pest species that have been introduced intercontinentally provide good subjects for investigation of the possible existence of biological differences among geographic populations. They also provide a rare opportunity for assessing the duration of time and the geographic distance required for the development of measurable biological variations in insects.

The Colorado potato beetle belongs to the genus *Leptinotarsa*, which consists of about forty-five species (Blackwelder, 1946; Jacques, 1972). The center of origin of this genus was southern Mexico. The Colorado potato beetle is the only species within the genus that has reached pest status. It is also the only species that has become distributed outside of North America because of its adaptation of feeding on cultivated potatoes. The Colorado potato beetle was first noticed as a pest on potatoes in North America in about 1860; by 1900 it had spread to all of the potato-growing areas in the eastern half of the United States and eastern Canada. Currently it is distributed between latitude 15° and 55° N

Journal paper No. 2524 Utah Agricultural Experiment Station, Logan, Utah.

throughout Mexico, the United States, and Canada, except that it has not been recorded for California, Nevada, or the Pacific Northwest coast. In continental Europe, the Colorado potato beetle was first reported in 1921 as established in the Bordeaux region of France. Over the next fifty years it spread rapidly and is now distributed between latitude 35° and 60° N, occupying an area similar in size to that in North America. Probably the original stock introduced to Europe from the eastern United States was of a rather small number. One would expect, therefore, that the Colorado potato beetle populations in Europe are less diverse, with less genetic variation, than are North American populations. In general, the species has one generation in the northern regions and two to three generations in the southern regions per year.

The remarkable adaptation of the Colorado potato beetle to vastly differing geographic regions has raised the question as to whether strains or races with dissimilar physiological and ecological characteristics exist or have evolved that adapt to local environmental conditions. Such variations, if they exist, could conceivably be useful in devising pest management strategies for controlling this pest in its varied regions. For the past three years, we have collaborated with J. de Wilde of the Agricultural University, Wageningen, Netherlands, on a study to assess the possible changes in physiological and ecological characteristics of Colorado potato beetles from differing geographic regions of North America and Europe. This chaper deals with the assessment of two important ecophysiological parameters--the host plant relationship and the photoperiodic response--of selected geographic populations of the Colorado potato beetle. The significance of these and other parameters in defining the occurrence of biotypes of the Colorado potato beetle will be discussed.

SPECIES ORIGIN AND PHYLOGENETIC IMPLICATIONS OF THE HOST PLANT RELATIONSHIP

In North America, the Colorado potato beetle infests about ten species of native and exotic solanaceous plants. With the exceptions of *Hyoscyamus niger* and certain varieties of the cultivated tomato, *Lycopersicon e culentum*, the hosts are of the genus *Solanum*, which contains an estimated seventeen hundred species (Whalen, 1979). Two species--the cultivated potato, *S. tuberosum*, and the common nightshade, *S. dulcamara*, originally imported from Europe--are of the subgenus *Potatoe* (Table 1). In the subgenus *Solanum*, *S. sarachoides* (=*S. villosum*) and *S. triflorum* are recorded as occasional hosts. The majority of the endemic hosts are of the subgenus *Leptostemonum*, which includes at least six host species: *S. carolinense*, *S. diversifolium*, *S. elaeagnifolium*, *S. melongena*, *S. sisymbriifolium*, and *S.*

Table 1

Food plant Spectrum of Four *Leptinotarsa* Species from the *Lineata* Group on Solanaceous Plants

Plant	*L. decemlineata*	*L. juncta*	*L. texana*	*L. defecta*
Genus *Solanum*				
Subgenus *Potatoe*				
S. tuberosum	+++*	-	+	-
S. dulcamara	+++*	+++*	++	++
Subgenus *Leptostemonum*				
S. rostratum	+++*	++	++	++*
S. carolinense	++*	+++*	++	+
S. elaeagnifolium	++*	-	+++*	+++*
S. melongena	+++*	++	++	+
Subgenus *Solanum*				
S. sarachoides	++*	-	-	-
Other genera				
Hyoscyamus niger	++*	+	-	-
Lycopersicon esculentum (certain varieties)	++*	-	-	-

Source: Hsiao, 1974; unpublished records.

Key: +++ = optimal growth and reproduction; ++ = moderate growth and reproduction; + = slight growth but no reproduction; - = no growth or reproduction.

*Known natural host for the species.

rostratum. The buffalo bur, S. rostratum, was considered to be the most important host prior to the introduction of the cultivated potato into the natural habitats of the Colorado potato beetle. This species belongs to the section Androceras, which consists of twelve species of prickly, largely annual herbs (Whalen, 1979).

Most of the species of Androceras are confined to Mexico and the southwestern United States. The buffalo bur, the most widely distributed member, is an aggressive weed. With human help, it has spread throughout North America and has found its way to every continent except Antarctica. Its original range probably included most of Mexico north to Mexico City, extending to the Great Plains of the United States (Whalen, 1979). Tower (1906) suggested that Spanish colonial travel and the movement of commerce might have resulted in the dispersal of the buffalo bur into Texas, New Mexico, and parts of Arizona. Such transport would have been initiated at the time of the Spanish conquest of Mexico in 1521. The subsequent dissemination of Androceras into the Great Plains conceivably could have been carried out by migratory mammals such as the bison. The northward distribution of the Colorado potato beetle from southern Mexico also could have been facilitated by these same or similar means. Judging from the rapid spread of the Colorado potato beetle from the foothills of the Rocky Mountains to the Atlantic coast in just twenty years, and from Bordeaux, France, to Eurasia within the past fifty years, Tower's view concerning the origin and early distribution of the Colorado potato beetle before its contact with the cultivated potato is worth considering.

A different approach to the question of the species origin of the Colorado potato beetle would be to compare its host affinity with related Leptinotarsa species. Tower (1906) divided Leptinotarsa on the basis of their morphological similarities into seven species groups, of which at least four groups include solanaceous feeders. Within the group he designated as lineata, four species occur in the United States: L. decemlineata, L. defecta, L. juncta, and L. texana. L. juncta is found in the eastern states and is the only Leptinotarsa species other than L. decemlineata distributed above latitude 40° N. L. defecta and L. texana are confined to southern Texas and adjacent states of Mexico. L. defecta is the least common of the four species.

Table 1 shows the food plant spectra of these three Leptinotarsa species and compares them with the known solanaceous hosts of the Colorado potato beetle. The hosts that are common to all four species are of the subgenus Leptostemonum. Outside this subgenus, only the Colorado potato beetle is capable of utilizing other plants as hosts. One exception is S. dulcamara, which is native to Europe and was introduced to North America. This species has now become a natural host of L. decemlineata and L. juncta in North America and is also suitable as a food plant for several other *Leptinotarsa* species (Hsiao, 1974). Within the subgenus *Leptostemonum*, *S. elaeagnifolium* is the most important plant host for *L. defecta* and *L. texana*, as well as for some geographical populations of *L. decemlineata*. As did *S. rostratum*, this species originated in Mexico and southwestern United States. A herbaceous perennial it has spread to Asia (Deb, 1979), Australia (Symon, 1979), South America, and South Africa (Goeden, 1971). The eggplant, *S. melongena*, originated in Asia (Khan, 1979) and is alien to the natural habitats of the four *lineata* species. Its general acceptance as a food plant for these beetle species resembles the finding for *S. dulcamara*. On the whole, the native host plants of the four *Leptinotarsa* species can be characterized as colonized and weedy species that thrive in sunny, often disturbed areas, such as roadsides, abandoned fields, and dry streambeds. With the production of large numbers of seeds per plant and dispersion aided by humans and animals, these species have spread widely over North America. Such effective dispersion of the plant also has enhanced the range of distribution of the phytophagous insects feeding on it. This is especially true in the case of the Colorado potato beetle.

The pattern of host plant relationships exhibited by these four *Leptinotarsa* species of the *lineata* group clearly reflects a close phylogenetic origin. Available evidence strongly suggests a coevolutionary relationship between plants of the subgenus *Leptostemonum* and these beetle species. Assuming the center of origin of the genus to have been in southern Mexico (Jacques, 1972), the four *Leptinotarsa* species appear to have radiated farthest north in the Western Hemisphere because of their adaptation to highly dispersible host species.

COMPARISON OF HOST PLANT RELATIONSHIPS OF VARIOUS GEOGRAPHIC POPULATIONS

The diverse host range of the Colorado potato beetle in North America has raised the question concerning the host plant affinity of beetle populations from various geographic regions. A recent study of four populations from southwestern United States on several solanaceous plants confirmed the existence of measurable differences in growth performance and reproductive success (Hsiao, 1978). Of particular interest is the finding that an Arizona beetle population is uniquely adapted to its natural host, *S. elaeagnifolium*. This plant is a typical unsuitable host for the Utah, New Mexico, and Texas beetle populations. The Arizona beetles also ranked most adaptable to the seven solanaceous hosts tested. The Utah beetles, which naturally fed on cultivated potato, were intermediate, whereas the New Mexico and Texas beetles, which were collected from *S. rostratum*, were the least adaptable. This finding strongly suggested that host-plant-adapted populations are developing among the Colorado potato beetles in North America.

In continental Europe, the Colorado potato beetle subsists mainly on cultivated plants--the potato and, to a lesser extent, the tomato and the eggplant. Among solanaceous plants native to Europe, only S. dulcamara has been reported to be infested occasionally. In the absence of several acceptable solanaceous hosts largely confined to North America, the food plant range of the Colorado potato beetle in Europe is rather limited. Considering that the European beetles have been isolated from the North American stock for about fifty years and have passed through an estimated one hundred generations in their new habitat, it would be interesting to assess whether changes in host plant affinity similar to those observed in North American populations have occurred.

In collaboration with J. de Wilde, a study was initiated at Wageningen in 1978 to assess the growth of the European Colorado potato beetles from six geographic localities on ten solanaceous plants under laboratory conditions. These geographic populations were from Wageningen, Netherlands; St. Clair, France; and Granada, Spain, in Western Europe; and Poznan, Poland; Budapest, Hungary; and Athens, Greece, in eastern Europe. For comparison, a North American population from Utah and a laboratory culture maintained at the Entomology Laboratory at Wageningen for more than twenty-four years were used.

Figure 1 presents data showing the rearing success of these European beetle populations on three hosts. The criteria used for comparison were the duration of development from larval hatching to adult emergence and the percentage of individuals successfully reared to adults. Data previously reported by Hsiao (1978) on four North American beetle populations are also included in Figure 1. On the cultivated potato, the percentage of rearing success among the European beetle populations varied between 50 and 65 percent, similar to the result obtained for Utah and Arizona beetles but somewhat higher than the 40 percent adult emergence recorded for the New Mexico and Texas beetles, which naturally fed on S. rostratum. The duration of development was not significantly different among all beetle populations tested. The results of rearings of these populations on S. rostratum were also similar for both criteria compared. The fact that the European beetle populations, which had not encountered this plant for an extended period of time, accepted S. rostratum as readily as did the North American beetle populations indicates that no measurable changes in host plant affinity have occurred with regard to this plant. None of the European populations grew well on S. elaeagnifolium. Only 5 to 15 percent were successfully reared to adults, and these were small in size. The duration of development was lengthened an average of two to three days. Only the Arizona beetle population developed well on S. elaeagnifolium, its natural host.

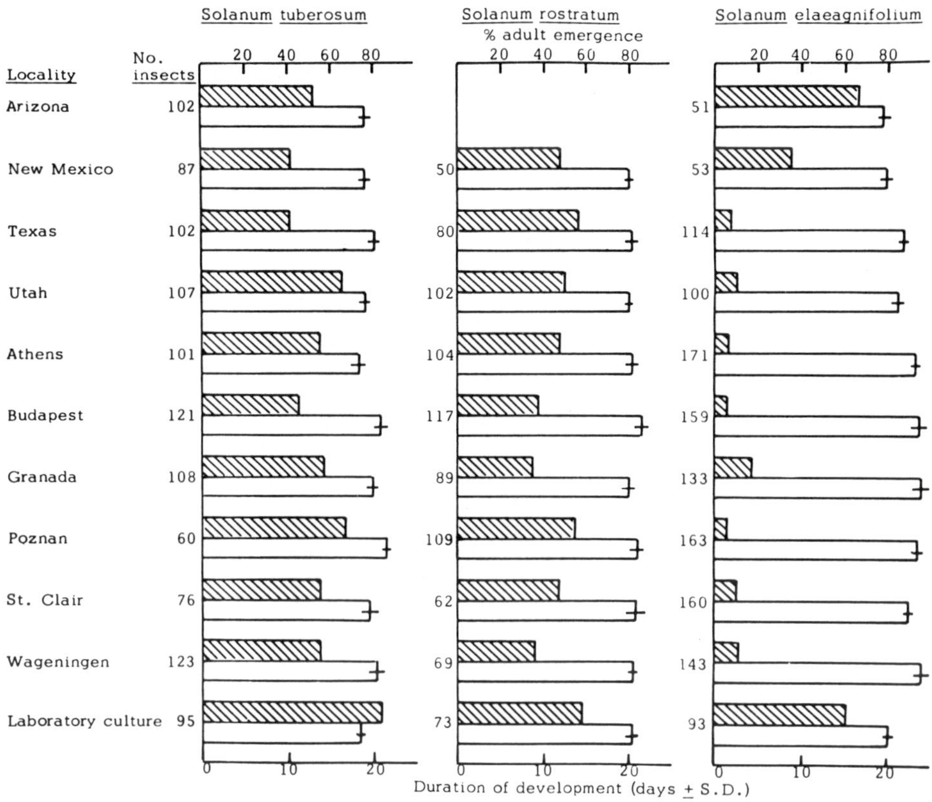

Figure 1
Comparison of rearing success of geographic populations of the Colorado potato beetle on three hosts. The shaded bar represents percentage adult emergence; the clear bar represents duration of development. Rearings were held at 25°C.

An unexpected finding was the high percentage of rearing success of the laboratory culture on S. elaeagnifolium. This laboratory culture also ranked the best in growth performance on S. tuberosum and S. rostratum (Figure 1) and survived better on several less acceptable plants than did any of the field populations tested. Adults of the laboratory culture were more fecund as well and had a shorter preoviposition period. Clearly this laboratory culture has undergone considerable changes as a result of interbreeding and continual manipulation during rearing. It has become a more vigorous strain and has developed a higher degree of tolerance to deterrency, which exists in various solanaceous plants tested. The precise mechanism involved in bringing about such changes in host plant response has not been analyzed. The existence of this phenomenon supports the view, however, that host plant adaptation can evolve in a relatively short period of time when the proper selection pressure is present. It also strengthens our contention that host-plant-adapted populations exist or are developing among geographic populations of beetles in North America. As yet, no such evidence is apparent for the European beetle populations.

GROWTH PERFORMANCE OF CROSSES BETWEEN GEOGRAPHIC POPULATIONS ON SOLANACEOUS HOSTS

The findings of differences in host plant affinity among various geographic populations of the Colorado potato beetle from southwestern United States led to the question as to what type of host relationship may result when these beetles hybridize. Such occurrence is unavoidable under natural conditions since none of these populations is completely isolated from others, and they can be dispersed by several means. The consequences of such hybridization are of fundamental importance to the understanding of host plant adaptations of phytophagous insects. A study of this problem also provides an opportunity for assessing the possible mechanism involved when the Colorado potato beetle shifted from its native host, *S. rostratum*, to the cultivated potato.

Two series of experiments were conducted to explore this problem, using the cultivated potato and *S. elaeagnifolium* as host plants. In the first series, three beetle populations were used. The natural hosts were the potato for the Utah beetles, *S. elaeagnifolium* for the Arizona beetles, and *S. rostratum* for the Texas beetles. To compare their growth performances, progeny from reciprocal crosses of the Utah beetles with either one of the other two populations were reared on the potato under identical conditions. The results (Table 2) show that all F1 progeny of these crosses yielded a percentage of adults at least equal to and in some cases better than that of the parent generation. Their duration of development from larval hatching to adult emergence was similar to that of the parents. The pupae of the hybrids, both male and female, were somewhat smaller than the parents in the cases of crosses between Utah and Arizona beetles but were larger than the parents in the cases of crosses between Utah and Texas beetles. The overall results show evidence of hybrid vigor when compared with the growth performance of the parents. These data clearly demonstrate that regardless of the original host plant of the parents, crossbred progeny of the three beetle populations readily accept the potato as a host plant. This finding helps to explain why the Colorado potato beetle was able to transfer without difficulty from its native hosts to the cultivated potato when the latter was introduced to the beetle's natural habitats.

The second series of experiments was carried out

Table 2
Growth and Development on *Solanum tuberosum* of Three Populations of Colorado Potato Beetle and Their Reciprocal Crosses

Locality	No. Insects	% Mortality: Larvae 1-4th	% Mortality: Prepupae and Pupae	Number and Mean Pupal Weight (mg ± S.D.): Female	Number and Mean Pupal Weight (mg ± S.D.): Male	Duration of Development (Days ± S.D.)	% Adult Emergence
Parent generations							
Utah	107	15.9	18.7	(33)136±11.6	(37)117±12.3	19.2±1.2	65.4
Texas	102	48.0	10.6	(19)103±11.1	(24) 93±12.0	20.3±1.1	41.4
Arizona	102	22.5	24.5	(21)142±12.9	(33)127±12.7	19.2±1.1	53.0
Crosses							
Utah female x Texas male	235	18.3	15.3	(76)114±12.8	(80)100±10.2	19.6±2.0	66.4
Texas female x Utah male	206	15.5	5.4	(82)116±12.2	(81)101± 9.4	20.9±1.3	79.1
Utah female x Arizona male	154	27.3	15.6	(37)129±16.9	(51)106±13.7	20.7±1.3	57.1
Arizona female x Utah male	164	25.0	12.2	(55)119±12.9	(48)102±13.2	20.7±1.4	62.8

Note: Reared at 25°C and sixteen-hour photophase.

with *S. elaeagnifolium* as a host. In contrast to the potato, this plant was not a suitable host for most beetle populations (Figure 1) but was a natural host for the Arizona beetles. Such a difference offers an ideal situation for assessing the outcome of hybridization among beetle populations. Progeny of reciprocal crosses between Arizona and Utah beetles were reared on *S. elaeagnifolium* under identical conditions to evaluate their growth performance. The results (Table 3) reveal that both F1 hybrids survived in greater proportions than did the Utah parents. They produced 33 percent of adults, about twice the percentage obtained with the Utah beetles but only about half of the adult yield (68 percent) of the Arizona parents. Nevertheless, the percentage of adult yield of these hybrids was significantly higher than that obtained from any of the other geographic populations (Figure 1). The duration of development of the hybrids lay intermediately between the two parents. The pupal weights were similar to those of the Utah parents but were lighter than those of the Arizona parents. A rather high percentage of prepupal and pupal mortality was observed in the two hybrids, suggesting a possible lack of genetic compatibility between the two parents. The overall improved growth performance of these hybrids on *S. elaeagnifolium* clearly indicates that genetic mechanism is involved in the host plant adaptation of these beetles. The nature of the inheritance determining such host relationship requires further elucidation. These data provide the first experimental crossing evidence, however, that host races exist among geographic populations of the Colorado potato beetle in North America.

SEASONAL ADAPTATION OF VARIOUS GEOGRAPHIC POPULATIONS

Seasonal adaptation of insects is governed by climatic conditions, of which photoperiod, temperature, and moisture are the most crucial factors. All of these undoubtedly play important roles in determining the life pattern of the Colorado potato beetle in the various geographic regions where it occurs. Because of its wide latitudinal distribution, the Colorado potato beetle is well suited for investigating the role of the photoperiod in relation to its seasonal adaptation. This species has long been regarded as a typical long-day insect. It continues to reproduce at day lengths longer than fifteen or sixteen hours and enters into adult diapause by burrowing into soil at shorter day lengths. The number of long days available during the summer months can be used to estimate the possible number of generations of the beetles in various geographic regions in the Northern Hemisphere (de Wilde, 1969). As yet, no experimental study has been conducted to determine the photoperiodic responses of various geographic populations.

Table 3

Growth and Development on *Solanum elaegnifolium* of Two Colorado Potato Beetle Populations and Their Reciprocal Crosses

Locality	No. Insects	% Mortality: Larvae 1-4th	% Mortality: Prepupae and Pupae	Number and Mean Pupal Weight (mg ± S.D.): Female	Number and Mean Pupal Weight (mg ± S.D.): Male	Duration of Development (Days ± S.D.)	% Adult Emergence
Utah	138	78.3	5.0	(11)112±11.4	(12)101± 7.1	22.7±2.0	16.7
Arizona	88	23.9	7.9	(28)137±20.1	(32)117±13.6	19.6±1.2	68.2
Arizona female x Utah male	81	45.7	21.0	(9)120±20.6	(18)106±11.8	20.8±1.6	33.3
Utah female x Arizona male	68	51.5	14.7	(12)113± 8.1	(11) 99±10.0	21.3±1.1	33.8

Note: Reared at 25° and sixteen-hour photophase.

We have now investigated four beetle populations from western United States, and, in collaboration with Professor de Wilde, eleven beetle populations from continental Europe (Hsiao and de Wilde, 1979). In addition, several related Leptinotarsa species from the United States were included for comparison. These studies were conducted in environment-controlled chambers set at 25°C and photoperiod regimes of LD = 18:6, 16:8, 14:10, 12:12, and 10:14. The F1 and F2 progeny of field-collected adults were used. The criteria for comparison were the incidence of adult diapause, the prediapause period, the preoviposition period, proportion of females reproducing, and fecundity. The photoperiodic response of the four North American beetle populations are compared in Figure 2. The data are expressed as percentage of nondiapausing adults of both sexes that did not enter soil regardless of whether they reproduced. The two northern populations--from Yakima, Washington (46°36' N), and Logan, Utah (41°44' N)--showed very few nondiapausing adults between the ten-fourteen hour photophases; they increased sharply in the sixteen-eighteen hour photophases. The critical photophase was

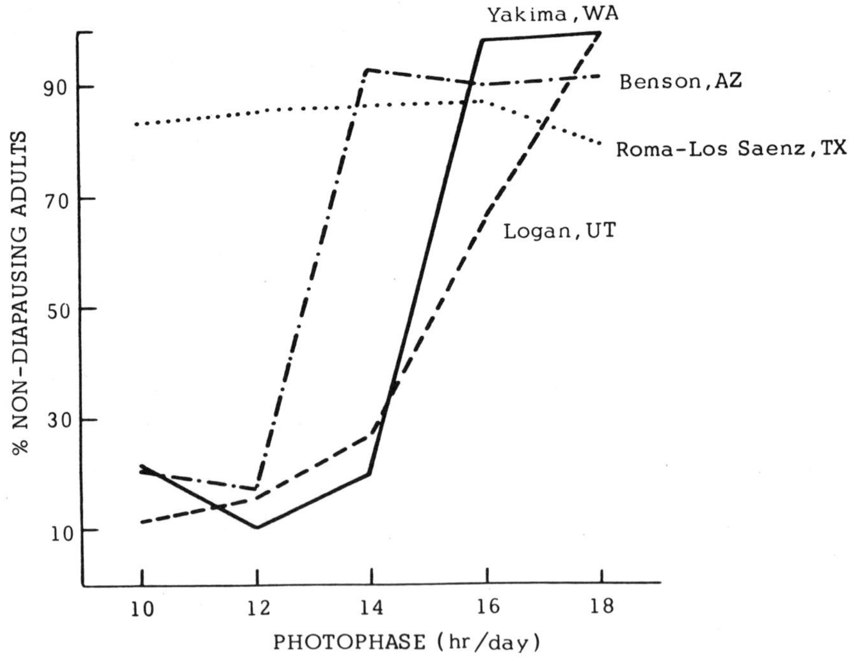

Figure 2
Effect of photoperiod on adult diapause of four geographic populations of the Colorado potato beetle from North America. Rearings were held at 25°C and on potato foliage.

about fifteen hours for the Utah population but was slightly lower for the Washington population. This finding is consistent with field data because the Colorado potato beetle has two generations per year in Yakima, Washington, and only one per year in Logan, Utah, even though Yakima is five degrees farther north in latitude, but lower in elevation, than Logan. Of the two southern populations, the beetles from Benson, Arizona (31°58' N), were mostly in diapause at the ten-twelve hour photophases and were nondiapause between the fourteen-eighteen hour photophases. This population has a critical photophase of less than thirteen hours. The beetles from Roma-Los Saenz, Texas (26°24'N), were the most unusual; they did not respond to photoperiod. Most adults remained nondiapause and reproduced at the ten-eighteen hour photophases tested. On the whole, the photoperiodic response curves of the North American beetle populations follow a predictable pattern, with a gradual increase in critical photophase as the species is distributed northward. Such responses reflect a seasonal adaptation well synchronized with the environment. The lack of photoperiodic response of the Texas beetle population provides an interesting example for further investigation as to the origin of photoperiodism in the Colorado potato beetle.

In general, the photoperiodic response curves of the European beetle populations are not as markedly different from each other, indicating that they have not undergone significant changes concerning this ecological factor. Considering that the climate of southern Europe is warm, a greater divergence in photoperiodic response may be anticipated if the patterns found among the North American beetle populations are extrapolated.

The results obtained thus far on photoperiodic responses of several related _Leptinotarsa_ are significant. _L. haldemani_ and _L. tumamoca_ from southern Arizona and _L. juncta_ from Virginia exhibited a typical long-day response, whereas _L. texana_ and _L. defecta_ from southern Texas did not show a definite photoperiodic response. The last two species were from the same geographic region as the Texas population of the Colorado potato beetle, which also exhibited no photoperiodic response. Considering this experimental evidence and the subtropical origin of the genus _Leptinotarsa_, it is apparent that the photoperiodic-induced adult diapause is an adaptive strategy that has evolved in several species distributed in the more northerly latitudes.

PHOTOPERIODIC RESPONSES OF CROSSES BETWEEN GEOGRAPHIC POPULATIONS

The marked differences in photoperiodic responses that exist among beetle populations in North America provide an opportunity to assess the possible mechanism involved. For this study the Texas beetles, which do not respond to photoperiod, were crossed with the Utah beetles, which have a typical long-day response. Progeny of these crosses and their inbred progeny were reared under a short-day (LD = 12:12) regime at 25°C to determine their photoperiodic responses. Figure 3 presents the results obtained for adult diapause and reproduction of the two populations and their reciprocal crosses. The incidence of adult diapause of the F1 hybrids of the two crosses was rather high compared with the Texas parents but was lower than that of the Utah parents. That more than 20 percent of the females of both crosses entered

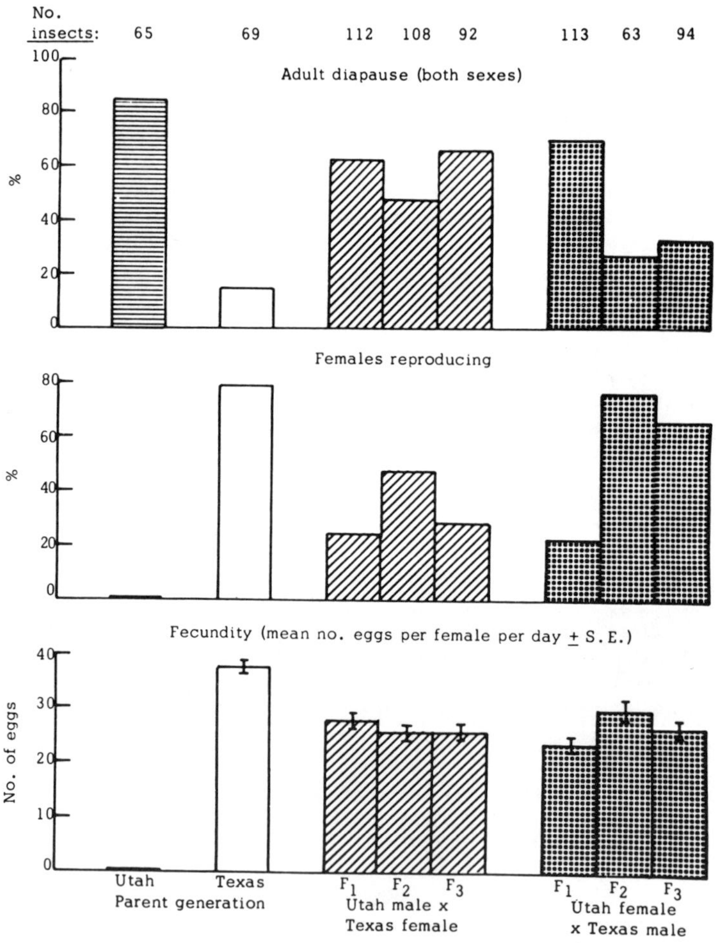

Figure 3
Effects of a short-day photoperiod (LD/12:12) on adult diapause and reproduction of two populations of the Colorado potato beetle and their reciprocal crosses. Rearings were held at 25°C and on potato foliage.

reproduction is of particular significance. By contrast, the Utah parents were unable to reproduce under this short-day photoperiod. The percentage of adult diapause remained high (48 and 66 percent, respectively) in the F2 and F3 progeny of the Utah male and Texas female cross but was considerably reduced (27 and 33 percent) in the Utah female x Texas male cross. Consequently more females in the last group than in the former groups entered reproduction. Fecundity of all progeny of the two crosses was remarkably similar, ranging between an average of twenty-four and thirty eggs per female per day. This figure is somewhat lower than that of the Texas parents, which produced an average of thirty-seven eggs per female per day. The low fecundity of these hybrids is probably due to their maladaptation to the short-day photoperiod. On the whole, these experimental crosses clearly show that the characteristic of photoperiod-induced adult diapause is inheritable in the Colorado potato beetle. The fact that the hybrids exhibited an intermediate photoperiodic response, when compared with their parents, would indicate that the diapause characteristic is neither dominant nor recessive. Since no clear segregation was apparent within the three generations tested, it may be assumed that polygenic control is involved. Further experimental crossing is needed to define the genetic nature of the photoperiodic response. These data support our view that strains with different photoperiodic responses exist among geographic populations of the Colorado potato beetle.

CONCLUSION

Current awareness of the importance of biological variations among insect populations is influenced by the finding of changes in quality of mass-reared colonies such as the screwworm, the European corn borer, and the codling moth (Bush, 1979; Boller and Chambers, 1977). These discoveries encouraged the development of methods for monitoring various biological characteristics of insects to detect the possible changes occurring during mass rearing (Huettel, 1976; Chambers, 1977). Even more influential are ecological theories of population regulation. The life-system concept proposed by Geier (1966) and Clark et al. (1967) stressed the interaction between subject populations of a species and its effective environment as codeterminants of abundance. This concept takes into account the ecological and physiological, as well as the genetic, characteristics of insect populations. It has profoundly revolutionized the thinking about pest control because of the recognition of the existence of different life systems of a pest in various regions where it occurs. Implementation of pest management strategy, therefore, requires an investigation of the various characteristics of a pest species that contribute to differences among life systems. The identification of such differences in ecological, physiological, and genetic characteristics among geographic populations of a pest is prerequisite to sound pest management practices.

Our study has focused on two important ecophysiological characteristics--host plant adaptation and seasonal adaptation as exemplified by photoperiodic responses--of various geographic populations of the Colorado potato beetle from North America and continental Europe. The results obtained thus far have demonstrated the existence of races among North American beetle populations that exhibit differing host plant affinity and photoperiodic response. Such evidence was not found among European beetle populations, possibly because of the relatively short time span--about fifty-five years--of the beetle's expansion there. Cytogenetic analyses currently being conducted in our laboratory strongly suggest the existence of chromosomal variations among the various geographic populations, indicating considerable genetic variability within the species. Many more attributes may be considered in measuring variations among geographic populations of an insect pest, including: ecological parameters such as temperature adaptations, moisture adaptations, and dispersal patterns; biochemical parameters such as protein and isozyme variations and insecticide resistance; morphological characteristics; and behavioral traits. As more data are accumulated on these attributes, it will be possible to determine their relative importance.

Information gathered from these studies may be directly incorporated into pest management programs that focus on the development of varietal resistance, genetic manipulation of pest species, and effective use of insecticide in pest control. This information is also invaluable for the understanding of the process of speciation and the phylogenetic relationships of insects.

REFERENCES

Blackwelder, R. E., 1946, Checklist of the Coleopterous Insects of Mexico, Central America, the West Indies, and South America, Bull. U.S. natn. Mus. 185:551-763.

Boller, E. F., and D. L. Chambers, 1977, Quality Aspects of Mass-Reared Insects, in Biological Control by Augmentation of Natural Enemies, ed. R. L. Ridgway and S. B. Vinson, pp. 219-236. Plenum Press, New York.

Bush, G. L., 1979, Planning a Rational Quality Control Program for the Screwworm Fly, in The Screwworm Problem: Evolution of Resistance to Biological Control, ed. R. H. Richardson, University of Texas Press, Austin.

Chambers, D. L., 1977, Quality Control in Mass Rearing, Annu. Rev. Entomol. 22:289-308.

Clark, L. R., P. W. Geier, R. D. Hughes, and R. F. Morris, 1967, The Ecology of Insect Populations in Theory and Practice, Methuen & Co., London.

Deb, D. B., 1979, Solanaceae in India, in The Biology and Taxonomy of the Solanaceae, ed. J. G. Hawkes, R. N. Lester, and A. D. Skelding, Academic Press, New York.

de Wilde, J., 1969, Diapause and Seasonal Synchronization in the Adult Colorado Beetle (Leptinotarsa decemlineata Say), Symp. Soc. Exp. Biol. 23:263-284.

Geier, W. P., 1966, Management of Insect Pests, Annu. Rev. Entomol. 11:471-490.

Goeden, R. D., 1971, Insect Ecology of Silverleaf Nightshade, Weed Sci. 19:45-51.

Hsiao, T. H., 1974, Chemical Influence on Feeding Behavior of Leptinotarsa Beetles, in Experimental Analysis of Insect Behaviour, ed. L. B. Browne, Springer-Verlag, New York.

Hsiao, T. H., 1978, Host Plant Adaptations among Geographic Populations of the Colorado Potato Beetle, Ent. Exp. & Appl. 24:437-447.

Hsiao, T. H., and J. de Wilde, 1979, Ecophysiological Variations among Geographic Populations of the Colorado Potato Beetle, in IX Intl. Cong. Plant Protection and 71st Annual Meeting of the American Phytophathological Society, Washington, D.C., Aug. 5-11 (Abstr.).

Huettel, M. D., 1976, Monitoring the Quality of Laboratory-Reared Insects: A Biological and Behavioral Perspective, Environ. Entomol. 5:807-814.

Jacques, R., Jr., 1972, Taxonomic Revision of the Genus Leptinotarsa (Coleoptera: Chrysomelidae) of North America, Xerox University Microfilms, Ann. Arbor, Michigan.

Khan, R., 1979, Solanum melongena and Its Ancestral Forms, in The Biology and Taxonomy of the Solanaceae, ed. J. G. Hawkes, R. N. Lester, and A. D. Skelding, Academic Press, New York.

Symon, D. E., 1979, The Genus Solanum in Australia, in The Biology and Taxonomy of the Solanaceae, ed. J. G. Hawkes, R. N. Lester, and A. D. Skelding, Academic Press, New York.

Tower, W. L., 1906, An Investigation of Evolution in Chrysomelid Beetles of the Genus Leptinotarsa, Carnegie Institution, Washington, D.C.

Whalen, M. D., 1979, Speciation in Solanum, Section Androceras, in The Biology and Taxonomy of the Solanaceae, ed. J. G. Hawkes, R. N. Lester, and A. D. Skelding, Academic Press, New York.

Role of Body Temperature and Thermoregulation in the Biology of the Colorado Potato Beetle

Michael L. May

The realization that insects may often experience body temperatures (T_b) well above the temperature of the surrounding air (T_a) is by no means new (Newport, 1837). Such elevation of T_b usually results either from absorption of solar radiation (ecothermy) or retention of metabolic heat generated by the flight muscles (endothermy). In addition, many insects are able to control the amount of heat they absorb, produce and/or retain and thereby maintain T_b within narrower limits than would be predicted for a body in passive thermal equilibrium with its surroundings; thus they thermoregulate. At least three extensive reviews of insect thermoregulation have appeared within the last decade (Cloudsley-Thompson, 1970; Heinrich, 1974; May, 1979).

The possible application of this expanded knowledge to problems of insect population biology and pest management has not been investigated systematically, partly because most insects that have been fruitful subjects of investigation of T_b are not major economic pests and because of the technical difficulty of making accurate determinations of insect T_b. The research described here is aimed at partially bridging the gap between thermoregulation studies and pest management.

INSECT THERMOREGULATION AND ITS IMPLICATIONS

Thermoregulation is a relative term as applied to any animal, especially insects. No organism maintains an absolutely constant T_b. Because of the small size of insects, their T_b usually is strongly influenced by T_a (Heinrich, 1974; May, 1978) despite adaptations that modify this influence. Comparatively small departures of T_b from T_a may be significant, however. The known effects of temperature on the survival, reproduction, growth rate, feeding, and other life processes of insects are often dramatic (Bursell, 1974). Many biological functions have well-marked temperature optima, and the benefits certainly must accrue to insects that maintain T_b as close as possible to these optima.

Uvarov (1977) has correctly pointed out that in many acridids (and other insects) the optima are not congruent; for example, growth commonly is maximal at higher T_b than is fecundity. In addition, T_b of locusts in the field may be regulated to some extent but apparently diverges considerably from some physiological optima. In some cases such divergence may be more apparent than real since the possibility remains that different life stages thermoregulate at different levels of T_b (Sherman and Watt, 1973). Nevertheless in many cases T_b in the field must be a compromise between what is possible in terms of environmental constraints and what is desirable in terms of physiology. Regardless of the specific nature of a given physiological response to T_b, however, any attempt to predict the response from data on T_a should take into account the sometimes complex relationship of T_b to T_a.

Adults of several beetle species are known or strongly suspected to thermoregulate behaviorally; these include gyrinids (M. Heath, pers. comm.) and desert-dwelling tenebrionids (Edney, 1971; Holm and Edney, 1973; Hamilton, 1971; Henwood, 1975). In addition endothermy occurs in some large dytiscids (Leston et al., 1965), cerambycids, and scarabaeids (Bartholomew and Casey, 1977; Bartholomew and Heinrich, 1978).

Apparently no beetle larvae have been studied previously from this standpoint, but a number of caterpillars maintain an elevated T_b and/or regulate T_b ectothermically (Sherman and Watt, 1973; Sullivan and Wellington, 1953; Wellington et al., 1954; Shephard, 1958), and a sawfly larva avoids overheating by evaporative cooling (Seymour, 1974). Casey (1976) showed that sphingid caterpillars thermoregulate by varying their position on their food plant and orientation relative to the sun and at the same time that T_b is an important determinant of feeding rate.

THERMOREGULATION IN LEPTINOTARSA

Adults and larvae of the Colorado potato beetle regulate their body temperature at least in a rudimentary fashion, and this capability may have marked effects on their food consumption, growth, and reproduction.

The relation of T_b to environmental conditions was studied in beetles feeding on potato near New Brunswick, New Jersey, from late June to mid-August. Potato stems were selected arbitrarily. T_b was determined for all larvae judged large enough for probe insertion and for all adults on the chosen stem. Body temperatures were measured using a copper-constantan thermocouple embedded in a 1 centimeter (cm), 26 gauge hypodermic needle and read with a Bailey Bat-4® portable electronic thermometer. Other environmental parameters measured were air temperature about 25 cm above ground in the potato crop (T_a), air temperature 1 meter (m) above the ground in the shade at the edge of the field (T'_a), leaf surface temperature (T_s), relative humidity (RH), wind speed above (W_1) and within (W_2) the crop canopy, and solar radiation intensity (SR). In addition, data were gathered for each specimen concerning whether it was feeding, whether it was in sun or shade, orientation of the long axis of the body relative to the sun, and position on the plant (upper, middle, or lower third; inner or outer layer of foliage; upper or lower surface of leaf, or on stem). Each individual beetle was collected for determination of mass (larvae) or sex (adults). Separate observations measured the biting rate of individual larvae and adults; immediately after these were determined T_b was ascertained.

The results of body temperature studies are summarized in Figures 1 and 2 for adults and larvae, respectively. In each case T_b is plotted against a measure of ambient temperature and a linear regression of T_b on ambient temperature calculated. A slope of zero suggests complete independence of T_b, and thus excellent thermoregulation, and a slope of 1.0 suggests the absence of thermoregulatory ability (May, 1976). The validity of the conclusions depends, however, on the appropriateness of the independent variable. The use of T_a as the standard to which T_b is compared is conservative because the effective environmental temperature to which an organism is exposed (T_e; Bakken and Gates, 1975) also depends on the radiant environment, drying power of the air, and other factors and generally is more variable than T_a.

For both adults and larvae $T_b > T'_a$, so that the assumption that T_b may be equated with standard screen or weather station temperatures can lead to serious underestimation of T_b. The slopes of T_b on T'_a give slight indication of thermoregulation; the air temperature (T_a) to which the beetles were actually exposed is, however, substantially higher and more variable than T'_a.

The best predictor of T_b is T_s. This has also been found in some other insects (Waloff, 1963). Examination of the regression lines suggests that T_b may not be completely dependent on T_s in adults because the slope is significantly less than 1 ($F = 14.9$, $p < 0.001$), but the effect, if any, is slight. Nevertheless, thermoregulation with respect to air temperature or other thermal parameters may still be possible by selection of a thermally favorable substrate. Other evidence gathered shows that this is indeed the case.

The plots of T_b versus T_a are interesting because, for adults at least, a degree of control of T_b is definitely suggested. Beetles experienced only about three-fourths as much variation in T_b as might be

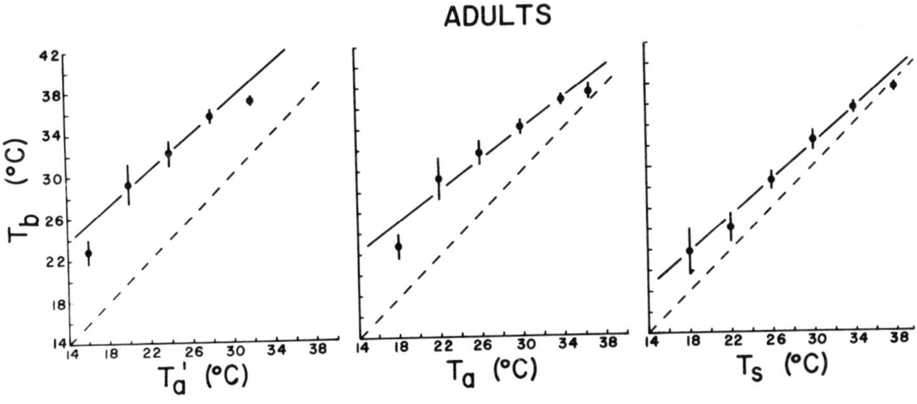

Figure 1
Body temperature of adult potato beetles (L. decemlineata) as function of various measures of environmental temperature, T_a', T_a, and T_s. The points and vertical lines indicate mean $T_b \pm 2$ SE within intervals of environmental temperature. Solid lines are least-square linear regressions of T_b on environmental temperature; dashed lines are isothermal lines. Regression equations are: $T_b = 12.16 + 0.84T_a'$, $r^2 = 0.778$, $n = 69$; $T_b = 12.09 + 0.73T_a$, $r^2 = 0.815$, $n = 75$; $T_b = 6.16 + 0.86\ T_s$, $r^2 = 0.912$, $n = 69$.

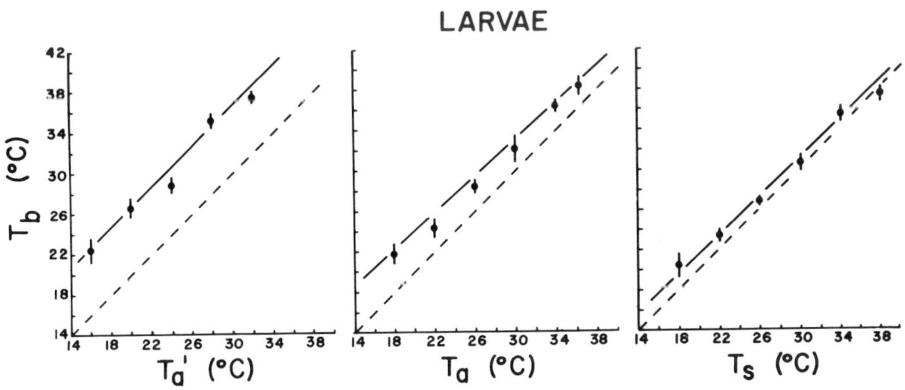

Figure 2
Body temperature of larval potato beetles as a function of environmental temperature. Symbols as in Figure 1. Regression equations are $T_b = 6.54 + 1.00T_a'$, $r^2 = 0.854$, $n = 115$; $T_b = 5.72 + 0.91T_a$, $r^2 = 0.875$, $n = 127$; $t_b = 2.34 + 0.96T_s$, $r^2 = 0.958$, $n = 121$.

expected from T_a. The mean temperature excess (T_b - T_a) was 4.59°C. If T_b values when solar radiation was less than 0.55 $kW.m^{-2}$ (occurring early in the morning and during cloudy periods) are eliminated from the regression, then $T_b = 14.3 + 0.67T_a$, $r^2 = 0.816$, so regulatory capabilities may be slightly further enhanced under favorable conditions. For larvae, plots of T_b versus T_a hint at a small degree of thermoregulation but do not allow a firm conclusion. Over the range of conditions covered here, expected T_b was elevated by 2° to 4°C above T_a; mean temperature excess was 3.36°C, maximum was 9°C.

Clearly, even for adults, variation in T_a accounted for the preponderance of variation in T_b. It is of interest to consider how other environmental parameters, as well as physical and behavioral properties of the insects, may have affected the difference between T_b and T_a. Thus I calculated single and multiple regression of $T_b - T_a$ on several independent variables. For adult beetles, the best fit was obtained by the multiple regression, $T_b - T_a = 5.45 + 6.90SR + 0.051RH - 0.406W_2 - 0.298T_a$, $r^2 = 0.557$, $n = 75$. Considering the effects of each independent variable separately, the slopes are significant on SR ($F = 26.4$, $p<0.0001$), T_a ($F = 50.6$, $p<0.0001$), and RH ($F = 4.67$, $p = 0.034$) but not on W_2 ($F = 1.93$ $p = 0.17$). This suggests that solar radiation was responsible for a considerable proportion of the temperature excess but that the latter may also have been reduced at low humidity by evaporative cooling. Increasing wind speed should bring T_b closer to T_a, but the effect was evidently slight in this case. The negative effect of T_a probably reflects behavorial changes on the part of the beetles, primarily shade seeking at high temperature. Behavior determines the effect of SR on $T_b - T_a$. A significant correlation between $T_b - T_a$ and SR exists only for those beetles with $T_b<30°C$ ($r = 0.636$). Above that temperature a comparatively high percentage of beetles were in the shade, and most of these had T_b relatively near T_a even at high SR, so $r = 0.099$.

The temperature excess of larvae is less easy to account for. It is best described by $T_b - T_a = 1.84 + 1.09SR + 0.057RH - 0.270W_1 - 0.069T_a$, $r^2 = 0.206$ $n = 125$. The low r^2 may be attributable to inaccuracies of measurement (especially of T_b), action of physical factors not accounted for, or behavior that affects $T_b - T_a$ but is not closely correlated with T_a; probably all three of these alternatives contribute. Among the parameters that contributed to $T_b - T_a$ in larvae, RH ($F = 6.06$, $p = 0.015$) and W_1 ($F = 8.81$, $p = 0.036$) were significant, T_a marginally so ($F = 3.84$, $p = 0.052$) and SR not significant ($F = 2.40$, $p = 0.12$).

Temperature excess of larvae and adults varied throughout the day in a pattern consistent with thermoregulation; that is, $T_b - T_a$ increased to a maximum by mid-to late morning and then decreased until mid-afternoon (Figure 3). No data are available after

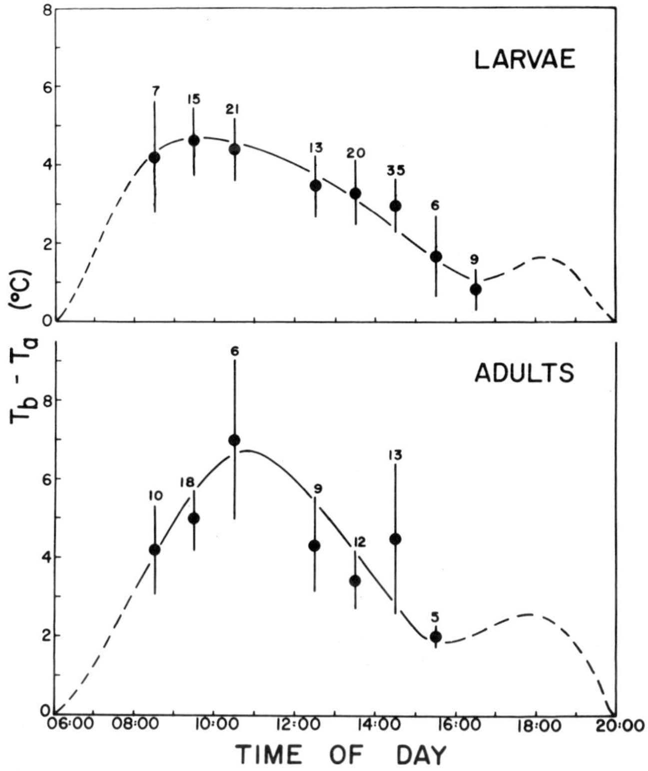

Figure 3
Temperature excess of larval and adult potato beetles as a function of time of day (EDT). Points and vertical lines represent mean $T_b - T_a \pm 2$ SE within one-hour intervals. Numerals above the lines are numbers in each sample. The curved line was drawn by eye through the data and projected to earlier and later times.

about 16:30 hours. I have drawn a curve through the available data by eye and have extended this on a purely hypothetical basis to approximate sunrise and sunset during July at this latitude. The late afternoon increase is based on the assumption that the beetles would have responded to falling T_a in such a way as to increase $T_b - T_a$ above the afternoon minimum. As *SR* became negligible toward sunset, $T_b - T_a$ must have fallen to nearly zero. The data mostly apply to rather hot midsummer days. Adult beetles newly emerged from diapause, as well as early first-generation larvae, might be exposed to much lower T_a at midday in May and early June. If these individual beetles thermoregulate, variation of $T_b - T_a$ might be rather different. Nevertheless, with these rather important reservations, it is possible to calculate from Figure 3 a mean excess of T_b over T_a by graphical integration. For larvae this turns out to be about 2.5°C averaged over the daylight hours or 1.5°C averaged over twenty-four hours; corresponding figures for adults are 6.0°C and 3.5°C. Such estimates may be useful in trying to determine the effects of elevated T_b.

Figures 4 and 5 indicate several behavioral changes in adults and larvae, respectively, that may have thermoregulatory significance. Both groups, particularly larvae, displayed an increasing tendency to rest and feed on the upper surface of leaves at low temperatures, probably increasing their exposure to solar radiation. Heinrich (1979) has shown that apparently unpalatable caterpillars are much more likely than cryptically colored species to feed continuously in exposed positions, and he suggested that this is possible only because of reduced predation on the former group. Also, brightly colored Hyles caterpillars thermoregulate more

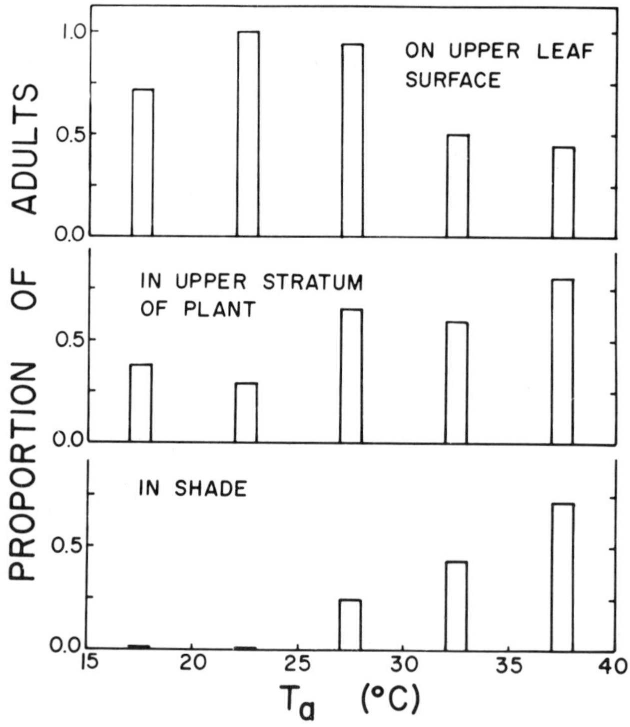

Figure 4
Proportion of adult potato beetles found in field on morphological upper surface of leaves (upper panel; specimens on stems were excluded from calculations), in upper third of plants (middle panel), and in shade (lower panel), within indicated ranges of T_a.

effectively than do cryptic Manduca (Casey, 1976). Colorado potato beetles are certainly aposematic and toxic to some predators (Hsiao and Fraenkel, 1969), properties that may have been important in the evolution of their thermal strategies.

Larvae were found more often in the inner layers of foliage as T_a increased (Figure 5), perhaps another adaptation for reducing exposure. Adults showed no such tendency but apparently moved into the upper third of the plant at higher T_a (Figure 4). This behavior may reflect a diurnal movement unrelated to thermoregulation, but it could also serve to increase exposure to wind and low humidity, as similar behavior does in locusts (Waloff, 1963). Finally, larvae, and especially adults, were much more likely to be found in the shade at high T_a, apparently also an adaptation to reduce heat load. Unlike many other insects (May, 1979) there was no clear relation between body orientation and solar azimuth.

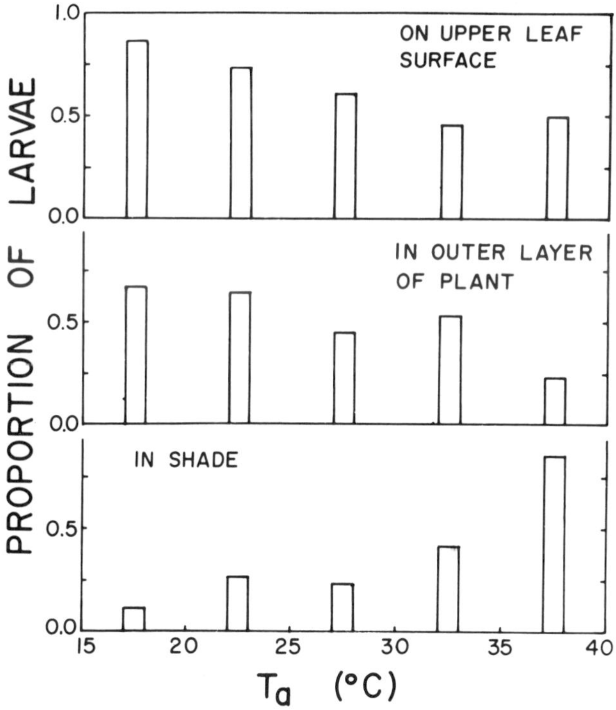

Figure 5
Proportion of larval potato beetles found in field on upper surface of leaves (upper panel), outer layer of foliage (middle panel), and shade (lower panel), within indicated ranges of T_a.

TEMPERATURE, POPULATION DYNAMICS, AND FOOD CONSUMPTION

The biological role of thermoregulation is still uncertain, but it is possible to point out some of the areas where significant effects might be expected and to indicate very roughly the direction and magnitude of those effects. Among the temperature-sensitive characteristics of most insects are those that bear most directly on population dynamics, survivorship and longevity, growth, and fecundity (Bursell, 1974). These in turn depend on a variety of behavioral and physiological processes, including food consumption and respiration, that may be independently influenced by temperature. From a management perspective consumption is of additional importance because it is the direct cause of damage to the host plant.

Data relating temperature to survival are comparatively scant. Figure 6 shows survival at various stages of development in northern European larvae studied by de Wilde (1950) at several constant temperatures. The apparent optimum for complete development is about 25°C.

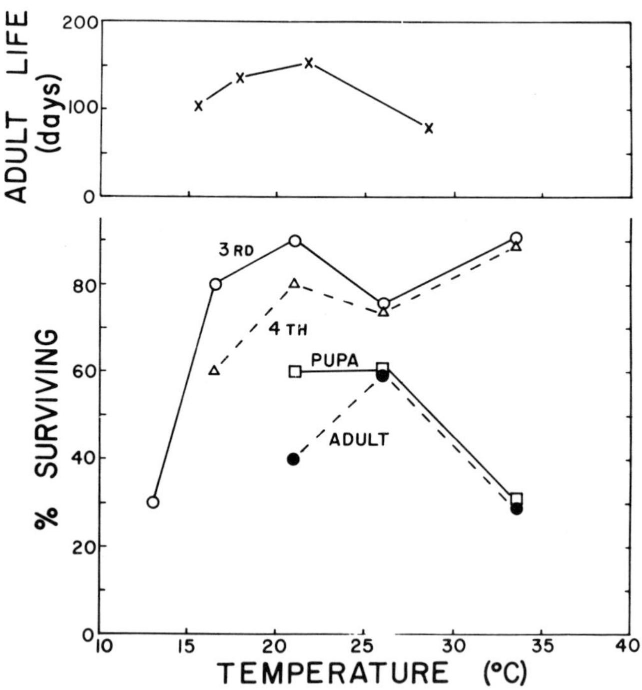

Figure 6
Average life span of female potato beetles as a function of temperature (upper panel) and percentage of larvae surviving to instar indicated as a function of temperature (lower panel) (data from Grison, 1950, and de Wilde, 1950).

Adult longevity (Figure 6; from Grison, 1950) has a rather broad maximum, probably around 20°C to 22°C. Both results suggest that elevation of T_b over T_a might be deleterious over much of the range experienced in New Jersey in midsummer because daily mean T_a must often exceed 20°C to 25°C. This conclusion may be misleading, however, since exposure to very high T_b occurs only for a few hours. The main effect of thermoregulation on survival may be an avoidance of acute exposure to lethal extremes, especially to high temperature.

Some information is available on larval growth and development, due largely to the work of Chlodny (1975) and de Wilde (1950). I have reanalyzed some of their data on larval development rate, growth, and consumption (Figures 7 and 8). Although these refer entirely to

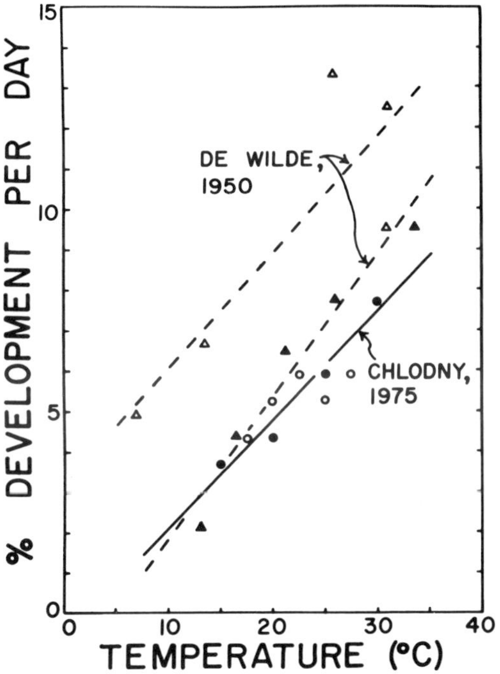

Figure 7
Developmental rate of larvae as a function of temperature. Data are from Chlodny (1975; circles and solid line) and de Wilde (1950; triangles and dashed lines). Closed symbols indicate development at constant temperature, open symbols average temperature of development under various regimes of alternating temperatures. Least-squares linear regression are: % dev. = 0.68 - 0.27T, r^2 = 0.962 (Chlodny, 1975; all data); % dev. = 1.73 - 0.35T, r^2 = 0.949 (de Wilde, 1950; constant temperatures); % dev. = 3.19 + 0.28T, r^2 = 0.732 (de Wilde, 1950; alternating temperatures).

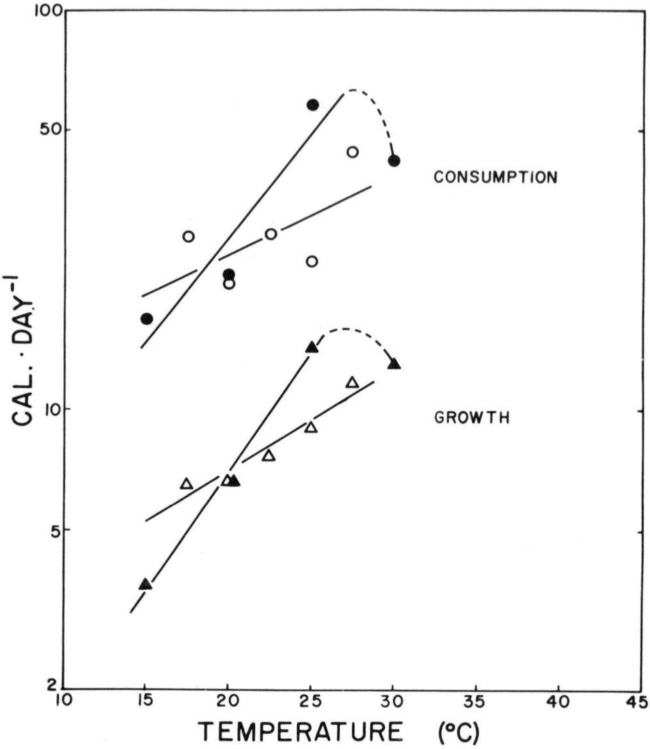

Figure 8
Potato leaf consumption (circles) and growth (triangles) by larval potato beetles. Closed symbols signify constant temperature regims, open symbols the means of 5°C or 10°C alternating temperatures. For both consumption and growth at constant temperatures, the lines drawn below 25°C are least-squares regressions on temperature; for consumption, Q_{10} *= 3.33, for growth,* Q_{10} *= 3.97. Above 25°C the lines are arbitrary. Least-squares regressions were drawn through all alternating temperature data; for consumption,* Q_{10} *= 1.57, for growth,* Q_{10} *= 1.70 (data from Chlodny, 1975).*

northern European beetles and sensitivity and temperature optima may vary geographically, it is a reasonable first assumption that the basic patterns of response are similar among populations. In brief, these patterns are: a linear increase in developmental rate with temperature; at constant temperatures a logarithmic increase of both production and consumption ($Q_{10}>3$)[1] up to maximum

[1] Q_{10} is the ratio of rates of some activity at a 10°C interval. More generally, $\log Q_{10} = [10 \log (R_2/R_1)]/(T_2-T_1)$, where R_1 and R_2 are rates at temperatures T_1 and T_2 respectively (Prosser, 1973).

somewhere between 25°C and 30°C, with a drop at higher temperatures; and at alternating temperatures a less rapid logarithmic increase (Q_{10} = 1.5-2.0) with possibly a somewhat higher optimum temperature. Additional data on leaf area consumed by adults held at constant temperatures (Grison 1950) indicate a similar pattern, with a sharp decline above the optimum (Figure 9). By contrast, de Wilde et al. (1969) reported maximal consumption at 30°C during four-hour tests when beetles were fed on potato and *Solanum dulcamara* and at least as high as 35°C on *S. dulcamara* alone. These authors also showed that food preference is temperature dependent; when given a choice between potato and *S. dulcamara* the beetle chose potato more often at low temperature but *S. dulcamara* above 30°C

Two sorts of data suggest that food consumption by New Jersey beetles may increase up to body temperatures of at least 35°C. First, the biting rate of both adults and larvae increased throughout the range of 20°C to 35°C (Figure 10). Second, the proportion of larvae feeding increased markedly with temperature to well above 35°C (Figure 11). Taken together these facts strongly imply that short-term consumption rates may increase up to T_b = 35°C.

Unfortunately most of the data on larval temperature responses fall considerably short of encompassing the full range of temperature or of variation in temperature

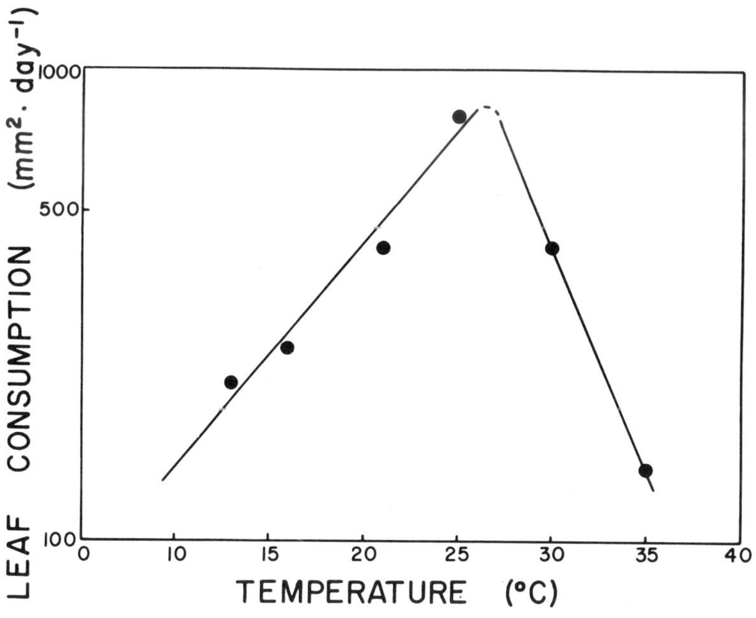

Figure 9
Potato leaf consumption by adult potato beetles. Line drawn as described in Figure 9; Q_{10} = 2.94 below 25°C (data from Grison, 1950).

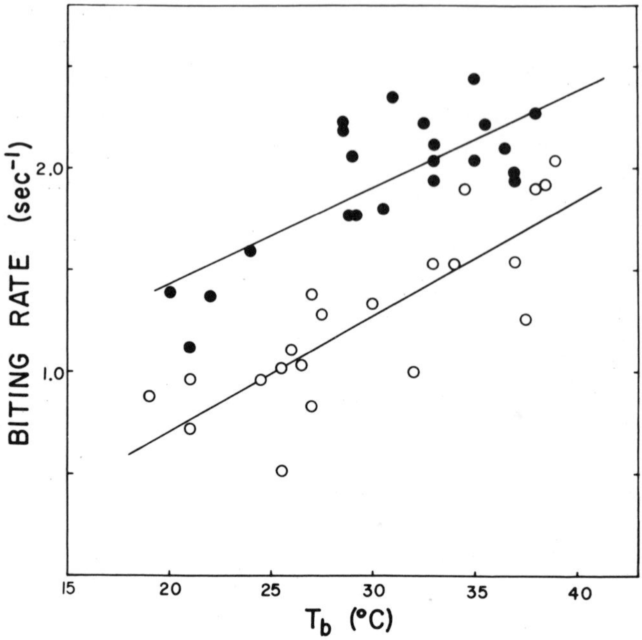

Figure 10
Biting rate of adult (lower line, open circles) and larval (upper line, closed circles) potato beetles, feeding on potato in field, as a function of T_b. Time required for ten or twenty mandible movements in continuously chewing individuals was converted to bites per second. Least-squares linear regression are: adults, rate =-0.434 - 0.057 T_b, r^2 = 0.702, n = 22; larvae, rate = 0.487 + 0.047T_b, r^2 ≐ 0.604, n = 22.

likely to be experienced by individual beetles in the field in New Jersey. Thus it is impossible to assess the effect of temperatures in the upper range to T_b or the possible benefits of thermoregulation in avoiding potentially harmful high T_b. I will assume, therefore, that only average T_b over the whole day is of consequence and consider only cases where this is below optimum. If larvae maintain the average temperature differentials hypothesized from Figure 3, the effects may be estimated from figures 7 through 9. A 1.5°C elevation of T_b by larvae should increase developmental rate by 0.41 to 0.55 percent per day, depending on the data set used; that is, duration of larval period could be reduced by about 1.5 days near T_a = 20°C. Food consumption of larvae could increase 7 to 20 percent, based on alternating and constant temperatures, respectively; the former corresponds more closely to natural conditions. Larval growth rate might increase 8 to 23 percent under the same circumstances. Adult consumption could be augmented by

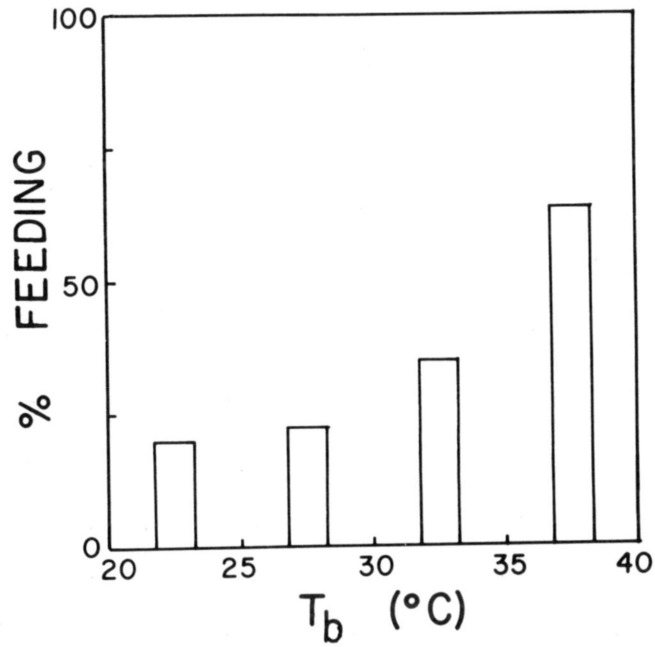

Figure 11
Proportion of larval potato beetles found feeding in field during T_b determinations, within indicated ranges of T_b.

as much as 46 percent. The estimate of mean $T_b - T_a$ might well be unrealistically low under cool, ambient conditions with sunlight available. Thus it appears that the temperature excess of beetles in the field has a potentially significant effect on the parameters considered. The significance of the ability to control T_b, as opposed to merely maintaining $T_b - T_a$, is still unclear, but sharp optima in growth and consumption rates must encourage selection for thermoregulatory behavior.

Another important component of the energy budget of L. decemlineata is respiratory metabolism. Gromadzca (1968) showed that oxygen consumption increased with T_b from 15°C to 30°C, with Q_{10}'s of 2.0 to 2.6 for various larval instars and 2.7 adults (Figure 12), values typical of many other insects (Keister and Buck, 1974). Such Q_{10}'s imply increases of 11 to 15 percent for larvae and 42 percent for adults, assuming mean temperature excesses at 1.5°C and 3.5°C. This is not an unmixed blessing. Increased resting metabolism may imply greater capability for activity, but it also means more rapid loss of usable energy. An important potential benefit of thermoregulation could be to keep metabolism within acceptable limits, but at present we have no way of knowing what these limits might be.

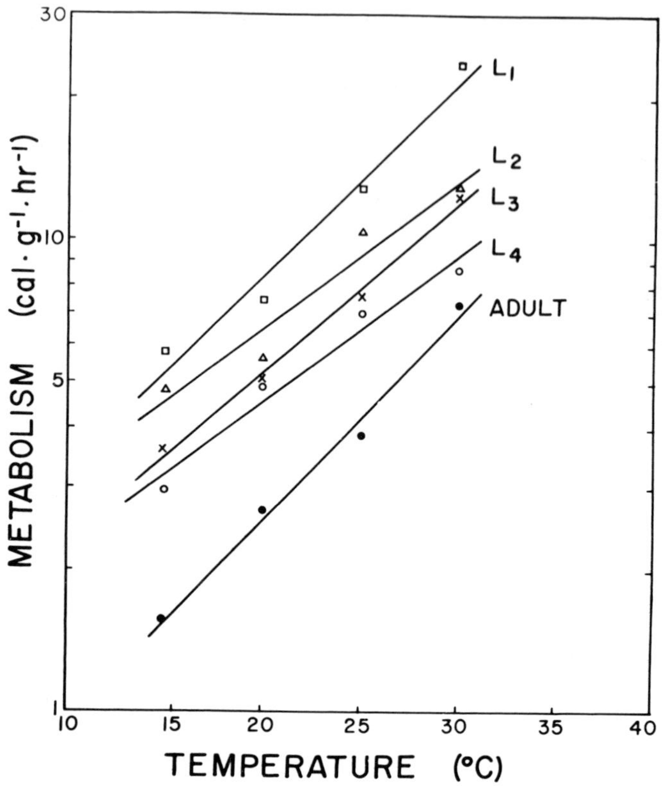

Figure 12
Energy metabolism of various larval stages and of adult potato beetles. Values converted from oxygen consumption measurements assuming a conversion factor of 4.7 cal. = 1.0 ml O_2. The Q_{10} values for each stage are: L_1, 2.60; L_2, 2.05; L_3, 2.29; L_4, 2.04; adult, 2.73 (data from Gromadzka, 1968).

Finally, temperature apparently has an extremely marked effect on fecundity. Figure 13 shows the total number of eggs produced during the lifetime of a female, as well as the number of eggs per day, at various constant temperatures (data from Grison, 1950). A very steep logarithmic increase occurs in both quantities below approximately 25°C (eggs per day may be maximal at an appreciably higher temperature; the curve above 22°C is drawn arbitrarily). Thus total and daily egg production could be increased as much as 175 percent and 123 percent, respectively, by maintenance of an average temperature excess of 3.5°C. In addition, onset of oviposition might be considerably accelerated at higher temperature. In preliminary experiments conducted in our laboratory, oviposition had not begun twenty-three days

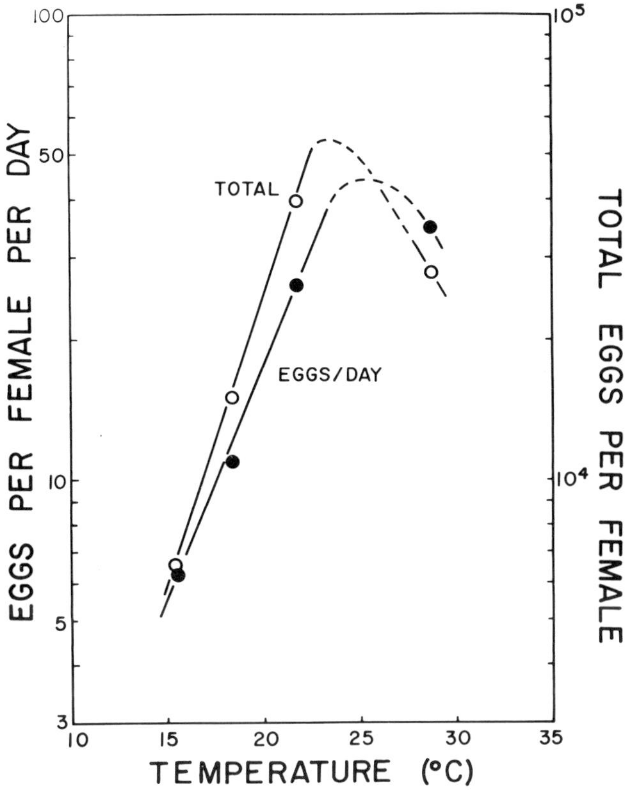

Figure 13
Total fecundity (eggs per female per lifetime; open symbols) and reproductive rate (eggs per female per day) in potato beetles at various constant temperatures. Lines drawn as described in Figure 9. For total fecundity, Q_{10} = 18.12 below 22°C; for reproductive rate, Q_{10} = 9.90 below 22°C (data from Grison, 1950).

after emergence at 15°C, although the same beetles commenced oviposition two days after being shifted to 25°C. At 20°C oviposition began after nine days, at 25°C after five days, and at 30°C after four days. These same experiments resulted in average rates of oviposition of zero eggs per day at 15°C, 8.8 at 20°C, 22.9 at 25°C, and 13.7 at 30°C. The 30°C data may have been depressed by the extremely rapid depletion and deterioration of the potato leaves at this temperature. de Wilde (1950) noted no oviposition at 17°C or below, but onset occurred after five, four and three days at 21°C, 25°C, and 30°C, respectively, slightly more rapidly than ours, again suggesting that New Jersey beetles may be adapted to higher T_b than are those in Europe.

It appears that oviposition rate and time of onset

may have maxima and minima, respectively, at slightly higher T_b than the optimum for total fecundity. This is consistent with findings in several other insects (Bursell, 1974). It is of interest to consider briefly the possible significance of this difference. In some regions climatic conditions may be such that Leptinotarsa can readily produce two, but never more than two, generations in a season. In that case it may be to the advantage of a female to produce the maximum number of offspring during her lifetime, with little regard for how long it takes. On the other hand, if a partial third generation is possible (as in New Jersey), selection may favor rapid reproduction at the cost of some diminution in total immediate offspring; that is, females that could reproduce fast enough for their late-season offspring to mature and overwinter might make a greater ultimate contribution to the gene pool than females that produced a larger number of eggs but some of whose offspring had not transformed by the end of the season. Differences in thermoregulatory behavior could be of crucial importance in different reproductive strategies of this sort since an increase or decrease of a few °C in average T_b could make the difference between operating at maximum reproductive rate or maximum total fecundity.

CONCLUSIONS

The body temperature of both adult and larval potato beetles is often elevated several degrees above air temperatures to which they are immediately exposed. In addition, there is clear evidence that adults exert a degree of control over T_b and somewhat less certain evidence that larvae, too, may thermoregulate weakly. The extent of elevation of T_b is sufficient to make potentially substantial differences in rates of food consumption, larval growth, and reproduction of the beetles. The sharp temperature optima exhibited by some of these biological characteristics make the possible selective advantages of thermoregulation obvious, but whether the degree of regulation exercised is of major quantitative importance remains to be seen.

Several types of additional information are needed for a more thorough analysis of the role of body temperature and its control. First, field measurements of T_b under a greater variety of environmental conditions, particularly early in the season, should be undertaken. Second, more data on the influence of both constant and fluctuating temperatures on consumption and population parameters are required; these should encompass the entire range of body temperatures and temperature variation to be expected in the field. Finally, the interaction of temperature with other physical and biotic factors, especially the amount and quality of food (House, 1966), should be investigated.

REFERENCES

Bakken, G. S., and D. M. Gates, 1975, Heat-Transer Analysis of Animals: Some Implications for Field Ecology, Physiology, and Evolution, in D. M. Gates and R. B. Schmerl, eds., Perspectives of Biophysical Ecology, pp. 255-290, Springer-Verlag, Berlin.

Bartholomew, G. A., and T. M. Casey, 1977, Endothermy during Terrestrial Activity in Large Beetles, Science 195:882-883.

Bartholomew, G. A., and B. Heinrich, 1978, Endothermy in African Dung Beetles during Flight, Ball Making and Ball Rolling, J. Exp. Biol. 73:65-83.

Bursell, E., 1974, Environmental Aspects--Temperature, in M. Rockstein, ed., The Physiology of Insecta, 2:1-41, Academic Press, New York.

de Wilde, J., 1950, Développement embryonnaire et postembryonnaire du Doryphore (Leptinotarsa decemlineata Say) en fonction de la température, 8th Int. Congr. Entomol., Stockholm, pp. 310-321.

de Wilde, J., W. Bongers, and H. Schooneveld, 1969, Effects of Hostplant Age on Phytophagous Insects, Ent. Exp. Appl. 12:714-720.

Casey, T. M., 1976, Activity Patterns, Body Temperature, and Thermal Ecology in Two Desert Caterpillars, Ecology 57:485-497.

Chlodny, J., 1975, Bioenergetics of the Larval Development of the Colorado Beetle, Leptinotarsa decemlineata (Say) in Relation to Temperature Conditions, Ann. Zool., Pol. Akad. Nauk. 33:149-187.

Cloudsley-Thompson, J. L., 1970, Terrestrial Invertebrates, in G. C., Whittow, ed., Comparative Physiology of Thermoregulation, pp. 15-77, Academic Press, New York.

Edney, E. B., 1971, The Body Temperatures of Tenebrionid Beetles in the Namib Desert of Southern Africa, J. Exp. Biol. 55:253-272.

Grison, P., 1950, Influence de la température sur l'activité du Doryphore (Leptinotarsa decemlineata Say) au stade imaginal, 8th Int. Congr. Entomol., Stockholm, pp. 226-234.

Gromadzka, J., 1968, Respiratory Metabolism of the Colorado Beetle (Leptinotarsa decemlineata Say), Ekol. Pol. A 16:375-383.

Hamilton, W. J., 1971, Competition and Thermoregulatory Behavior of the Namib Desert Tenebroinid Genus Cardiosis, Ecology 52:810-822.

Heinrich, B., 1974, Thermoregulation in Endothermic Insects, Science 185:747-756.

Heinrich, B., 1979, Foraging Strategies of Caterpillars: Leaf Damage and Possible Predator Avoidance Strategies, Oecologia 42:325-337.

Henwood, K., 1975, A Field Tested Thermoregulation Model for Two Diurnal Namib Desert Tenebrionid Beetles, Ecology 56:1329-1342.

Holm, E., and E. B. Edney, 1973, Daily Activity of Namib Desert Arthropods in Relation to Climate, Ecology 54:45-56.

House, H. L., 1966, Effects of Temperature on the Nutritional Requirements of an Insect, *Pseudosarcophaga affinis* Auct. nec Fallen (Diptera: Sarcophagidae), and Its Probable Ecological Significance, Ann. Ent. Soc. Amer. 59:1263-1267.

Hsiao, T., and G. Fraenkel, 1969, Properties of Leptinotarsin: A Toxic Hemolymph Protein from the Colorado Potato Beetle, Toxicon 7:119-130.

Keister, M., and J. Buck, 1974, Respiration: Some Exogenous and Endogenous Effects on Rate of Respiration, in M. Rockstein, ed., The Physiology of Insecta, vol. 6, Academic Press, New York.

Leston, D., J. W. S. Pringle, and D. C. S. White, 1965, Muscular Activity during Preparation for Flight in a Beetle, J. Exp. Biol. 42:409-414.

May, M. L. 1976, Thermoregulation and Adaptation to Temperature in Dragonflies (Odonata: Anisoptera), Ecol. Monogr. 46:1-32.

May, M. L., 1978, Thermal Adaptations of Dragonflies, Odonatologica 7:27-47.

May, M. L., 1979, Insect Thermoregulation, Ann. Rev. Entomol. 24:313-349.

Newport, G., 1837, On the Temperature of Insects and Its Connexion with the Functions of Respiration and Circulation, Proc. Roy. Soc. Lond. 3:479-482.

Prosser, C. L., 1973, Comparative Animal Physiology, W. B. Saunders, Philadelphia.

Seymour, R. S., 1974, Convective and Evaporative Cooling in Sawfly Larvae, J. Insect Physiol. 20:2447-2457.

Shephard, R. F., 1958, Factors Controlling the Internal Temperature of Spruce Budworm Larvae, *Choristoneura fumiferana* (Clem.), Can. J. Zool. 36:779-786.

Sherman, P. W., and W. B. Watt, 1973, The Thermal Ecology of some *Colias* Butterfly Larvae, J. Comp. Physiol. 83:25-40.

Sullivan, C. R., and W. G. Wellington, 1953, The Light Reaction of Larvae of the Tent Caterpillars, *Malacosoma disstria* Hbn., *M. americanum* (Fab.), and *M. pluviale* (Dyar), Can Entomol. 85:297-310.

Uvarov, B., 1977, Grasshoppers and Locusts: A Handbook of General Acridology, vol. 2, Centre for Overseas Pest Research, London, 588p.

Waloff, Z., 1963, Field Studies on Solitary and *transiens* Desert Locusts in the Red Sea Area, Anti-Locust Bull. 40:1-93.

Wellington, W. G., 1950, Effects of Radiation on the Temperature of Insectan Habitats, Sci. Agric. 30:209-234.

Wellington, W. G., C. R. Sullivan, and W. R. Henson, 1954, The Light Reactions of the Spotless Fall Webworm, *Hyphantria textor*, Can Entomol. 86:529-542.

Note: This work was performed as part of NJAES Proj. No. 08194 supported by the N.J. Agricultural Experiment Station and Hatch Act Funds.

Estimating and Projecting Colorado Potato Beetle Density and Potato Yield Loss

Patrick A. Logan

The Colorado potato beetle (CPB), *Leptinotarsa decemlineata* Say, is the familiar yellow and black chrysomelid pest of potatoes and other solanaceous hosts. Extensive resistance to insecticides and a tendency toward continual monoculture make the CPB the major potato defoliator throughout most of its range. Development of a quantitative pest management program for the CPB on potatoes will be a stepwise process, eventually culminating in a recommendation algorithm of sufficient complexity to provide reasonably accurate projections of insect density and crop response under an assortment of management options.

This chapter is a first step in that process, an empirical consideration of three questions that arise when making a CPB management decision: how to estimate CPB densities, how to predict future densities, and how to predict the effect of these densities on yield. The information presented on density estimation corroborates and expands previous work by Harcourt (1963, 1964). The density and yield models are original attempts to summarize gross aspects of system behavior, prior to careful investigation of the components of that behavior. The models may also serve to guide initial spray decisions on early maturing varieties.

METHODS

This study was conducted in 1978 and 1979 at the University of Rhode Island (URI) Agronomy Farm, Kingston. The early-maturing potato variety, Superior, was planted May 3 both years in 4 row x 4 meter (m) plots with a tilled soil buffer of 6.5 m between plots. A gradient of densities was established by collecting and counting all adults twice weekly and then redistributing them in numbers based on a reference plot. In 1978, densities of 3, 1, 0.5, and 0.05 times reference plot density were established. In 1979, ratios were 1, 0.3, 0.1, and 0.03

times reference. The sedentary behavior of adults plus the wide buffer zones helped maintain the gradient, affording a range of subsequent egg and larval densities.

Twice weekly, eggs and larvae were counted on thirty stems in each plot (about 17 percent of the stems in an average plot). For adults, the entire plot was the sample unit. Yield was assessed by grading, counting, and weighing the tubers in each of the four rows in each plot. Yield estimates in 1979 were adjusted prior to analysis to account for unusually poor stand establishment.

DENSITY ESTIMATION

A pest management decision begins with an estimate of pest density. The reliability of the estimate may be described in terms of sample variance, and generally it may be increased by increasing sample size. Use of density estimates in modeling may require normalizing or variance stabilizing transforms derived from knowledge of the underlying count frequency distribution. For the CPB, all of these topics have been discussed by Harcourt (1963, 1964). Because of the importance of reliable CPB sampling and because of the historically significant role taken by the potato beetle in the literature on statistical treatment of insect count data, it is appropriate to include a brief review here and to offer some refinements made possible by our field studies.

Variance to Mean Relation

The variance (s^2) to mean (m) relation provides a useful index of insect spatial pattern and tendency to aggregate. The relations for CPB egg mass and larval counts are illustrated in Figure 1, representing 481 egg and 471 larval samples (thirty stems per sample, zero-density samples ignored), taken in 1979. Harcourt (1963) used an exponential function (Taylor, 1961) to characterize this relation as

$$s^2 = am^b \tag{1}$$

where a and b are derived from

$$\log(s^2) = a_o + b \log(m) \tag{2a}$$

and

$$a = 10^{a_o}. \tag{2b}$$

Application of Equation 1 to the Figure 1 data yields

$$s^2 = 1.31\ m^{1.07} \tag{3a}$$

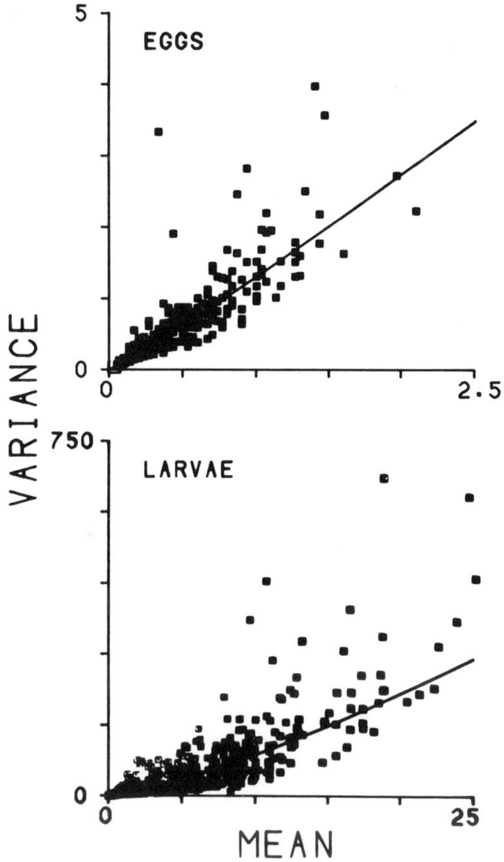

Figure 1
Variance to mean relation for CPB egg masses and larvae. Taylor's power function, $s^2 = am^b$, is applied.

for egg masses and

$$s^2 = 4.32\ m^{1.306} \tag{3b}$$

for larvae. (Values of r^2 for the log-log regressions were 0.93 and 0.85, respectively.)

Equation 1 is appealing for its simplicity and apparently general applicability. In using it to describe CPB aggregation, however, a note of caution is advised. Although Equation 1 provides a reasonable fit to the Figure 1 data, the s^2 to m relation actually changes faster than is indicated by the b of the equation. This is visible in Figure 1, but it is easier to see if the data are stratified, with Equation 1 fit within each strata. For the larval data, for example, and using five strata ($0 < m \leq 5$, $5 < m \leq 10$, . . . , $m > 20$), Equation 1 yields b's of 1.32, 1.12, 1.87, 2.6, and 4.4, respectively. Groups of samples from lower-density populations tend to produce lower estimates of b. This accounts for the disparity between Harcourt's (1963) estimate of $b = 1.45$ for larval stem samples (consistent with URI estimates of $b = 1.45$ and 1.41 for 1977 and 1978 studies)[1] and the estimate of $b = 1.306$ (SE = 0.025) of the 1979 URI study. The lower value of the latter estimate reflects the fitting of a collection of samples from a lower-density population and is not a sufficient indication of any fundamental population change (that is, toward less aggregated). This shortcoming may apply to other insects (Logan, 1980) and should be kept in mind when comparing sets of population statistics.

Negative Binomial

Another useful way to describe the frequency distribution of CPB larval counts is by fitting the negative binomial. This is a skewed distribution with a variance-to-mean relation of

$$s^2 = m + m^2/k. \tag{4a}$$

Beall (1942) used the similar

$$s^2 = m + Km^2 \tag{4b}$$

to describe the CPB variance to mean as a prelude to developing normalizing transforms. Evans (1953) fit the negative binomial to counts of CPB larvae, leaving open, according to Harcourt (1963), the question of the validity of using data that were not separated by instar. Using such data permits the possibility that populations with differing spatial patterns (depending on instar-specific behavior) were being superimposed.

There are several ways of estimating the negative binomial parameter, k, including use of Equation 4a (Anscombe, 1949), but the statistically most efficient is Fisher's (1953) maximum likelihood estimate (MLE). To test whether the negative binomial adequately represents CPB count data, the 1979 egg and larval samples were individually fit to expected negative binomial frequencies derived from sample mean and sample MLE k, and goodness of fit was evaluated using chi-square (all computations according to Davies, 1971). For many samples it was impossible to estimate an MLE k, as for example when variance and mean are equal (see Davies, 1971, for other criteria). This eliminated 202 egg mass and 21 larval samples from the analysis. Many of the remaining samples had inadequate degrees of freedom for testing, dropping another 191 egg and 53 larval samples.

---

[1] 1977: 247 samples, 12 stems per sample, $r^2 = 0.89$ (log-log);
1978: 317 samples, 30 stems per sample, $r^2 = 0.94$ (log-log).

This was so even though the usual expected cell size minimum of $n = 5$ (Steel and Torrie, 1960, p. 350) was replaced by $n = 1$. Of 88 testable egg samples, 77.3 percent were not significantly different from the negative binomial($\alpha = .1$), a weak indication of agreement with this distribution. Egg masses apparently are more uniformly distributed than would be expected under the negative binomial. Of the 397 testable larval samples, 88.2 percent were not different from the negative binomial, a more substantial indication of agreement with this distribution.

Harcourt (1963) determined that negative binomial distributions of larval samples did not have a common k. He demonstrated that k changed with density for egg masses, four larval instars, and adults by simply stratifying the density levels, the higher density strata having a higher common k in each of the six cases (see Anscombe, 1949; Bliss and Owen, 1958, for methods of calculating common k). For the 450 URI larval samples for which it was possible to estimate MLE k's, the relation

$$\text{MLE } k = .689 + 1.017 \log(m) \tag{5}$$

($r^2 = 0.15$, $n = 448$ (two outliers dropped), F significant at $p < 0.001$) supports Harcourt's claim for dependency of k on m.

Finally, the question of whether changes in the underlying distribution were attributable to age class was partially resolved by Harcourt (1963), who showed slightly lower common k estimates for first instars but no consistent trend across all four instars, over two density strata. The URI data were not divided into age classes, but there was no significance to change in first-generation MLE sample k's over time, supporting a hypothesis of no change in spatial pattern between age classes. Thus although it may be of interest to quantify the change in aggregation of first instars dispersing away from the egg mass cluster, there appears to be little reason to look much further for meaningful changes in pattern due to changes in age structure.

Normalizing Transforms

Negative binomially distributed CPB larval count data may be normalized by any of several transforms. Harcourt (1963) recommended $z = \log(y_i + 1)$ as best for CPB larvae (y_i = stem count), and this is usually advised prior to analysis of variance (Beall, 1942). Even when the overall correlation between variance and mean is reduced by the transforms, however, there is likely to remain a residual correlation at low densities (Taylor, 1970; Logan, 1980). The implications of this problem have not yet been resolved. Conclusions about highly significant effects in the analysis of variance of untransformed data are unlikely to be altered by changes introduced by transformation, but conclusions about less significant effects may be altered (Beall, 1942).

Sample Size

Although variance increases with sample mean, the coefficient of variation (CV) of CPB egg mass and larval samples,

$$CV = 100 \cdot s/m \tag{6}$$

(where s is sample standard deviation), declines with the mean, indicating that fewer samples will be needed for a given degree of accuracy (that is, standard error within a preset percentage of the mean) when the samples are taken from a higher-density population. As a useful approximation (ignoring biases introduced by the skewness of the underlying distribution), the necessary sample size (n) for SE within p percent of m is

$$n = (CV \cdot m/100 \cdot SE)^2 = (CV/p)^2 \tag{7}$$

as illustrated in Figure 2.

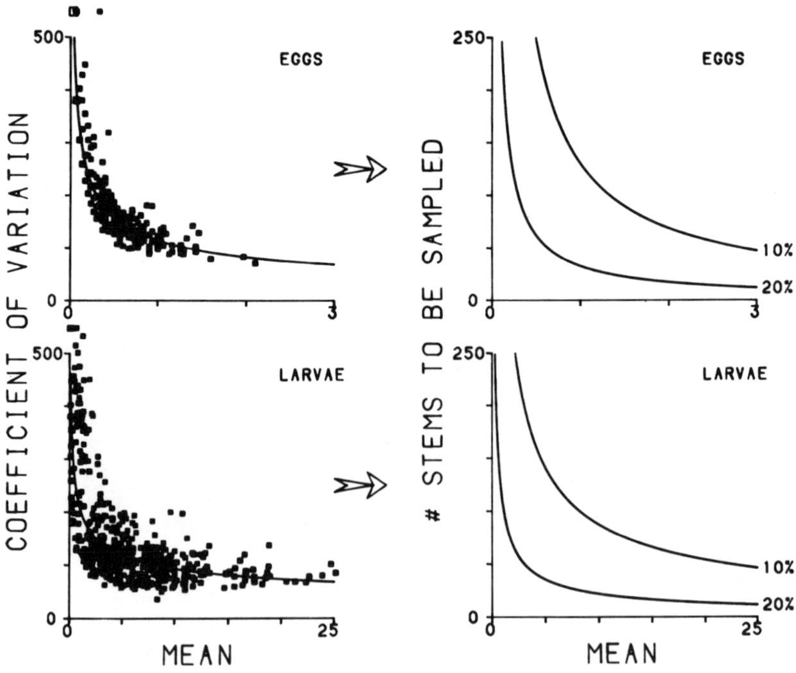

Figure 2
Relation between coefficient of variation and mean CPB egg masses and larvae and estimated number of stems to be sampled for standard error to be within 10 or 20 percent of mean.

PREDICTING DENSITY

Predictions of insect population dynamics require an accurate estimate of initial density and a predictive model that closely approximates actual conditions. This section discusses the second part of the predictive problem, building models.

There are two quantitative approaches to modeling: the component, or mechanistic approach, and the empirical approach. The component model focuses on individual parts of the system, describing system behavior as a combined action of these parts. The empirical model overlooks component-level details, focusing only on the net system response to key variables, which in practice involves merely fitting equations to data. Empirical models may be used in turn to describe subcomponents, and science is frequently a mixture of the two types of modeling.

At this stage in worldwide research on the CPB, it is useful to develop an empirical model of system behavior that will summarize gross aspects of the numerical behavior of larval density over time and the effect of CPB density on potato yield. The result should serve as a prelude to further research into the subcomponents of this system.

Incidence Curves

CPB larval density changes in an orderly relation to accumulated degree days (DD), as seen in twelve 1978 URI plots (Figure 3). The changes can be represented by a series of bell-shaped, skewed incidence curves (Figure 3, dashed lines) of the form

$$y = e^{b_0 + b_1 DD + b_2 DD^2 + b_3 DD^3} - 1 \qquad (8)$$

where y is count per stem and the coefficients, b_i , are estimated from log-log regressions. This series is marked by changes in shape and a shift in time of peak density, changes that are related to the integral of the curve (shown as the area under it). Possible reasons for this density-related shift include changes in oviposition rate, egg cannibalism, and starvation, all associated with density-related changes in defoliation rates.

The orderly change in peak densities and curve shapes in Figure 3 is the result of changes in the curve coefficients, which are plotted over the integral of their respective curves (A), expressed as CPB DD, in Figure 4. For curves to be skewed to the right, signs of the four coefficients must be −, +, −, and +, respectively. Although some of the curves in Figure 3 are skewed to the left (signs +, −, +, −), this is probably an artifact of shortcomings in the adult manipulation technique.

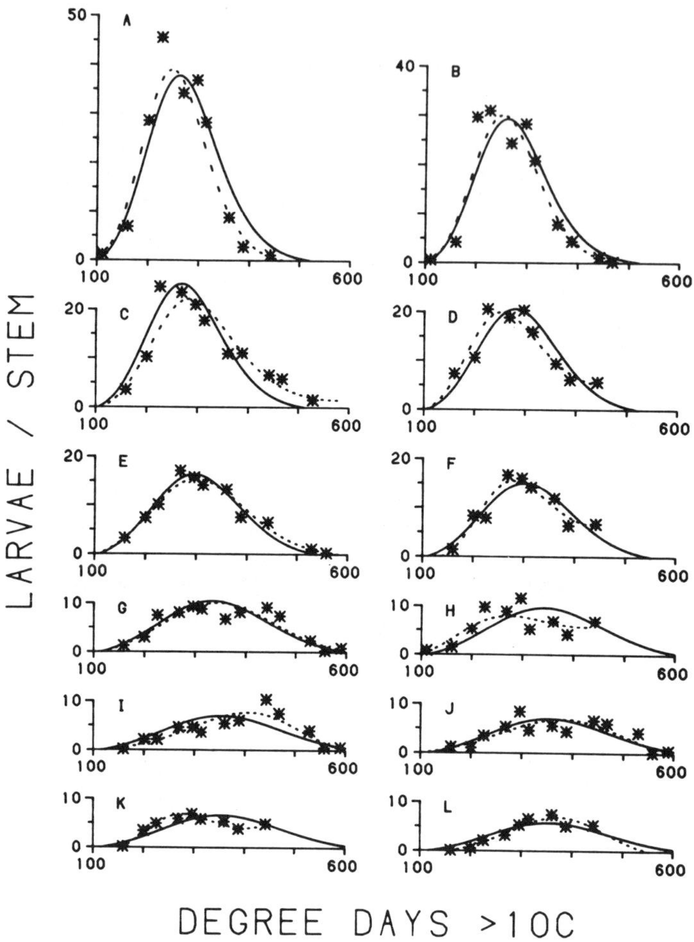

Figure 3
Application of model (solid line) to 1978 data. Regression, Equation 8, is shown as dotted line. Plots D, F, H, K, and L were not included in model development.

Four power curves

$$b_i = a_i A^{c_i}, \quad i = 1, \ldots, 4, \tag{9}$$

represented by the solid curves in Figure 4, were fit to the coefficients of the five complete curves of Figure 3 that had coefficients with proper signs. (Five of the incidence curves--D, F, H, K, L--were terminated prematurely with insecticides and are not included in the analysis.) A log transform using absolute values of the b's was used to estimate a_i and c_i. Equation 9 provided the required shifts in peak and shape, but application of the power curve over all A's led to complications: time of peak density approached infinity as A approached zero, and curves failed to descend to zero at high densities.

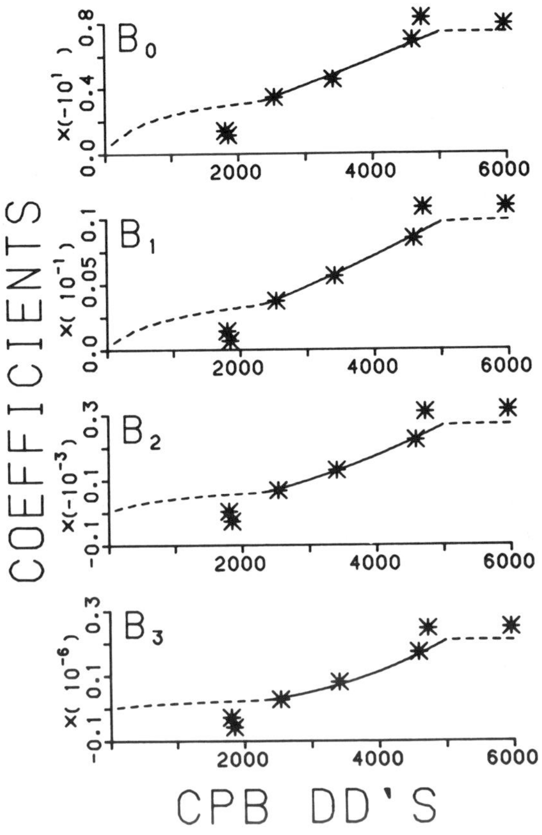

Figure 4
Relation between curve (Equation 8) coefficients and area under incidence curve.

Also the actual area of the curve (A'), defined by the integral of Equation 8 using the *b*'s of Equation 9, began to deviate from the intended area (A) at very low and very high densities. For these reasons, the power curves in Equation 9 were bounded by

$$A_l \leq A \leq A_u, \quad A_l = 2324, \; A_u = 5024. \tag{10}$$

A_l and A_u were chosen to limit the range of peak densities to 260 and 360 *DD* (the approximate range in Figure 3) with $A_l' = A_l$ and $A_u' = A_u$. Outside of these bounds, the family of curves is defined by

$$\hat{y} = \delta \hat{y}_B \tag{11a}$$

where $\hat{y}_B$ is the curve at either A_l or A_u and

$$\delta \leq 1, \text{ if } A \leq A_l \tag{11b}$$

and

$$f \geqq 1, \text{ if } A \geqq A_u. \tag{11c}$$

Several sets of points on curves with area $0 < A \leqq A_1$ or $A_u \leqq A < 6000$ were generated using Equations 8 through 11 and were subjected to regression, yielding the dotted curve sections of Figure 4.

The curves of Figure 3 thus may be represented as three subfamilies. At lowest densities, density-dependent effects are ignored. Since all low-density curves are proportioned to a single reference curve, they have equivalent peak time and skew characteristics. The high-density group of curves is similar, under the assumption that further density-related changes in peak time and skew have little practical significance.

A Search Technique

One or more sets of field counts, each associated with $DD > 10°C$ since crop emergence, may be used to form a seasonal projection of the incidence curve through proper selection of one of the curves described previously. The selection process consists of an iterative fitting procedure. First, the appropriate subfamily of curves--$A \leqq A_1$, $A_1 \leqq A \leqq A_u$, or $A > A_u$--is determined. A trial curve of $A = A_1$ (2324 CPB DD), with coefficients defined by Equation 9 and the curve itself defined by Equation 8, is evaluated first. For each observed density, p_i (here, the mean of thirty stem counts), taken on DD_i, a corresponding point on the trial curve, $\hat{y}_i$, is computed. Errors, or differences between observed and trial curve densities,

$$\varepsilon = p_i - \hat{y}_i \tag{12}$$

are classified as positive $(p_i \geqq \hat{y}_i)$ or negative $(p_i < \hat{y}_i)$. These errors are compared using the test statistic[2]

$$\gamma = (\Sigma\varepsilon+ - \Sigma\varepsilon-)/(\Sigma\varepsilon+ + \Sigma\varepsilon-). \tag{13a}$$

A solution is sought that will yield

$$/\gamma/ \leqq \lambda \tag{13b}$$

(where, arbitrarily, $\lambda = 0.0001$). If the trial curve ($A = 2324$) satisfies Equation 13b, it is accepted as the minimum error solution. If not, and the curve is too

---

[2]Minimizing error sum of squares, $SSE = \Sigma\varepsilon^2$, by setting $dSSE/df = 0$ yields $f = \Sigma p_i\hat{y}_i/\Sigma\hat{y}_i^2$, a minimum variance but biased solution.

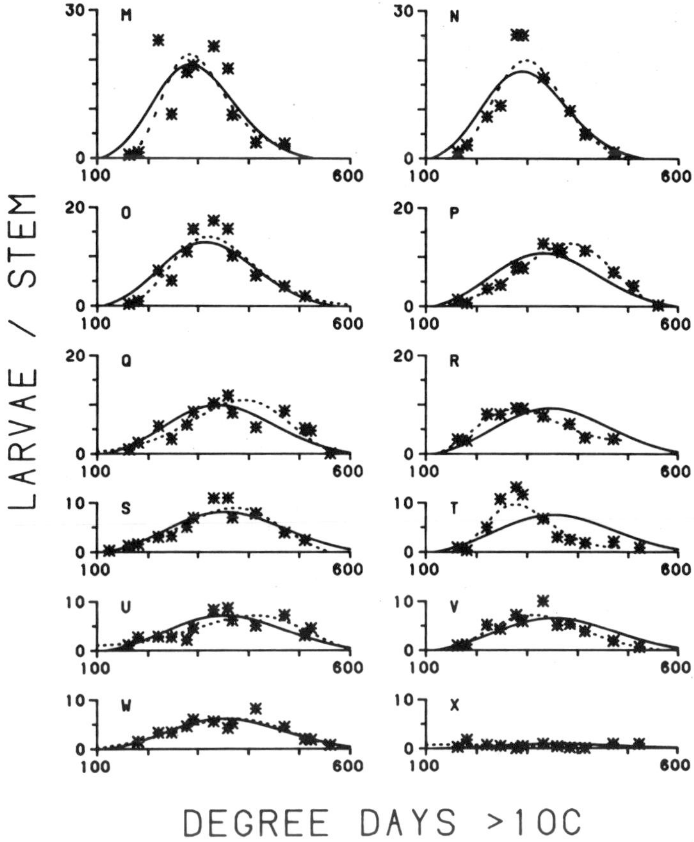

Figure 5
Model (solid line) and regression (Equation 8) applied to 1979 larval incidence data.

high, and the correct solution lies in the subfamily of curves with area in the range $0 < A \leq A_1$. This range is iteratively divided and new trial curves, Equations 11a and 11b, are evaluated, using γ to determine whether the curves are too high or too low. Convergence to a solution-satisfying Equation 13b is generally rapid.

Alternatively, using the trial curve of area A_1 could yield $\gamma < \lambda$; making the curve too low. A second trial curve of area A_u would then be evaluated, using γ and if Equation 13b is satisfied, this trial curve is taken as the desired solution. If $\gamma < \lambda$, this curve is still too low, and an iterative search is performed over the range $A_u < A < 2A_u$,[3] using trial curves defined by Equations 11a and 11c.

---

[3] In practice, a curve of $A > 10048$ CPB DD is highly improbable.

Finally, if the trial curve of $A = A_u$, if $\gamma > \lambda$, the trial curve is too high, the solution must lie in the range of curves with $A_l < A < A_u$. Again, this range of trial curves is divided, using curves defined by Equation 8 with coefficients from Equation 9, and the test statistic γ is used to converge on a solution satisfying equation 13b.

Application of the model to the original 1978 data, Figure 3 (solid line), indicates good agreement with the data from plots D, F, H, K, and L, which were not used for derivation. Figure 5 illustrates application to twelve 1979 plots, again indicating adequate agreement except for plot T. The shape of this plot was largely due to the influence of adjacent high-density plots from another experiment. Comparing the model to regressions (Equation 8) through the same data (dashed line in Figures 3 and 5) shows that although the model will be unable to improve on the reduction in sum of squared errors possible with the regression, in several cases the model makes more sense than the regression. In Figure 5, plot M, for example, the least-squares regression fails to return to zero at the end of the first generation, a problem corrected by the restrictions of the model.

Predicting Yield Loss

Factors affecting yield will vary among growing seasons. Thus although maximum expected yield in any field in any year may always be expressed as 100 percent, prediction of the absolute yield will involve extrapolation from previous field history, adjustments for weather, and other factors.

To derive a predictor of percentage yield, the area (A) under the seven complete 1978 incidence curves (regressions) of Figure 3 was integrated numerically (Euler approximation to within 0.01 percent) and compared to the plot yield (Y), in hundredweight per acre, of U.S. No. 1 tubers, giving

$$Y = 381.6 - 0.0642A \quad (r^2 = 0.95) \tag{14a}$$

(yield of seven beetle-free control plots was 397 ± 39(sd) hundredweight per acre). Using the density model to compute area instead of the regression gave

$$Y = 357.3 - 0.055A \quad (r^2 = 0.96). \tag{14b}$$

By comparison, 1979 yields, aided by favorable late season conditions and additional fertilizer, related to area under the model curves as

$$Y = 501.4 - 0.089A \quad (r^2 = 0.90,\ n = 8).^{4} \tag{14c}$$

---

[4] Four plots were planted one month early in 1979 and are not included in this analysis.

Thus, to permit comparisons, expected yield should be estimated from

$$Y = Y_m(1 - 0.000168A), \quad (15)$$

where Y_m is the grower's estimate of maximum potential yield. The rest of the right-hand side of Equation 15 is derived from Equation 14a, converting Y at $A = 0$ to 1.

DISCUSSION

In practice, density projections are extremely sensitive to errors in the original density estimates. The determination of proper sample size is critical to accurate forecasts. It is appealing to take samples as early as possible, before densities have a chance to build, but the necessary sample sizes may be quite large. A benefit of the approach given here, though, is that errors in estimating density one day are likely to be offset by subsequent samples because all samples can be used in the forecast.

The empirical model described here is presented as a first step in a continuing modeling process. It may be applied to guide control decisions until the first spray, and it is expected that a similar approach will prove useful in developing guides for subsequent decisions. The intention of the modelers, however, and greatest use of the model will continue to be a summary of overall system behavior, a system that eventually will be better understood through continual refinement of component models.

REFERENCES

Anscombe, F. J., 1949, The Statistical Analysis of Insect Counts Based on the Negative Binomial Distribution, Biometrics 5:163-173.

Beall, G., 1942, The Transformation of Data from Entomological Field Experiments So That the Analysis of Variance Becomes Applicable, Biometrika 32:243-262.

Bliss, C. I., and A. R. G. Owen, 1958, Negative Binomial Distributions with a Common k, Biometrika 45:37-58.

Davies, R. G., 1971, Computer Programming in Quantitative Biology, Academic Press, New York, 485p.

Evans, D. A., 1953, Experimental Evidence Concerning Contagious Distributions in Ecology, Biometrika 40:186-211.

Fisher, R. A., 1953, Note on the Efficient Fitting of the Negative Binomial, Biometrics 9:197-200.

Harcourt, D. G., 1963, Population Dynamics of Leptinotarsa decemlineata Say in Eastern Ontario. I. Spatial Pattern and Transformation of Field Counts, Canad. Entomol. 95:813-820.

Harcourt, D. G., 1964, Population Dynamics of Leptinotarsa decemlineata Say in Eastern Ontario. II. Population and Mortality Estimation during Six Age Intervals, Canad. Entomol. 96:1190-1198.

Logan, P. A., 1980, Spatial Distribution of Cereal Leaf Beetle (Oulema melanopus (L)) Eggs and Larvae and Treatment of Count Data, Environ. Entomol. 9:186-189.

Steel, R. G. D., and J. H. Torrie, 1960, Principles and Procedures of Statistics, McGraw-Hill, New York, 481p.

Taylor, L. R., 1961, Aggregation, Variance and the Mean, Nature, Lond. 189:732-735.

Taylor, L. R., 1970, Aggregation and the Transformation of Counts of Aphis fabae (Scop.) on Beans, J. Appl. Biol. 65:181-189.

Control of the Colorado Potato Beetle with Fungi

Donald W. Roberts, Roger A. LeBrun,
and *Maurie Semel*

Entomopathogenic fungi have played major roles in the development of insect pathology. The first pathogen of an insect to be identified as truly pathogenic was a fungus, and the first insect pathogen to be utilized in an attempt to control an insect pest was another fungus. The former was *Beauveria bassiana*, reported to be in silkworm larvae in 1835, and the latter was *Metarhizium anisopliae* utilized in the mid-to late 1880s to control cockchafers in Russia (Steinhaus, 1956, 1975). This was followed by two energetic but poorly developed efforts to control pests in the United States in the late nineteenth and

---

The research described here was supported in part by cooperative agreements between Agricultural Research, Science and Education Administration, U.S. Department of Agriculture, and Boyce Thompson Institute (58-32U4-8-6) and University of Rhode Island (58-32U4-9-89). The able technical assistance of Donna Palermo, Lori Longo, Holly Beermann, Joan Parker, and Russell Campbell is gratefully acknowledged. The support of Richard S. Soper, Agricultural Research, Science and Education Administration, U.S. Department of Agriculture Insect Pathology Research Unit at Boyce Thompson Institute, also is very much appreciated. Information provided to Donald Roberts in discussions with N. V. Lappa, O. A. Alyoshina, and A. A. Yevlakhova on the use of *Beauveria* in the Soviet Union for insect control, is also gratefully acknowledged, as is the gift of Boverin by O. A. Alyoshina and her colleagues at the All-Union Scientific Research Institute of Microbial Means of Plant Protection and Bacterial Preparations as part of the U.S./U.S.S.R. Joint Commission on Science and Technology Working Group on the Production of Substances by Microbial Means. (University of Rhode Island Experimental Station #1989.)

early twentieth centuries. These were the use of B. bassiana against chinch bugs and a group of fungi, eventually known as the friendly fungi, against scale insects in citrus. Both efforts apparently failed. After a period of little developmental work with fungi for insect control, several projects are currently in progress. Several recent reviews provide details and references to specific literature on the subject (Roberts and Yendol, 1971; Roberts and Campbell, 1977; Ferron, 1978; Roberts and Humber, 1980; Burges, 1981; and Roberts, 1981).

FUNGI UNDER DEVELOPMENT FOR INSECT PEST CONTROL

Over 750 fungi in approximately 100 fungal genera infect insects. Of these genera only about a dozen currently are under development as insect control agents (Table 1). Two (Coelomomyces and Lagenidium) are aquatic

Table 1
Fungal Genera with Entomopathogenic Species Currently Considered Promising for Insect or Mite Control

Genus	Habitat	Target Hosts
Coelomomyces	Aquatic	Mosquito (larvacide)
Lagenidium	Aquatic	Mosquito (larvacide)
Entomophthora	Foliage	Lepidoptera (caterpillars), aphids, grasshoppers
Beauveria	Foliage, soil	Coleoptera (beetles), lepidoptera
Metarhizium	Foliage, soil, aquatic	Coleoptera, mosquito, spittlebugs
Verticillium	Greenhouse	Aphid
Hirsutella	Citrus	Mites
Paecilomyces	Foliage	Coleoptera, Lepidoptera
Nomuraea	Foliage	Lepidoptera
Culicinomyces	Aquatic	Mosquito (larvacide)

fungi. Coelomomyces has been of interest for approximately thirty years as a control agent of mosquitoes, but significant development has not yet been attained. Recently it was demonstrated that one of the major problems was that the fungus has a very complicated life cycle, including an alternate host. The problem of utilization is further complicated by the fact that mass production in artificial media is not possible at present. Nevertheless, there is continued interest in several species of this genus, and projects to develop these for mosquito control are currently underway.

Lagenidium for insect control is restricted to L. giganteum. Some strains of this fungus have a very wide host range among larval mosquitoes (Culicidae).

Currently there are few field studies with this organism, but it could be of considerable promise if the problems of production, storage, and susceptibility to water pollution are solved.

Those species of Entomophthorales with promise for insect control have traditionally been placed in one genus, *Entomophthora*. Current taxonomic thinking, however, indicates that there are several genera within this so-called one genus. The Entomophthorales are Zygomycetes. The fungi involved are quite similar in many respects, but they vary widely in ability to grow on artificial media and in host range. The Entomophythorales are the fungi most commonly encountered in natural fungal epizootics in insect populations. They produce short-lived conidia, which usually are readily airborne. Some species produce resting spores that are very long lived but in most cases are difficult to germinate in the laboratory. In recent years progress has been made in producing resting spores in fermenters and in germinating the resting spores of some very interesting species, particularly one, *E. obscura*, that might be used for aphid control. Because of the new technology with resting spores, the Entomophthorales are now of considerable interest for insect control (Soper, 1978).

A number of deuteromycetous (Fungi Imperfecti) fungi are currently being developed or being seriously considered for development for insect pest management. The most widely used of these are the two major species of *Beauveria*: *B. bassiana*, usually used against foliage pests and *B. brongniartii*, used against soil insects, particularly beetles. These fungi are fairly easily produced in artificial culture.

Metarhizium anisopliae is another fungus that has been studied since the nineteenth century. Rather easily produced on artificial media, it shows considerable promise for larval mosquito control using an approach similar to that used in making insecticidal applications. The fungus does not produce conidia on submerged aquatic insects and therefore is unlikely to recycle in the mosquito population. It is most commonly isolated from soil insects, particularly Coleoptera, and it may prove useful against this increasingly important group of insect pests. The largest current project is the use of *M. anisopliae* versus spittlebugs (Cercopidae: Homoptera) in sugarcane in Brazil. The fungus is produced by grower associations with very simple technology: on sterile rice in autoclavable plastic bags. Approximately 150,000 hectares (ha) were treated in the 1979-1980 growing season. The fungus is now under development for use in spittlebug management programs being developed for the 150 million ha of pasture in Brazil. Registration has been granted to two commercial firms by the Brazilian government (Roberts and Humber, 1980).

Verticillium lecanii, until recently known as *Cephalosporium lecanii*, does not infect any type of plant, in spite of its generic name. It has been reported from scale insects and whiteflies and appears to be very effective against aphids, particularly in greenhouses (Hall, 1975). Registration of this fungus for use against greenhouse aphids in England is expected soon. (S. Lisansky, pers. comm.).

Another fungus quite far along in its commercial development is *Hirsutella thompsonii*, used against the citrus rust mite, *Phyllocoptrutta oleivora*, in the southern United States, particularly Florida (McCoy, 1980). This fungus occurs naturally in epizootics in rust mite populations, but these outbreaks generally occur too late to protect the citrus fruit. Artificial introductions (such as by sprays or dusts) will be made to initiate earlier epizootics. Methods for commercial production are under development. Safety tests have been completed, and submission for registration is possible in 1981 (T. Couch, pers. comm.).

Paecilomyces is being investigated, particularly in Eastern Europe, for the same types of environments in which *Beauveria bassiana* is used (Bajan, 1973).

Nomuraea rileyi (= *Spicaria rileyi*) occurs annually virtually throughout the world in noctuid (Lepidoptera) pests of soybeans in devastating epizootics. In some cases its presence obviates the need for insecticide use. The epizootics, however, usually occur after some damage to the plants has resulted. There is interest at present in introducing the fungus in high levels in the field a few weeks before epizootics naturally occur to hasten their onset sufficiently to protect the plants (Ignoffo et al., 1976). *N. rileyi* is not as easily mass produced as the other deuteromycetous fungi listed here, and innovative production technology is needed if this fungus is to be commercialized.

Culicinomyces clavosporus is an unusual deuteromycetous fungus in that it produces conidia readily on submerged aquatic hosts (mosquito larvae) and in submerged artificial cultures. Two isolates are known: one from Australia and one from North Carolina. Small-scale outdoor tests indicate the Australian isolate may have promise for mosquito control. A major technological problem at present is difficulty in stabilizing conidia for storage (A. W. Sweeney, pers. comm.).

INFECTION CYCLE

Nearly all of the entomogenous fungi enter their hosts by penetrating the insect's cuticle. The fungi are unique in this respect because other pathogens enter the alimentary tract. The infective unit, usually a conidium, therefore must sometimes germinate under adverse conditions in the environment. Once the conidium germinates on the host exocuticle, a resulting germ tube in some fungal species penetrates the cuticle directly and in others produces an adhesive button called an appressorium. A penetrating peg grows from this appressorium and, with some help from enzymes, forces itself into the cuticle. Although penetration is often rapid, it can take several days in wireworm larvae. Once within the hemocoel, most fungi convert from mycelial growth to a yeast phase. Toxins and enzymes can kill the host, or the host dies by depletion of nutrients or physical disruption. Under certain conditions, thick-walled long-lived chlamydospores or resting spores are produced in the hemocoel. If moist conditions continue, however, massive numbers of shorter-lived conidia are produced on the cadaver.

CONCEPTS OF MICROBIAL CONTROL

Colonization

The introduction of natural insect enemies to control arthropod pests is well documented by state and federal organizations. Few biological control projects in the past, however, have considered introduction of imported beneficial microorganisms. Most introductions and colonizations have been relegated to a redistribution of indigenous groups or those fungi accidentally introduced with the insect host. Permanent establishment of new fungal pathogens in disease-free areas allows them to spread naturally and serve as a source for greater distribution. Milner and Soper (1981) successfully introduced *Zoophthora radicans* in Australia against the recently introduced spotted alfalfa aphid. The fungus has since become established and has caused epizootics in the aphid populations (R. S. Milner and R. S. Soper, pers. comm.).

Mycoinsecticides

The preferred means of control is colonization, which is cheap and not laborious. This is not always possible, however, and a mycoinsecticide approach is needed. Successful instances of control with *Beauveria*, for example, have required massive field inoculations. Some failures with fungi in pest control experiments apparently occurred because only the colonization concept was used and too little inoculum was applied. In the past, timing for low-target population and use of high inoculum levels has been successful. Reapplication of the fungus is made each time the pest population increases to economically significant numbers. A single mass introduction each season to hasten the initiation of epizootics that normally occur but too late to protect the crop has been proposed for *Hirsutella* and *Nomuraea*. This approach is a combination of colonization with a mycoinsecticide.

Integrated Control

The integrated pest management approach is designed to take advantage of existing mortality factors and to utilize chemical pesticides in emergency situations only. In general, economic thresholds for pest insects are such that 99 percent control, such as was normally sought with chemicals, is not needed. Accordingly the integrated pest management approach has afforded an opportunity for widespread use of biological control agents because even though they may be slower acting and reduce the pest population less, they often will afford adequate population reduction while causing less devastating effects than do chemicals on the ecosystem. The fungi utilized may be naturally occurring ones encouraged by manipulation of chemicals and other field conditions, or they may be introduced. Comprehensive knowledge is needed concerning the fungus involved (including susceptibility to agricultural chemicals), the host's ecology, and epizootiology of the disease.

BEAUVERIA VERSUS THE COLORADO POTATO BEETLE

The *B. bassiana*-Colorado potato beetle (CPB) association is interesting because of the importance of both the pest and the crops it damages and because promising research results from the Soviet Union, Poland, and France indicate a high probability of success in using the fungus elsewhere. Preparations of *B. bassiana* conidia are distributed under the name Boverin in the Soviet Union. Production data are unavailable, but interest (and government support) is high. One facility distributed twenty-two tons in 1977 (Yevlakova, 1978), and a large factory that will produce the substance is under construction in Minsk. Polish workers have published widely on the *B. bassiana*-CPB association, and they report long-term as well as immediate detrimental effects on CPB populations. Results of field tests conducted in Normandy in 1978 indicated that 10^{14} lyophilized, clay-treated blastospores per hectare could afford plant protection (Fargues et al., 1980).

Strain Selection and Characterization

Isolates of *B. bassiana* from various localities and hosts are known to differ in virulence to codling moth and CPB (Lappa and Goral, 1981). In general, Lappa (1978) found isolates from the target host in the target locality to be from ten to a hundred times less virulent to the target population than exotic isolates. It is obvious that examination of a wide variety of isolates should allow selection of strains much more efficient in causing CPB mortality than naturally occurring fungi. In addition, isolates can be obtained by mutagenesis (Pekrul and Grula, 1979) or by culturing from insects that are surface sterilized several hours after exposure to fungal conidia (Al-Aidroos and Roberts, 1978).

Characterization of the most promising isolates should be carried out for unequivocal identification and to try to correlate virulence with specific traits. The latter information may facilitate the selection process. The methodologies include isozyme analysis by starch gel electrophoresis (May et al., 1979), polypeptide analysis by polyacrylimide gel electrophoresis, immunology by immunodiffusion and by enzyme-linked immunosorbant assay, and toxin production, particularly oxalic acid (Roberts, 1981). In addition, strains show morphological differences, pigment production variation, media preferences, and variation in viability retention on storage.

Fungus Production

Four methods are available for production of infective units of *B. bassiana* (Soper and Ward, 1981). First, the fungus may be grown on solid media (agar media or heat-sterilized grains such as rice, barley, or bran). In this case, conidia are produced on aerial conidiophores. Second, semisolid media production is accomplished by using wheat bran or a similar substance. The substance with the necessary nutrients is placed either in a bin or a rotating drum through which sterile air is passed. This seems to be the method of choice at present by Western industry for production of conidia of entomopathogenic deuteromycetes. Third, cultures are incubated in fermenters to a high yield of mycelia, and then the mycelia are removed to shallow trays for production of aerial conidiophores and conidia. This approach considerably hastens production; mycelium is produced in approximately two days and conidia in another five days. This is the preferred method in the Soviet Union (N. Lappa, pers. comm.) and will be utilized in the factory under construction in Minsk. Fourth, using a special medium, slightly elevated temperature, neutral pH, and low aeration, conidia of *B. bassiana* can be produced in submerged (fermenter) culture (Goral, 1972). Previously only blastospores (essentially fragments of mycelium without the thick walls of conidia) were produced in submerged culture. Blastospores have been used in field trials, but they are short-lived in relation to conidia. Longevity has been improved by drying washed blastospores in certain clays (Fargues et al., 1978). The addition of bentonite clay permitted spray drying of *B. brongniartii* conidia at 150°C (Fargues et al., 1979). Longevity of conidia is greatly improved by storing at reduced temperature (5°C or less) (Roberts and Campbell, 1977; Fargues et al., 1979). Unfortunately conidia produced in submerged culture to date have been less virulent and have reduced shelf life in comparison with aerially produced conidia (Goral, 1978; Lappa and Goral, 1981). Accordingly even though the submerged method obviates the space-consuming and physically awkward step of surface incubation, the method of choice for small-scale production is a submerged-surface (diphasic) technique.

The medium on which spores are produced can influence their virulence and longevity (Lappa and Goral, 1981; Roberts, 1974). Therefore conidia produced on other media should be compared to those produced on a standard medium to determine the optimal nutrients for mass production of isolates.

Formulation and Application

Formulation of entomopathogenic fungi for field application is still in the experimental stage. Recent studies of the effects on *Entomophthora* resting spores of several dilutents, dispersants, and wetting agents used in the chemical pesticide industry indicate that several of them may be used in formulating this fungus (Mietkiewski and Soper, 1979). These and numerous other compounds are being tested at the Boyce Thompson Institute for their suitability for formulating *B. bassiana* conidia as a wettable powder.

Foliage application protocols developed in the Soviet Union for control of CPB with *B. bassiana* apparently afford satisfactory plant protection (Lappa, 1978; Lappa and Goral, 1981). Careful population monitoring is carried out; and because young larvae are considered the most susceptible, the fungus is applied when 70 to 80 percent of the first (spring) generation is still in the first or second instar. A second application is made ten days later. The cycle is repeated for the second generation. In the Ukraine this generation produces the overwintering adults. The recommended dosage is 1.6 kilograms (kg) per ha of a fungus preparation containing 30×10^9 conidia per gram (g), or 0.4 kg of fungus plus one-fifth the normal dose of chemical insecticide. *B. bassiana* is sufficiently compatible with some insecticides used routinely in the Soviet Union in conjunction with several such chemicals (Lappa, 1978; Lappa and Goral, 1981). The soviet scientists now agree with the findings of Fargues (1975) that the effect of the combined agents is additive rather than synergistic.

Long-Term Effects

Soil is an essential environment in the developmental cycle of the CPB. Within this substrate, the insect is protected during vulnerable, inactive periods of pupation and diapause. During protracted periods of cold soil temperatures, hibernating adults have been known to spend more than one year in the soil. Because of the high probability of physical conditions appropriate to disease initiation, such as high moisture, moderate temperature, and protection from ultraviolet light, soil may prove to be one of the most rewarding habitats for use of fungal entomopathogens (Roberts and Campbell, 1977).

Frequent observation in the soil of CPB adults naturally infected by *B. bassiana* has prompted European and Soviet researchers to evaluate soil application of this fungus as a management tool. Encouraging results have been reported by Bajan et al. (1975) in Poland, Ferron (1978) and Reisinger et al. (1977) in France, and Lappa and Goral (1981) in the Soviet Union.

Long-term persistence of conidia may be the most important factor influencing the effectiveness of soil-applied *B. bassiana* on CPB populations. Wartenberg and Freud (1961) reported that the mycostatic substances excreted by various soil fungi, actinomycetes, and soil bacteria inhibited germination of *B. bassiana*. The presence of the host insects stimulated germination of spores. Exposure to some soil fungi reduced the pathogenicity of *B. bassiana* spores (Kmitowa, 1973). Clerk (1969) pointed out, however, that extracts of deep soil layers were less inhibitory than were extracts of layers near the surface. Possibly spores injected into the soil persist longer than those applied to the soil surface. In Poland, the cumulative pathogenic effects of *B. bassiana* on CPB populations in the soil persisted over two years (Wojciechowska et al., 1977).

Safety

Safety data on *B. bassiana* gathered in the Soviet Union, over more than a decade of small animal laboratory tests, examination of production personnel, and field application experience indicate that this fungus is not hazardous to warm-blooded vertebrates, and governmental permission to produce and utilize *B. bassiana* at the commercial level has been granted in that nation (Fargues and Remaudiere, 1981; Lappa, pers. comm.). The only two reports of infected vertebrates involved reptiles held in zoos (Austwick and Keymer, 1977; Fromtling et al., 1980). A number of fungi were isolated from the infected animals in one of these studies so it is presumed that the animals were stressed and the fungi were facultative parasites. Based on current information, it is anticipated that safety clearance of *B. bassiana* for registration by the U.S. Environmental Protection Agency is likely after completion of the tests this organization currently is outlining. The first step will be the granting of an experimental use permit for large acreage use, based on preliminary safety tests.

Field Trials on Long Island

Experiments in 1978

As part of the collaborative research portion of the U.S./U.S.S.R. Agreement on Science and Technology, 3 kg of *B. bassiana* conidia in clay were provided by the laboratory of O. A. Alyoshina in Moscow (All-Union Scientific Research Institute of Microbiological Means of Plant Protection and Bacterial Preparations). The conidia were produced by a methodology still under development, involving submerged (fermentation) culture. The kaolin-conidia formulation that was obtained was severely aggregated. The first generation of CPB on Katahdin potatoes was treated when 80 percent of the population was in the first and second larval instars and again one week later. Tween 80® was used as a wetting agent, and sprays were applied at the rate of seventy-five and one hundred gallons per acre in the first and second application, respectively. Treatments were replicated four times and included high and low dosages of fungus and a combination of the low dose of fungus and a dose of the insecticide endosulfan at one-quarter the commercially recommended rate. Unfortunately, CPB on Long Island were found to be very tolerant to this chemical in 1978. The Boverin proved difficult to use because of rapid settling of the suspended formulated particles and the need for large-size nozzles. Coverage was essentially on the upper leaf surfaces only. There were no detectable larval population reductions resulting from field applications of up to 8 kg per ha (5×10^{13} colony-forming units per ha).

Ten leaflets from each plot were placed in sterile plastic bags (Whirl-paks) at the time intervals after spray application indicated in Table 2. Ten milliliters (ml) of sterile water were introduced into each bag, and the bag was shaken for one minute. Ten-fold dilutions were streaked (0.2 ml per plate) on YpSs medium (Emerson, 1941) containing 350 milligrams (mg) rose bengal per liter. The plates were incubated at room temperature (∿25°C) for six days and then obvious (by colony morphology) *B. bassiana* colonies counted. The number of colony-forming units was reduced at least 90 percent in twenty-four hours and 99 percent in forty-eight hours (Table 2). The higher counts observed with low doses of endosulfan are presumed to be associated with the disassociation effects of the formulation ingredients of the endosulfan concentrate on the aggregates of Boverin. In laboratory tests, spray-tank strength endosulfan (0.25 pound active ingredient per 100 gallons) in agar medium was found to reduce germination of laboratory-produced conidia after eighteen hours of incubation from 66 percent with no insecticide to 15 percent with the low field dose. The normal full-strength level of

Table 2

Survival of *Beauveria bassiana* Conidia (Boverin) on Potato Leaves Following Two Field Applications, Long Island, 1978

Treatment	*Beauveria bassiana* Colony-Forming Units per Leaf				
	Day				
	0[e]	1	2	3	4
Exp. I[a]					
Control	15	---	0	---	5
Low dose B[b]	34,000	---	325	---	305
Low dose B + low dose F[c]	99,500	---	1,000	---	665
High dose B	364,000	---	21,125	---	8,875
Exp. II[d]					
Control	0	5	0	---	---
Low dose B	59,065	4,965	145	---	---
Low dose B + low dose F	191,900	14,625	135	---	---
High dose B	865,000	61,690	2,465	---	---

[a] Field application: June 22, 1978.
[b] B = Boverin; high dose = 8 kg/ha; low dose = 2 kg/ha.
[c] F = Furadan (4 Flow.), 0.25 lb active ingredient (endosulfan)/acre.
[d] Field application: June 28, 1978.
[e] Leaf samples were collected approximately two to three hours after spraying in exp. I and approximately one to two hours after spraying in exp. II. Exp. I was sprayed in the morning and exp. II, in the evening.

insecticide permitted only 2.7 percent germination, whereas one-tenth of this amount permitted 44 percent. In contrast TH 60-41 [1-(4 chlorophenyl-carbamoyl)-3-(4 chlorophenyl)-2 pyrazoline] did not inhibit germination even at full tank strength (230 mg per 100 ml). Similar lack of inhibition has been noted with synthetic pyrethroids tested against a U.S. isolate of *B. bassiana* (Roberta Clarke, pers. comm.).

Experiments in 1979

In 1979 approximately forty *B. bassiana* strains from many locations worldwide were evaluated for production of conidia on solid (Sabouraud dextrose agar) medium and for virulence to first instar CPB larvae. The former was based on milligrams of spores per petri dish and the latter on the time of 50 percent mortality after exposure to spores on potato-leaf discs. The stock solutions consisted of 50 mg of pure conidia suspended in 10 ml of 0.1 percent Tween 80®. Clumps were broken with a glass-Teflon homogenizer and tenfold dilutions prepared. Normally 10^{-1} and 10^{-2} dilutions were tested. One-inch circles of potato leaf were cut with a punch and sprayed with 0.5 percent Tween 80® on both sides and air dried. Spraying was necessary to permit spreading of the spore suspensions over the hydrophobic leaves. Spreading was accomplished by pipetting 0.1 ml of suspension on each side of leaf with drying before the opposite side was treated and before introduction of five freshly hatched (not previously fed) CPB larvae onto each disc. The discs were placed individually on moist filter paper in 60 millimeter (mm) petri dishes and incubated at 25°C. The discs were replaced with untreated leaves at forty-eight hours, and mortality counts were taken at twenty-four-hour intervals for ten days.

An isolate (RS 252) from diseased CPB collected in Maine consistently performed well in both the production and virulence tests and was therefore selected as the standard for further studies, including field application. Production was accomplished by autoclaving 300 ml of half-strength Sabouraud dextrose medium in nylon oven bags. On removal from the autoclave, the bags were placed on a flat surface and folded in half until the agar had jelled to give a cake in each bag approximately 6 mm thick. The bags were inflated with sterile air, and a suspension of three-day-old mycelia was sprayed onto the medium with a plastic hand-pumped household sprayer. They were then closed with wire ties and incubated at room temperature until heavy sporulation had occurred, usually six to ten days. The nylon bags allowed considerable escape of moisture, and after conidiation spores were vacuumed from the surface of the almost-dry medium with a plastic tube attached to a miniature glass cyclone collector fashioned, with the help of a glass blower, from a 500 ml Erlenmeyer flask. A 0.45μ filter was placed between the vacuum source and the cyclone collector to prevent contamination of the vacuum source with conidia. The spores were stored at 4°C. Spore counts of RS 252 averaged ∿2 x 10^{11} per g by hemocytometer and ∿1 x 10^{11} per g by dilution plate. Spore viability ranged from 85 to 95 percent.

Immediately before spraying, a 5 percent formulation of conidia was prepared in mineral oil containing 10 percent Agrimul 26B. This was applied in the evening with one-gallon hand sprayers to Katahdin potatoes in 0.00526 ha plots replicated four times in randomized blocks. The first application was made when the first generation was still 80 percent first and second instar larvae, and the second application was approximately one week later. Water volume was 100 gallons per acre. The treatments were (1) mineral oil-Agrimul 26B - H_2O control; (2) isolate RS 252 at 10^{14} spores per ha per application applied on a fixed weekly schedule; (3) isolate RS 252 at 10^{14} spores per ha per application applied two times per generation of CPB; (4) isolate RS 252 at 5 x 10^{13} spores per ha per application applied two times per generation, with a high molecular weight polymer, Pro-Tec, containing a sunscreen (PABA), added at 0.75 percent in the tank; and (6) an isolate of _B. bassiana_ from Japan, RS 302, at 10^{14} spores per ha per application applied two times per generation.

Field evaluation of CPB populations was done by counting all egg clusters, small larvae (L_1 and L_2), large larvae (L_3 and L_4), and adults on central ten-foot lengths of each of the two rows in each of the twenty-five-foot-long plots. Yield data were based on weight of new potatoes mechanically harvested from the entire plots.

B. bassiana-treated plots had significantly ($p \leq 0.05$) smaller first-generation larval and adult CPB populations than control plots (Table 3), and all _B. bassiana_ treatments afforded obvious foliage protection. The latter was not evaluated objectively, but it is reflected in the significantly higher yield of new potatoes harvested in late July from _B. bassiana_-treated plots (Table 3). The field study was terminated and potatoes harvested in late July because exceedingly large numbers of CPB adults migrated into the experimental area. The adults and ensuing larvae in the fungus-treated plots were present in such large numbers that _Beauveria_ was not successful in controlling them, apparently because the fungus permitted excessive plant damage prior to insect mortality.

Table 3
Colorado Potato Beetle Control through the First Generation Using *Beauveria bassiana*, Long Island, 1979

Fungal Isolated	Treatment		Mean Counts per 10 Foot Row[a]			Yield[b]
	Spores per ha	Application Schedule	L_1-L_2	L_3-L_4	Adults	(ton/ha)
Control	0	Weekly	48.2 ab	27.1 a	13.5 a	12.7 c
RS-302[c]	10^{14}	Twice/gen.	56.1 a	27.1 a	5.5 b	15.9 b
RS-252[d]	10^{14}	Weekly	41.8 bc	9.9 b	1.1 b	17.3 ab
RS-252	10^{14}	Twice/gen.	34.7 c	12.1 b	1.8 b	18.1 ab
RS-252	5×10^{13}	Twice/gen.	33.1 c	14.1 b	2.1 b	19.7 a
RS-252	10^{14}	Twice/gen. + Pro-Tec	42.8 bc	13.2 b	2.2 b	19.8 a

Note: Means within columns followed by the same letter are not significantly different at the 0.05 level (Duncan's multiple range test).

[a] Means of 11 counts taken during the first generation starting on day of first field application.

[b] Harvested as "new" potatoes.

[c] RS-302: *B. bassiana* strain originally isolated in Japan.

[d] RS-252: *B. bassiana* strain originally isolated in Maine.

Future Research Needs

The research needed to develop *B. bassiana* as a control agent for CPB closely parallels that needed currently to develop virtually all entomopathogenic fungi for commercial-scale use:

1. Elucidation of modes of disease induction and development. This includes studies of spore germination; chemicals on the cuticle that enhance or inhibit germination; penetration of the cuticle; development within the host; and discovery, isolation, characterization, and evaluation of significance of toxins.
2. Genetic manipulation. The objectives will be to increase and maintain virulence, facilitate mass production, and develop strains of *B. bassiana* with pesticide tolerance.
3. Standardization of the fungus product. In addition to unequivocal identification of the fungus strain and quantification of spore viability and number, development of reproducible bioassays to evaluate virulence is needed.
4. Epizootiology. Information will be required on environmental effects; host effects, including the genetic state of the host in relation to susceptibility to the pathogen and population density; and pathogen effects, including the virulence and density of the pathogen.
5. Elucidation of the long-term effects of introduced *B. bassiana* on insect populations.
6. Mass production. Current methods of production and harvesting of conidia need improvement.
7. Formulation. These studies will be directed toward preserving viability, preserving or enhancing virulence, and reducing the possibility of allergenic responses.
8. Combinations of fungi with other mortality agents. This will include combining more than one species of fungus and using *B. bassiana* with low doses of insecticides.
9. Field application. The most appropriate application equipment, timing, and dosage must be determined for local conditions.
10. Safety of entomopathogenic fungi. The effects of *B. bassiana* on invertebrates other than CPB and on vertebrates, including humans and domestic animals, must be determined.

CONCLUSIONS

Based on the 1979 experiments reported here and on literature from Europe, the control of CPB with B. bassiana appears possible. This insect is increasing in importance in the United States at the present time, and alternatives and supplements to synthetic insecticides are urgently needed. It is anticipated that further basic and applied studies will afford effective products containing fungi, especially B. bassiana, for CPB control.

REFERENCES

Al-Aidroos, K., and D. W. Roberts, 1978, Mutants of Metarhizium anisopliae with Increased Virulence toward Mosquito Larvae, Can. J. Genet. Cytol. 20:211-219.

Anonymous, 1979, Microbial Processes: Promising Technologies for Developing Countries. U.S. National Academy of Science Washington, D.C. p. 198.

Austwick, P. K., C., and I. F. Keymer, 1977, Reptilian Mycoses, Abstracts of Second International Mycological Congress, 1:28.

Bajan, C., 1973, Paecilomyces fumoso-roseus (Wize): Pathogenic Agent of the Colorado Beetle (Leptinotarsa decemlineata Say), Ekol. Pol. 21:705-713.

Bajan, C., T. Bilewicz-Pawiska, A. Fedorko, and K. Kmitowa, 1975, Natural Reduction of Leptinotarsa decemlineata Say on Potato Crops, Rep. 8th Int. Plant Prot. Cong., Moscow, Sec. V, pp. 33-40.

Burges, H. D., ed., 1981, Microbial Control of Pests and Plant Diseases: 1970-1980 Academic Press, London.

Clerk, G. C., 1969, Influence of Soil Extracts on the Germination of Conidia of the Fungi Beauveria bassiana and Paecilomyces farinosus, J. Inv. Path. 13:120-124.

Emerson, R., 1941, An Experimental Study of the Life Cycles and Taxonomy of Allomyces, Lloydia 4:77-144.

Fargues, J., 1975, Etude expérimentale dans la nature de l'utilisation combinée de Beauveria bassiana et d'insecticides à dose reduite contre Leptinotarsa decemlineata, Ann. Zool. Ecol. Anim. 7:247-264.

Fargues J., J. P. Cugier, and P. Van de Weghe, 1980, Expérimentation en parcelles du champignon Beauveria bassiana (Hyphomycete) contre Leptinotarsa decemlineata (Col., Chrysomelidae), Acta Ecologica 1:49-61.

Fargues, J., O. Reisinger, and P. H. Robert, 1978, Use of Argil-Coating to Increase the Resistance of Spores of Entomopathogenic Fungi to Biodegradation, Proc. Internat. Collog. On Invertebrate Pathol., Prague.

Fargues, J., and G. Remaudière, 1981, Host Specificity and Virulence of Entomopathogenic Fungi, in D. W. Roberts, ed., Entomopathogenic Fungi, Allanheld, Osmun, and Co., Montclair, N.J.

Fargues, J., P.-H. Robert, and O. Reisinger, 1979, Formulation des productions de masse de l'hyphomycete entomopathogene Beauveria en vue des applications phytosanitaires, Ann. Zool. Ecol. Anim. 11:247-257.

Ferron, P., 1978, Biological Control of Insect Pests by Entomogenous Fungi, Ann. Rev. Entomol. 23:409-442.

Fromtling, R. A., S. D. Kosanke, J. M. Jensen, and G. S. Bulmer, 1980, Fatal Beauveria bassiana Infection in a Captive American Alligator, Jour. Amer. Vet. Med. Assoc. 175(9):934-936.

Goral, V. M., 1972, The Submerged Method of Obtaining Beauverine, Zashchita Rastenif, no. 3:23.

Goral, V. M., 1978, Effect of Cultivation Conditions on the Entomopathogenic Properties of Muscardine Fungi, in Proc. 1st Joint US/USSR Conf. Prod. Selec. Stand. Entomopath. Fungi, Jurmula (Riga) Latvia SSR 20-21 May 1978, pp. 217-228.

Hall, R. A., 1975, Aphid Control by a Fungus Verticillium lecanii within an Integrated Programme for Chrysanthemum Pests and Diseases, Proc. 8th Br. Insect. Fungic. Conf., pp. 93-99.

Ignoffo, C. M., N. L. Marston, D. L. Hostetter, B. Puttler, and J. V. Bell, 1976, National and Induced Epizootics of Nomuraea rileyi in Soybean Caterpillars, J. Invertebr. Pathol. 27:191-198.

Kmitowa, K., 1973, Interaction of a Few Species of Soil Fungi and Two Species of Insect Pathogenic Fungi on the Changes of Their Parasitic Activity, Ekol. Pol. 21:687-697.

Lappa, N. V., 1978, Practical Applications of Entomopathogenic Muscardine Fungi, in Proc. 1st Joint US/USSR Conf. Prod. Selec. Stand. Entomopath. Fungi, Jurmula, (Riga) Latvia SSR 20-21, May 1978, pp. 51-61.

Lappa, N. V., and V. M. Goral, 1981, Insect Control with Fungi in Eastern Europe, in D. W. Roberts, ed., Entomopathogenic Fungi, Allanheld, Osmun and Co., Montclair, N.J.

McCoy, C. W., 1980, Pest Control by the Fungus Hirsutella thompsonii, in H. D. Burges, ed., Microbial Control of Pests and Plant Diseases, 1970-1980, Academic Press, New York.

Madelin, M. F., 1968, Fungal Parasites of Invertebrates. 1. Entomogenous Fungi, in The Fungi, ed., G. C. Ainsworth and A. S. Sussman, 3:227-238, Academic Press, New York.

May, B., D. W. Roberts, and R. S. Soper, 1979, Intraspecific Genetic Variability in Laboratory Strains of Entomophthora as Determined by Enzyme Electrophoresis, Exp. Mycology 3:289-297.

Mietkiewski, R., and R. S. Soper, 1979, Progress toward Commercial Formulation of Entomophthora virulenta Resting Spores, Proc. Internat. Colloquium on Invertebrate Pathol., Prague, Czech., 1978, pp. 137-138.

Milner, R. S., and R. S. Soper, 1980, Bioassay of Entomophthora against the Spotted Alfalfa Aphid Therioaphis trifolii f. maculata, J. Invertebr. Pathol. 37(2):168-173.

Pekrul, S., and E. A. Grula, 1979, Mode of Infection of the Corn Earworm (Heliothis zea) by Beauveria bassiana as Revealed by Scanning Electron Microscopy, J. Inv. Pathol. 34:238-247.

Reisinger, O., J. Fargues, P. Robert, and M. F. Arnould, 1977, Effet de l'argile sur la conservation des microorganismes. I. Etude ultrastructurale de la biodégradation dans le sol de l'Hyphomycète entomopathogène Beauveria bassiana (Bals.) Vuill. Ann. Microbio. (Inst. Pasteur). 128:271-287.

Roberts, D. W., 1974, Fungal Parasites of Mosquitoes, in A. Aubin, S. Belloncik, J. P. Bourassa, E. Lacouisiere, and M. Pellissier, eds. Le controle des moustiques, pp. 143-193, University of Quebec Press, Montreal.

Roberts, D. W., 1980, Fungi: Toxins, in H. D. Burges, ed., Microbial Control of Pests and Plant Diseases: 1970-1980 Academic Press, London.

Roberts, D. W., 1981, Entomopathogenic Fungi, Allanheld, Osmun & Co., Montclair, N.J.

Roberts, D. W., and A. S. Campbell, 1977, Stability of Entomopathogenic Fungi, Misc. Publ. Entomol. Soc. Am. 10:19-76.

Roberts, D. W., and R. A. Humber, 1980, Entomogenous Fungi, in G. T. Cole and B. Kendrick, eds., Biology of Conidial Fungi, Academic Press, New York.

Roberts, D. W., and W. G. Yendol, 1971, Use of Fungi for Microbial Control of Insects, in H. D. Burges and N. W. Hussey, eds., Microbial Control of Insects and Mites, pp. 125-149, Academic Press, London.

Soper, R. S., 1978, Development of Entomophthora Species as Possible Microbial Insecticides, in Proc. 1st Joint US/USSR Conf. Prod. Selec. Stand. Entomopath. Fungi, Jurmula (Riga) Latvia SSR 20-21 May 1978, pp. 270-282.

Soper, R. S., and M. G. Ward, 1981, Production, Formulation and Application of Fungi for Insect Control, in G. C. Papavizas, ed., Biological Control in Crop Production, BARC Symp. V. Allanheld, Osmum and Co., Montclair, N.J.

Steinhaus, E. A., 1956, Microbial Control--The Emergence of an Idea. A Brief History of Insect Pathology through the Nineteenth Century, Hilgardia 26:107-160.

Steinhaus, E. A., 1975, Disease in a Minor Chord, Ohio State University Press, Columbus, 488p.

Wartenberg, H., and K. Freud, 1961, Der Konservierungseffekt antibiotischer Mikro-organismen an Konidien von *Beauveria bassiana* (Bals.) Vuill. *Zentralbl. Bakterol. Parasitenkd. Infekt. Hyg.* Abt. 2 Naturviss. 114:718-724.

Wojciechowska, M., K. Kmitowa, A. Fedorko, and C. Bajan, 1977, Duration of Activity of Entomopathogenic Microorganisms Introduced into the Soil, *Pol. Ecol. Stud.* 3:141-148.

Yevlakova, A. A., 1978, Basic Trends in the Use of Entomopathogenic Fungi in the Soviet Union, In *Proc. 1st Joint US/USSR Conf. Prod. Selec. Stand. Entomopath. Fungi, Jurmula (Riga) Latvia SSR 20-21 May 1978,* pp. 35-50.

Insects and Diseases in the Production of Seed Potatoes

Richard H. Storch

Stringent pest control is needed in seed potato production because the object is to produce seed essentially free of disease. Knowledge of the sources of the pathogen, size of the pathogen population, and dissemination of the pathogen is important for disease control. If a pathogen is spread by active vectors, then the vector sources, their population size, and their spread must also be considered.

Potatoes are raised commercially in every state in the United States. It is estimated that in 1979 in the United States, 506,000 hectares of potatoes were harvested with an average yield of 30,376 kilograms (kg) per hectare (ha) (USDA, 1979). In Maine, 46,500 ha were planted to potatoes in 1979. Approximately 2800 kg of round white seed or 1800 kg of Russet Burbank seed are required to plant 1 ha of potatoes. The cost of round white seed to plant the 1979 crop in Maine was about $10.00 per 100 kg (Smith, 1979). The Maine Farm Planning Guide estimated the per-hectare cost of production of tablestock potatoes from planting to harvest (exclusive of land expenses, interest on loans, and taxes) to be about $1340 for round whites and $1050 for Russet Burbank. Seed potato production costs are generally higher than costs for tablestock production because more effort is expended to control potato pests and lower yields are obtained.

POTATO PESTS

The potato plant is affected by many pests that can reduce the quality and yield of tubers. These pests

---

I wish to thank F. E. Manzer, M. E. MacGillivary, G. W. Simpson, and K. E. Storch for reviewing this manuscript.

include invertebrates (primarily insects), pathogens, and weeds, which, like other organisms, have variants (strains, races, biotypes, and so forth). Simpson (1977) records 101 potato insect pests in North America, all but two of which are individual species. Insect pests damage plants in several ways: by defoliation, removal of plant sap, and boring into foliage, roots, and tubers. Pathogens are actively transmitted from plant to plant by some insects during feeding, and, in addition, any break in the plant epidermis provides a place for passive entry of certain potato pathogens. Weeds compete with potato plants for nutrients and moisture in the soil. When weed populations are high, the roots of some weeds can penetrate the tuber and provide an opportunity for pathogens to enter. Weeds, also, can serve as reservoirs for potato pathogens and potato invertebrate pests. Rich (1977) states that about 120 potato pathogens are recorded in the _Index of Plant Diseases in the United States_. The severity of symptom expression may vary with strains of a given pathogen, variety of potato plant, presence of other pathogens, and, usually, with growing conditions. Pathogens, found in or on the tissues of potato foliage and tubers, can be spread from diseased to healthy foliage or tubers mechanically by contact between plants, by contaminated animals and inanimate objects, or by infected animals. Some pathogens can overwinter in the tubers produced from diseased plants. It is important to have seed potatoes free of pathogens to ensure a high yield of quality tubers for the fresh market and processing because potatoes are propagated vegetatively to maintain the characteristics of a given variety.

PATHOGENS

The most important diseases to consider in seed potato production are ring rot, mosaics, virus leaf roll, and spindle tuber. Ring rot caused by _Corynebacterium sependonicum_ (Spieck. and Koth.) is spread mechanically. Some in-field spread from diseased to healthy foliage by equipment can occur, but the pathogen is spread to seed pieces primarily by contact with diseased seed pieces and with contaminated seed-cutting knives, picker-planters, and other equipment. Potato virus X (PVX) is spread from diseased to healthy plants in the field by contaminated cultivator and spray equipment (Manzer and Merriam, 1961). The virus can also be spread to healthy tubers by contact with infected seed pieces and with contaminated seed-cutting knives and picker-planters. Potato virus A (PVA) and potato virus Y (PVY) can be spread mechanically like PVX, but in nature they are spread primarily by aphids (Beemster and Rozendaal, 1972). Spindle tuber is caused by the spindle tuber viroid, which is transmitted mechanically and through true seed (Fernow et al., 1970). Leaf roll, caused by potato leaf roll virus (PLRV), is transmitted only by aphids.

One of the most important pathogen-host plant relationships to consider in seed potato production is the translocation of the pathogen from the foliage to the tuber, especially of those pathogens transmitted only by insects. Bradley and Ganong (1953) suggest that translocation of PLRV from vines to tubers requires from eight to sixteen days. This study, however, was conducted on plants in July. Regression analysis of field data collected in late August and September using minimum daily temperature and incidence of PLRV (Manzer and Storch, unpubl.) shows that in a three-day interval from inoculation to harvest, there will be about 1 percent increase in PLRV for each degree increase in temperature above 6.6°C for Katahdin and about 3.1°C for Russet Burbank (Figure 1). When the interval between inoculation and harvest is fifteen days, the increase in PLRV is 8.1 percent and 15.4 percent for Katahdin and Russet Burbank, respectively, for each degree increase in temperature above about 5.7°C. In addition, Beemster (1961) found that different strains of PVY are translocated at different rates.

APHID LIFE HISTORY

The potato-infesting aphids in northeastern North America are *Aphis nasturtii* Kaltenbach (the buckthorn aphid), *Acyrthosipon solani* (Kaltenbach) (the foxglove aphid), *Myzus persicae* (Sulzer) (the green peach aphid), and *Macrosiphum euphorbiae* (Thomas) (the potato aphid). In Maine these aphids overwinter as eggs on the primary host or as nymphs and adults on secondary hosts in buildings. In the spring after the eggs hatch and several generations of wingless aphids develop on the primary host, a generation of winged aphids is produced that flies to various secondary hosts. A given species of aphid usually has numerous secondary hosts, such as weeds and cultivated plants. Some of these hosts are more readily accepted than others. When aphid populations increase and crowding occurs, winged forms, which fly to secondary hosts, are produced. In late summer or early fall, alate aphids develop, which fly back to the primary host to produce overwintering eggs once again (Figure 2).

Each species and life stage of aphid differs in its ability to transmit potato pathogens; each also differs in seasonal abundance. *Rhamnus alnifolia* L'Her., the alder-leaved buckthorn, is the most important overwintering host of *A. nasturtii* in Maine (Shands and Simpson, 1971). *A. nasturtii* is a faily efficient vector of PVA and PVY but a poor vector of PLRV. Alate aphids infected with PVY are more efficient vectors and remain infective longer than do apterous forms (Kostiw, 1975). Few winged forms of this aphid are produced during the summer (MacGillivray, pers. comm.). The main overwintering host of *M. euphorbiae* is *Rosa palustris* (Marsh.) (the swamp rose), but some eggs are laid on

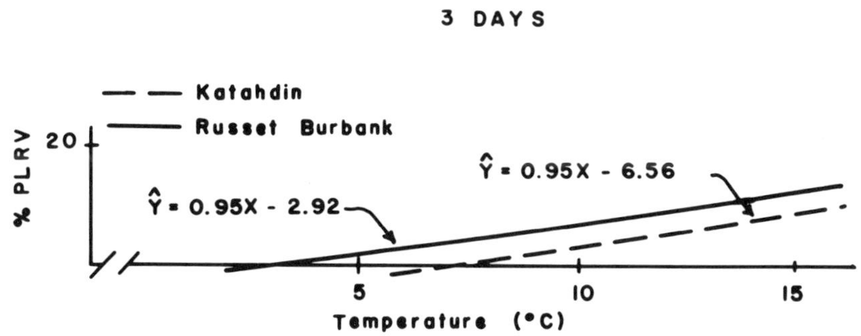

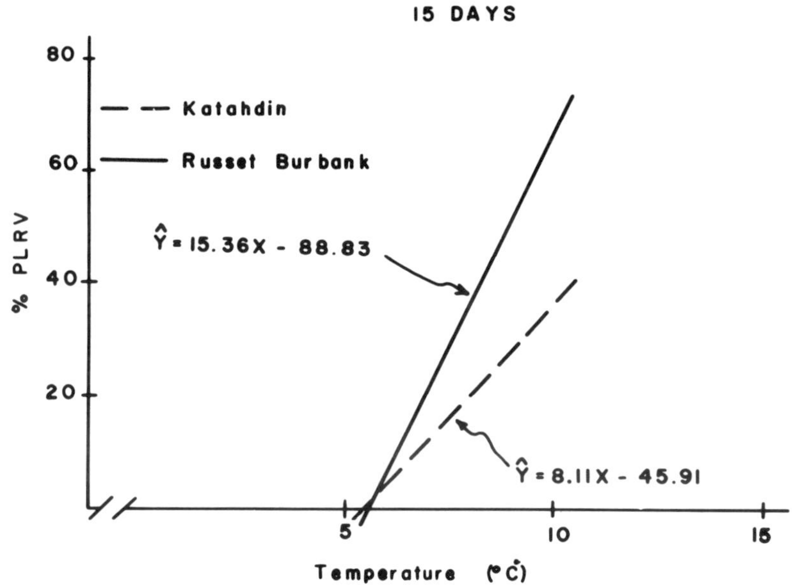

Figure 1
Linear regression of minimum daily temperature on percentage of PLRV from inoculation to top-kill. The experiment consisted of five replicates (five plants of each variety) with nine infection dates, three days apart beginning August 20. Vines were severed at the soil level three or fifteen days after infestation with viruliferous aphids. The experiment was repeated for three years. The analyses shown are preliminary. The top graph shows the three-day period from infestation to harvest, the bottom graph the fifteen-day period from infestation to harvest.

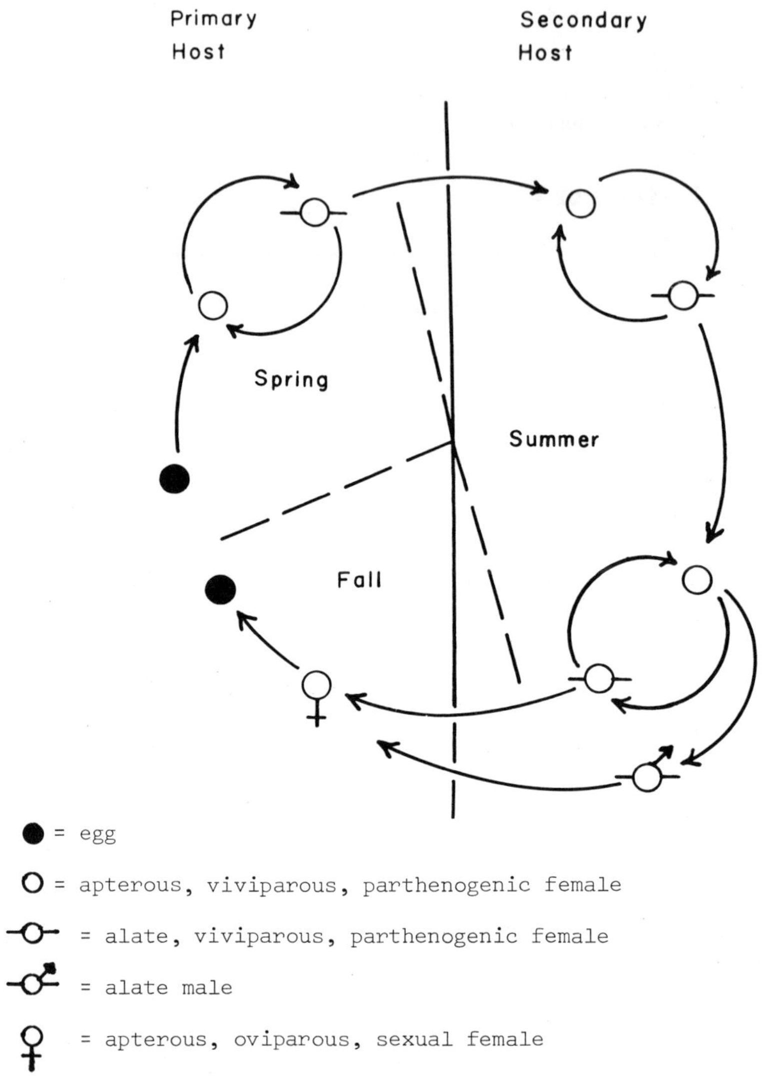

Figure 2
Typical life cycle of potato-infesting aphid in northeastern North America. Usually aphids overwinter in the egg stage on the primary host. In the spring the eggs hatch, and the resulting nymphs develop into wingless, viviparous females (stem mothers). A few generations of viviparous females develop on the secondary hosts. In late summer or early fall, winged males are produced, which fly back to the primary host. At the same time winged viviparous females fly from the secondary to the primary host and deposit nymphs, which develop into wingless, oviparous females. These females mate with the males and produce the overwintering eggs.

Prunus nigra Ait. (Canada plum) (Shands et al., 1972). Populations of *M. euphorbiae* usually decline in August due to the action of pathogenic fungi (*Entomophthora* spp.). In late August fall migrants move back to the primary hosts. This aphid is a faily efficient vector of PVA but an inefficient vector of PVY and PLRV. *P. nigra* is the primary host of *M. persicae* in Maine (Shands et al., 1972). *M. persicae* also overwinters on sprouted potatoes in potato storage buildings and on plants in homes and greenhouses. The fall migrants move back to the primary host in late August or September. This aphid is a very efficient vector of PVA, PVY, and PLRV.

Weeds are hosts for various insect pests and pathogens. *Chenopodium album* L. (lamb's quarters), *Brassica campestris* L. (yellow mustard), and *Amaranthus retroflexus* L. (red-rooted pigweed) are among the most common weed pests of potatoes in Maine and are also secondary hosts for all species of potato-infesting aphids (Patch, 1939). *A. retroflexus* is also a host for PVX and PLRV and *C. album* a host for PVX (Thornberry, 1966).

APHID FLIGHT

Behavior of alate aphids is important in the spread of some potato pathogens (see Kring, 1972). After molting to the adult stage, the aphid will take off when light intensity and quality, temperature, and wind conditions are correct. Two types of flight have been observed. Migrating aphids move upward, avoiding objects, and may be caught in low-level wind currents, and nonmigrating aphids fly to and among bushes and trees (Figure 3). When the urge to fly is satisfied, the aphid will alight. At this time aphids are especially attracted to yellow objects. After alighting on a plant, an aphid will probe and then make several short flights prior to settling down. Nymphs are usually deposited each time the aphid alights.

APHID TRANSMISSION OF POTATO VIRUSES

Aphid-transmitted pathogens can be separated by the way they are transported by the vector. Stylet-borne pathogens such as PVA and PVY are carried on the mouth parts (Figure 4). Infectivity of the vector is lost at molting. These pathogens can be acquired (acquisition period) from and transmitted (transmission period) to plants in probes of one minute or less. It is not known whether ingestion is necessary for stylet-borne pathogens to be acquired (Pirone and Harris, 1977). Circulative pathogens, like PLRV, are carried in the salivary gland(s), and infectivity is not lost at molting (Figure 5). The acquisition period for PLRV is several minutes because the stylets must enter the phloem, and ingestion must begin so that the pathogen will accumulate

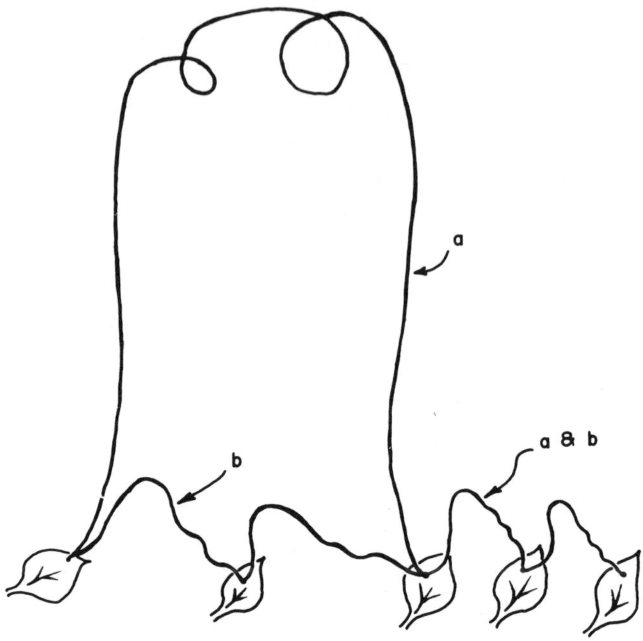

Figure 3
Flight behavior of winged aphids. (a) Flight path of a migratory aphid. (b) Flight path of a nonmigratory aphid.

in the digestive tract. The vector is not infective for several hours after acquisition because the pathogen must penetrate the wall of the digestive tract and be carried to the salivary gland(s) before transmission can occur. The minimum transmission period for PLRV is several minutes. Stegwee and Ponson (1958) reported that PLRV multiplied in the vector, but others, most recently Eskandari et al. (1979), have not been able to demonstrate propagation of the virus in the vector.

SPREAD OF APHID-BORNE PATHOGENS

Stylet-borne and circulative pathogens are spread at different rates within a potato field. Often several vector species transmit a given stylet-borne pathogen. The time required for the vector to become infective and transmit the pathogen is short (measured in minutes), so stylet-borne pathogens can be spread readily within a field. Circulative pathogens are spread less readily because there are usually only one or two vector species that can efficiently transmit a given circulative pathogen and the time required for the vector to become infective and transmit the pathogen is relatively long (measured in hours).

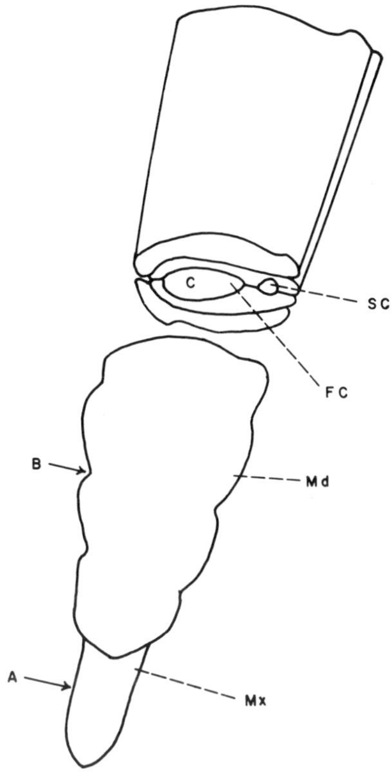

Figure 4
Diagram of aphid mandibular and maxillary stylets showing where stylet-borne virons may be located. (A) On smooth maxillary stylet (Mx). (B) In grooves on mandibular stylet (Md). (C) In the food canal (FC). Stylet-borne virons have not been found in the salivary canal (SC).

Circulative pathogens are readily spread between fields because the vectors remain infective for life. Aphids carrying stylet-borne pathogens do not remain infective after several probes. The reason for this is not known, but it is suggested that the stylets are cleansed in the process of probing. The time of infectivity of wingless aphids in the absence of probing is seldom more than a few hours.

The difference between the initial spread of a virus between fields and the spread within a field by aphids can be determined. If a viruliferous aphid enters a field of healthy potatoes, plants acquiring the virus are randomly distributed in the field because the winged aphids have probed several plants before settling down (Figure 6). The spread of the virus in a field already having diseased plants is indicated by a clumping of diseased plants around the originally infected plant (Figure 7). This localized spread is normally accompanied

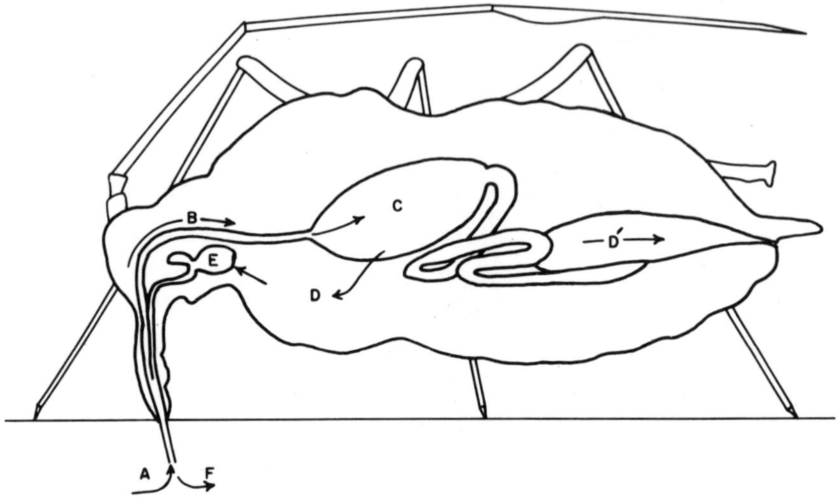

Figure 5
Diagrammatic cross-section of an aphid. The illustration shows the mode of acquisition and inoculation of circulative viruses. Virons are acquired from a viruliferous plant (A) and pass through the stomodaeum (B) to the mesenteron (C) along with food materials. Virons passed to the proctodaeum (D') and excreted with waste materials are not infective. Some virons pass through the cell wall of the mesenteron into the haemocoel (D) and are transported to the salivary glands (E). Virons passed from the salivary gland to another plant (F) are infective.

by wingless aphids that develop from the nymphs deposited by the migrant. Wingless aphids tend to remain sedentary unless disturbed. The spread of pathogens within a field is increased when infected aphids are moved throughout the field by humans, animals, and equipment. Virus spread within a field might be greater when the population of predators is high (Roitberg and Myers, 1978).

IPM PROGRAM FOR SEED PRODUCTION

An initial concern in establishing a pest management program for seed potatoes is the climate and location of seed potato fields. Cold winters and cool, damp summer climates are detrimental to most potato pests. Even though late-maturing varieties may not develop fully in these regions, it would be possible to produce high-quality seed potatoes of these and earlier maturing potato varieties. Tablestock and processing potatoes and other crops susceptible to potato pests should not be grown in seed-potato-producing areas because pest control requirements for them are not as stringent as for seed

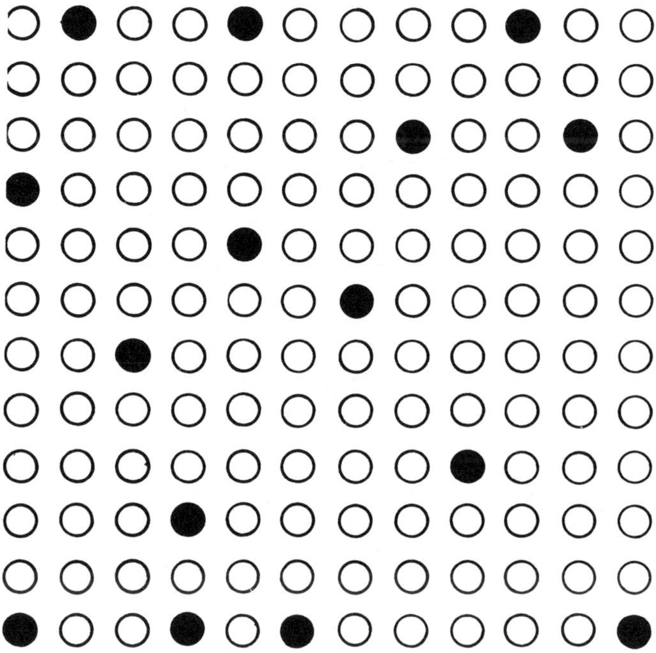

Figure 6
Diagram showing spread of PLRV into a field of initially healthy potatoes by winged, viruliferous aphids. Open circles represent healthy plants. Solid circles represent infected plants.

potato production. Seed potato plots should be located on high, windswept land not surrounded by windbreaks because aerial plankton (including potato pathogen vectors) are less likely to settle in these areas (Lewis, 1969).

The use of disease-free seed will reduce sources of ring rot, spindle tuber, mosaics, and leaf roll. Diseased plants should be rogued from the field early in the growing season; the entire plant, including the seed piece, should be taken out, placed in a bag, and destroyed. Pathogens and vectors can spread from diseased plants left in or near the field. Roguing should be done before plants are large because pathogens transmitted mechanically can be spread.

Burning vines after harvest destroys early blight, Rhizoctonia pathogens, and overwintering European corn borers and eliminates an overwintering site of flea beetles. The destruction of potato dump piles and volunteer potatoes will reduce the sources of early and late blight, ring rot, spindle tuber, mosaics, and virus leaf roll pathogens and sources of potato-infesting aphids. Vector sources can be reduced by keeping greenhouses free of pests, importing pest-free bedding

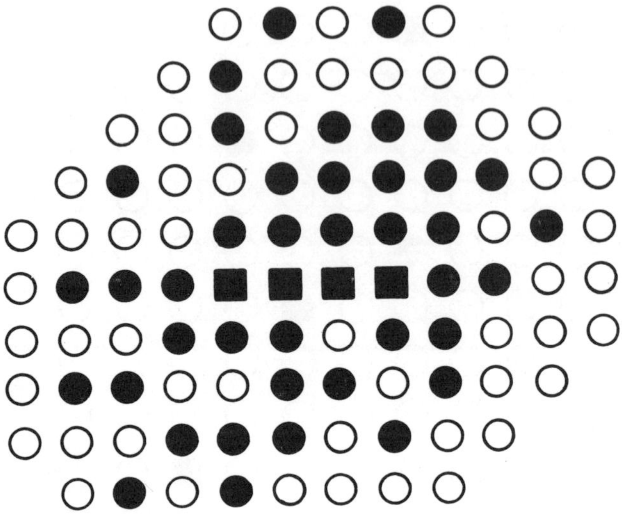

Figure 7
Diagram of part of a potato field showing spread of PLRV primarily by wingless aphids from diseased to healthy plants. Squares represent plants from diseased seed pieces, and all circles represent plants from healthy seed pieces. Open circles represent healthy plants. Solid circles represent infected plants (modified from Simpson, 1939).

plants, and removing overwintering hosts and sites.

A reduction in the spread of aphid-borne pathogens can be accomplished by a reduction of vector populations. A systemic insecticide therefore should be used to prevent the rapid increase of vector populations. Winged M. persicae do not appear in large numbers until late in the growing season. A reduction in the spread of circulative pathogens, such as PLRV, will result if potato tops are killed when winged vectors first appear. When the winged M. persicae occur, the average daily temperature is usually above 15°C, and PLRV is probably translocated from the foliage to tubers in two to four days. The use of insecticides applied to foliage to kill winged vectors will not prevent the spread of circulative or stylet-borne pathogens because no insecticide available will kill vectors before pathogens can be transmitted to the plant.

Other practices can prevent the spread of mechanically transmitted pathogens. Use of whole seed will greatly reduce the spread of pathogens such as ring rot, spindle tuber, and PVX by eliminating seed-cutting knives, contact of cut seed surfaces, and picker-planters, which could be replaced with cup planters if whole seed is used. When potato tops become large, cultivation and ground spray equipment should not be used.

Program Needs

In several places in this scheme, problems arise. A reduction in hosts for vectors is often difficult to achieve. It would be difficult, for example, to remove all R. alnifolia (primary host of A. nasturtii) because the plant is numerous and widespread in Maine. Volunteer potatoes continue to be difficult to control because they often occur in fields planted to small grains or other crops in the rotation cycle. Chemicals that can kill volunteers probably will kill the crop being grown as well.

It is also difficult to prevent aphids from coming into the area on low-level jet streams. Long-distance (several hundred miles) aphid migrations are known to occur (Berry and Taylor, 1968).

There is evidence that some spread of PLRV occurs early in the growing season. Manzer et al. (1977) suggest that stem-end browning (SEB) can occur when young plants become infected with PLRV. Net necrosis can result when older plants become infected with the virus. SEB symptoms were expressed when plants were inoculated experimentally with PLRV in late June or early July (Figure 8). The populations of M. persicae in Maine potato fields at this time are very small. Alate forms are not viruliferous if they come from primary hosts. It is possible, therefore, that aphids spreading PLRV at this time are arriving from areas several hundred miles away. M. persicae populations currently are being monitored in other areas to the south and west of Maine to determine when and from where migrations might occur. When aphids take off in those areas, weather data on low-level jet streams and other atmospheric conditions can be used to determine whether the migrating aphids will be deposited in Maine. If a particular area to the south or west is shown to produce aphids that migrate to seed-producing areas in Maine, then those areas should be provided with the cleanest seed available. It may also be possible to have those areas use control measures to keep aphid populations low so that the production of winged aphids is kept to a minimum.

Another problem is encountered in the spread of stylet-borne pathogens. Some species of aphids other than those normally infesting potatoes are able to spread these pathogens. These species could spread pathogens if migrants probed diseased and then healthy potatoes in the search for adequate secondary hosts. This problem could be overcome by identifying the vectors and then reducing their sources.

An Example of Success

The effectiveness of techniques of this nature is indicated by the success of the Maine-New Brunswick potato program (summarized from MacGillivray, 1979). In the early 1970s, an increase in PLRV was noted in seed potato

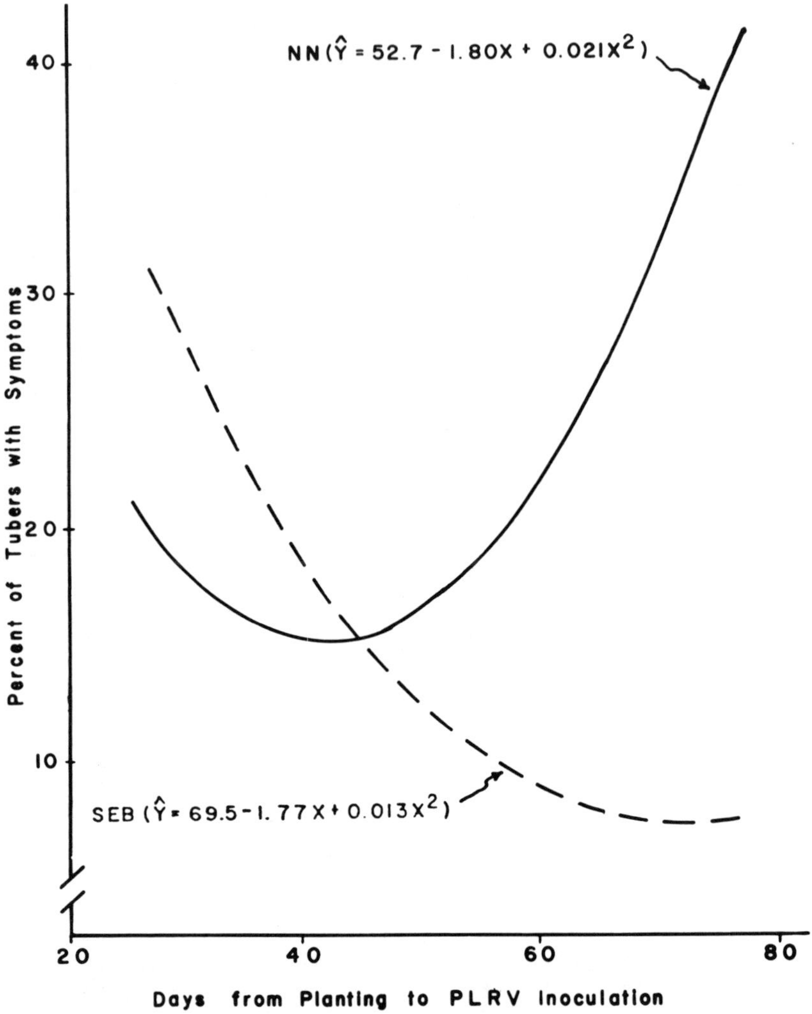

Figure 8
Regression of days from planting to inoculation with PLRV on percentage of tubers (Green Mountain var.) showing stem-end browning (SEB) and net necrosis (NN). Planting was done on May 26, June 4, 9, 16, 23, and 30. Inoculation with viruliferous aphids was done at four weekly intervals for each planting beginning July 21.

lots as indicated by the winter test program. A committee of scientists from Maine and New Brunswick was formed to develop a pest management program. The immediate objective was to decrease the injurious effects of potato pest insects and the amount of PLRV in seed lots by reducing the sources of the vector and the virus. The long-range objectives were to expand the program to include all potato pests and to determine areas where further information or research was needed to make the program most effective.

The original program was set up in 1973 using existing information about PLRV and *M. persicae*, the most efficient vector of leaf roll. Seven recommendations were made: use the cleanest seed possible, plant early, rogue, top-kill early, use a systemic insecticide at planting, reduce *M. persicae* populations in greenhouses, and import aphid-free bedding plants. The result was a reduction of PLRV in seed lots as indicated by the winter test program. Data collected from monitoring aphid populations and seed lots confirmed the four points. (1) The transmission of PLRV to healthy potatoes coincides with the flights of *M. persicae* into the potato fields. (2) A systemic insecticide at planting suppresses aphid populations but does not prevent the spread of PLRV into potato fields by winged aphids. (3) Fields that were not exposed to flights of *M. persicae* prior to top-killing had little or no PLRV. (4) PLRV spread occurred in fields that were not top-killed three days after winged *M. persicae* were caught in yellow pan traps in that field. The program is now being expanded to reduce other potato pests by studying the biology of each pest and using techniques to reduce sources, populations, and spread of the pest.

REFERENCES

Beemster, A. B. R., 1961, Translocation of Leaf Roll and Virus Y in the Potato, *Proc. 4th Conf. Potato Virus Dis. Braunschweig 1960*, pp. 60-70.

Beemster, A. B. R., and A. Rozendaal, 1972, Potato Viruses: Properties and Symptoms, in J. A. de Bokx, ed., *Viruses of Potatoes and Seed-potato Production*, pp. 115-143. Centre for Agricultural Publishing and Documentation, Wageningen, the Netherlands.

Berry, R. E., and L. R. Taylor, 1968, High Altitude Migration of Aphids in Maritime and Continental Climates, *J. Anim. Ecol*. 37:713-722.

de Bokx, J. A., ed., 1972, *Viruses of Potatoes and Seed-Potato Production*, Centre for Agricultural Publishing and Documentation, Wageningen, the Netherlands.

Bradley, R. H. E., and R. Y. Ganong, 1953, Note on the Time Potato Leafroll Virus Takes to Pass from Aphids on the Leaves into the Developing Tubers, *Can. J. Bot*. 31:143-144.

Eskandari, F., E. S. Sylvester, and J. Richardson, 1979, Evidence for Lack of Propagation of Potato Leaf Roll Virus in its Aphid Vector, *Myzus persicae*, *Phytopathology* 69:45-47.

Fernow, K. H., L. C. Peterson, and R. L. Plaisted, 1970, Spindle Tuber Virus in Seeds and Pollen of Infected Potato Plants, *Am. Potato J*. 47:75-80.

Kostiw, M., 1975, Investigation on the Retention of Potato Viruses M and Y in Two Species of Aphids (Myzus persicae Sulz. and Aphis nasturtii Kalt.), Potato Res. 18:637-640.

Kring, J. B., 1972, Flight Behavior of Aphids, Ann. Rev. Entomol. 17:461-492.

Lewis, T., 1969, Factors Affecting Primary Patterns of Infestation, Ann. Appl. Biol. 63:315-344.

MacGillivary, M. E., 1979, The Potato Industry of the Upper-Saint John River Valley: Potato Research--Pest Management. Paper presented to AIC Meeting, Fredericton, N. B., Aug. 20.

Manzer, F. E., and D. Merriam, 1961, Field Transmission of the Potato Spindle Tuber Virus and Virus X by Cultivating and Hilling Equipment, Am. Potato J. 38:346-352.

Manzer, F. E., D. C. Merriam, and R. H. Storch, 1977, Effects of Time of Inoculation with PLRV on Internal Tuber Necrosis, Am. Potato J. 54:476.

Patch, E. M., 1939, Food-Plant Catalog of the Aphids of the World Including Phylloxeridae, Maine Agr. Exp. Sta. Bull. 393:35-431.

Pirone, T. P., and K. F. Harris, 1977, Nonpersistent Transmission of Plant Viruses by Aphids, Ann. Rev. Phytopathol. 15:55-73.

Rich, A. E., 1977, Potato Diseases, in O. Smith, Potatoes: Production, Storing, Processing, pp. 506-549. AVI Publishing Co., Westport, Conn.

Roitberg, B. D., and J. H. Myers, 1978, Effect of Adult Coccinellidae on the Spread of a Plant Virus by an Aphid, J. Appl. Ecol. 15:775-779.

Shands, W. A., and G. W. Simpson, 1971, Seasonal History of the Buckthorn Aphid and Suitability of Alder-Leaved Buckthorn as a Primary Host in Northeastern Maine, Maine Life Sci. Agr. Exp. Sta. Tech. Bull. 51, 24 p.

Shands, W. A., G. W. Simpson, and H. E. Wave, 1969, Canada Plum, Prunus nigra Aiton, as a Primary Host of the Green Peach Aphid, Myzus persicae (Sulzer), in Northeastern Maine, Maine Agr. Exp. Sta. Tech. Bull. 39, 32 p.

Shands, W. A., G. W. Simpson, and H. E. Wave, 1972, Seasonal Population Trends and Productiveness of the Potato Aphid on Swamp Rose in Northeastern Maine, Maine Life Sci. Agr. Exp. Sta. Tech. Bull. 52, 35 p.

Simpson, G. W., 1939, Report for 1937-38, Maine Agric. Exp. Sta. Bull. 391, 301 p.

Simpson, G. W., 1977, Potato Insects and Their Control, in O. Smith, Potatoes: Production, Storing, Processing, pp. 550-605. AVI Publishing Co., Westport, Conn.

Smith, D., 1979, The Maine Idea, Potato Councillor 25:10.

Stegwee, D., and M. B. Ponson, 1958, Multiplication of Potato Leafroll Virus in the Aphid Myzus persicae (Sutz.), Entomol. Exp. Appl. 1:291-300.

Thornberry, H. H., 1966, Plant Pests of Importance to North American Agriculture, Index of Plant Virus Diseases, USDA ARS, Agr. Handbook 307, 446 p.

USDA, 1979, Statistical Reporting Service, CR-PR 2-2, October.

Role of Entomophthoran Fungi in Aphid Control for Potato Integrated Pest Management

Richard S. Soper

Aphids must be included as a major component of any potato integrated pest management (IPM) system. The green peach aphid, *Myzus persicae* (Sulzer), is important as a plant virus vector, and the potato aphid, *Macrosiphum euphorbiae* (Thomas), can build damaging populations. Other species that infest potatoes in the Northeast are the buckthorn aphid, *Aphis nasturtii* Kaltenbach, and the fox-glove aphid, *Acyrthsiphon solani* (Kaltenbach).

A complex of entomopathogenic fungi is found attacking these aphids. An understanding of the biology and epizootiology of these pathogens is essential to their utilization in IPM. An aphid monitoring system with which the prevalence of disease can be measured is needed for the prediction of impending epizootics and promotion of the efficient use of chemical control measures.

There are two basic approaches to exploiting the entomophthoran pathogen complex for aphid control. First, enhancement of the aphid mycoses can be achieved by utilizing compatible chemical pesticides, introducing fungicide-resistant strains, and establishing entomophthoran reservoirs in nonpotato aphids on adjacent intercrops. The second tactic, the use of these fungi as mycoinsecticides, would give IPM managers the greatest flexibility in using these fungi for aphid control.

Although integrated pest management strategies for potato crop protection vary regionally, the importance of aphid control is of common concern. The role of these insects as vectors of potato viruses makes aphid control

---

This chapter reports the results of research only. Mention of a pesticide does not constitute a recommendation for use by the USDA nor does it imply registration under FIFRA as amended. Also mention of a commercial product does not constitute an endorsement of this product by the USDA.

vital for seed production. Undoubtedly the early seasonal occurrence of aphids is a primary factor that governs the ability of an area to produce certified seed potatoes. In areas with a long growing season, aphid populations can move from crop to crop. Under these conditions, aphid infestations on potatoes can build up, causing yield loss and preventing the production of certified seed. Current control practices rely exclusively upon systemic and contact insecticides.

In Northern Maine, a concentrated effort to eliminate Canada plum, *Prunus nigra* Ait, the host of overwintering green peach aphids, *Myzus persicae* (Sulzer), has had limited success. The normal seasonal occurrence of this aphid has been retarded by approximately one to two weeks (Hammond and Holbrook, 1979). Generally, however, heavy reliance is made upon the use of chemical insecticides. The constant application of these substances in large amounts, although necessary for crop protection, is costly. Pesticides are generally not host specific and thus can also be detrimental to beneficial insects, such as parasites, predators, and pollinators. In addition, chemical insecticides may have high mammalian toxicity and pose a threat to human health. Finally, these pesticides are based on petrochemicals, which may make their use uneconomical as the cost of petroleum soars. It is prudent, therefore, to take advantage of alternative crop protection methodology that does not rely exclusively on chemical pesticides. The use of entomopathogenic fungi constitutes an important alternative control strategy (Engler and Rogoff, 1976; Franz, 1973).

PATHOBIOLOGY

The insect pathogenic fungi that attack potato aphids naturally throughout their range are an important component of the natural enemies complex. The importance of Entomophthorales in controlling the potato aphids in Maine was first recorded by Patch in 1907, followed by a long-term study by Shands et al. (1962, 1963, 1972). During these investigations, it was determined that a complex of entomophthoran species was playing an important role in the control of potato aphids. The most important species were *Conidiobolus obscurus* (Hall et Dunn) Remaudière et Keller (= *Entomophthora obscura* Hall et Dunn = *Entomophthora thaxterina* Petch = *Entomophthora ignobilis* Hall et Dunn) (Humber, 1978; Remaudière et al., 1979), *Erynia neoaphidis* Remaudière et Hennebert (= *Entomophthora aphidis* Hoffman), *Zoophthora radicans* (Bref.) Batko (= *Entomophthora sphaerosperma* Fresenius), and *Entomophthora planchoniana* Cornu. Subsequently (Soper et al., 1975; Soper, unpubl.), *Zoophthora occidentalis* (Thaxter) Batko and *Conidiobolus thromboides* Drechsler (= *Entomophthora virulenta* Hall et Dunn) have been isolated from potato aphids. Of these, *C. obscurus* was by far the most important pathogen, followed by *E. neoaphidis* and *E. planchoniana*. The potato aphid proved to be the most susceptible to mycosis, followed by the green peach aphid and the buckthorn aphid. A key to these entomophthoran species is found in Table 1.

Table 1
Key to Entomopthoran Species on Potato Aphids

1.	Conidia present,	2
	Resting spores present,	7
2.	Conidia bell-shaped (with pointed apex and truncate base), *Entomophthora planchoniana*	
	Conidia spherical to ovid,	3
3.	Rhizoids present,	5
	Rhizoids absent or only from mouth parts,	4
4.	Conidia spherical with broad rounded basal papilla, 18-32μm x 16-28μm, *Conidiobolus thromboides*	
	Conidia broad pyriform, 26-43μm x 24-34μm, *Conidiobolus obscurus*	
5.	Conidia long ellipsoid, capillispores produced; no cystidia produced,	6
	Conidia ovid, 28-30μm x 11-18μm; cystidia may be present; secondary spores like primary; no capillispores present, *Erynia neoaphidis*	
6.	Conidia 14-25μm x 5-8μm, *Zoophthora radicans*	
	Conidia 35-45μm x 10-12μm, *Zoophthora occidentalis*	
7.	Rhizoids present,	9
	Rhizoids absent or from mouth parts,	8
8.	Resting spores small, 15-31μm, *Conidiobolus thromboides*	
	Resting spores large, 23-50μm, *Conidiobolus obscurus*	
9.	Species not separable by resting spore characteristics: *Entomophthora planchoniana* (25-36μm), *Zoophthora occidentalis* (20-35μm), *Zoophthora radicans* (20-30μm)	

Before utilizing entomophthoran species in an integrated pest management system, one must understand the factors that control the pathogen life cycle (Figure 1). The infection process begins with the contact of a conidium with a susceptible aphid. The conidium germinates and produces a germ tube that can penetrate the cuticle at any point (Brobyn and Wilding, 1977; Keller, 1975). It is likely that the penetration process is accomplished by both enzymatic and mechanical means. The invading hyphae colonize the haemocoel and fragment into hyphal bodies, which are then circulated through the insect, attacking first the fat bodies and then the cephalic nerve ganglia. Following invasion of the muscle tissue, the aphid usually dies. Any rhizoids produced by an entomophthoran species usually form prior to death of the host; these structures serve to anchor the host to

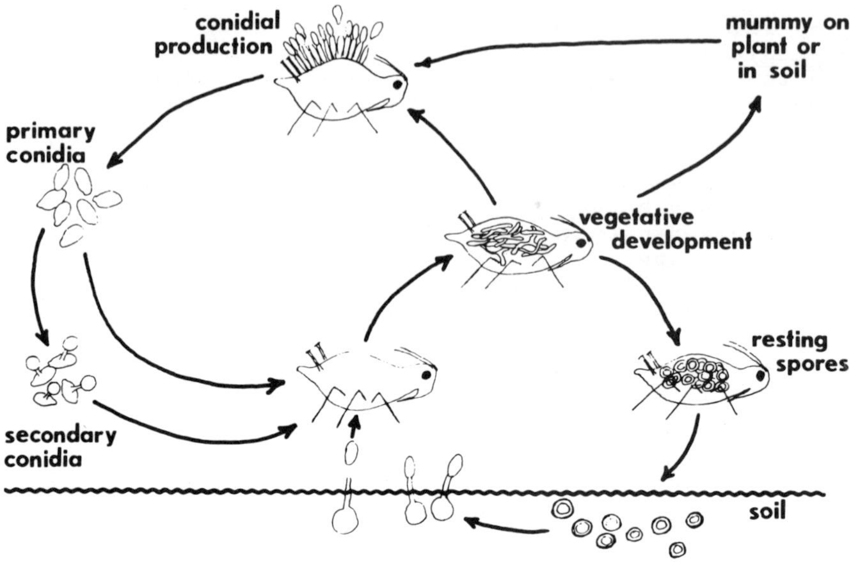

Figure 1 (Drawn by R. A. Humber)
Entomophthoran development in aphids. The development depends on rapid disease increase through conidial infections. Long-term survival is either through resting spores in the soil or dead aphids containing mycelium.

the substrate. On perennial plants, the resting spores may persist in mummified aphids until the next season (Soper, 1962).

In the case of E. neoaphidis, pseudocystidia--very large hyphal elements that break through the insect cuticle and assist in the emergence of the conidiophores--are formed. Developing conidiophores in C. obscurus and Z. radicans emerge in clusters through the host body wall without the benefit of pseudocystidia. It has been suggested that mechanical or enzymatic action on the cuticle is concentrated at these points and thus aids in the penetration of the conidiophores. Eventually however, the entire aphid surface becomes a sporulating hymenium.

The formation and discharge of conidia usually occur five hours after death of the host but may be delayed for a considerable period depending on temperature. C. obscurus, for example, was found by Wilding (1971) to discharge conidia between 5°C and 25°C but not at 0°C or 30°C. Although the earliest discharge of conidia occurred at 20°C, the maximum discharge rate was obtained at 25°C. Regardless of the temperature, however, the same number of conidia were produced.

Wilding (1971) also found that light played an important part in the discharge cycle of C. obscurus. With alternation of twelve hours of light with twelve hours of darkness, the rate of conidia discharge was substantially increased two hours after exposure to light. In the field, one would reasonably expect that the maximum rate of discharge would occur shortly after dawn. Spore trap data support this contention (Wilding, 1970).

The conidium is forcibly discharged, and if it fails to contact a susceptible host, it may produce a secondary conidium, which in turn is discharged. This process may be repeated several times until the protoplasm has been exhausted. Germination of conidia is dependent upon very high relative humidity (Wilding, 1969). In some species, such as Z. radicans, a second type of resporulation occurs. In this instance, the conidium produces a very slender capillary tube on which is formed an almond-shaped spore, generally referred to as a capillospore or an adhesive spore. This is formed perpendicular to the substrate and generally is picked up by the host as it moves over the plant surface.

Under conditions that are not well understood, resting spores may be formed. Although it has been argued that resting spores are formed in response to abiotic factors such as temperature (Wilding and Brobyn, 1973), this is not always the case. Soper and his coworkers (unpubl.) have shown that in the spruce budworm, Choristoneura fumiferana (Clemens), infected with Z. radicans, early instars 1 through 4 usually produce conidia, whereas instars 5 and 6, pupae, and adults, with extremely few exceptions, produce resting spores. This suggests that the spore state--conidia or resting spores--formed by these pathogens depends upon the physiological condition of the host.

The function of resting spores is to provide survival of the pathogen during periods of unfavorable climatic conditions. In some entomophthoran species, they are produced to the exclusion of conidia, whereas in others both conidia and resting spores are found in the same individual host (Soper, 1962; Soper and MacLeod, 1963). Resting spores are produced within the insect and are released to the environment only upon the disintegration of the dead insect. Ultimately they find their way into the soil, which serves as the reservoir of infection for the next season's aphid generations (Latteur, 1977a).

More than one mechanism may be responsible for triggering the germination of the spores. In the case of Zoophthora canadensis (MacLeod, Tyrrell et Soper) Remaudière et Hennebert on the woolly pine needle aphid, Schizolachnus piniradiatae (Davidson), field investigations have shown that the disease does not occur until early to mid-July, although aphids are present in May (Soper and MacLeod, in press). Recently Wallace et al. (1976, 1978) have shown that these spores may be germinated by exposure to fourteen hours day length or more. In the case of C. obscurus, Latgé et al. (1978a, 1979) obtained a high germination rate of resting spores by maintaining the spores at 7°C for two months in contact with soil. *C. thromboides* resting spores, on the other hand, can be induced to germinate by high-speed blending in a weak alcoholic solution (Soper et al., 1975). Kenneth and his coworkers (pers. comm.) in Israel have succeeded in germinating *Triplosporium fresenii* (Nowak.) Batko (= *Entomophthora fresenii* (Nowak.) Gustafs.) resting spores, a pathogen of aphids, by collecting resting spores that have been in the field for several months. In all of these cases, the resting spores produce another spore, which is forcibly ejected; this stage is thought to be the unit of infection. We have called these spores *germ conidia* to distinguish their origin from resting spores as contrasted to conidia with similar morphology from infected insects (Soper et al., 1975).

An alternative survival strategy has been developed by *E. neoaphidis*. There is some controversy in the literature as to whether this fungus produces resting spores. If in fact it does, they are extremely rare and probably are not the principal means used by this fungus to survive adverse conditions. Byford and Reeve (1969) have found *E. neoaphidis* in England infecting *Rhopalosiphoninus staphylae* Koch and *Myzus persicae* in marigold clumps in which the aphids had overwintered. Wilding (1973) could store entomophthoran fungi for long periods at low relative humidity and temperature within aphid mummies. Remaudière and Michel (1971) kept infected aphids in the laboratory at 2°C and -196°C; when they returned the aphids to moist conditions, the fungus sporulated, and its conidia were proven infective. Further evidence that entomophthorous pathogens may survive in this manner has been presented by Kenneth et al. (1972). They observed that *Triplosporium floridanum* (Weiser et Muma) Weiser, a pathogen of spider mites, could survive six weeks at 4°C and a further nine and a half months at room temperature. By returning the mites to a damp environment, *T. floridanum* was induced to sporulate.

Understanding the factors that control the development of the pathogen in its hosts is only a part of the story. Disease is a population phenomenon, and epizootiology delineates and isolates the key controlling factors of pathogen distribution in the population. Understanding the epizootiology is a key to manipulating the host-pathogen system.

EPIZOOTIOLOGY

The seasonal distribution of aphid populations in temperate climates tends to form definite patterns, as does the occurrence of the associated entomophthorous pathogens. Tingey and van de Klashorst (1976) recently devised a method to alter this distribution. By interplanting Chinese cabbage in areas of high populations of green peach aphids, an early buildup of this aphid could be obtained on potatoes. They were thus able to test insecticides before the usual late-season collapse of the green peach aphid population caused by entomophthoran infections. Typical examples are presented in the long-term study of Shands et al. (1963) for the potato aphid in northern Maine. The populations peaked between mid-July and mid-August, whereas the prevalence of entomophthoran pathogens lagged one to two weeks behind the highest aphid density.

The late seasonal occurrence of entomophthoran populations is common. In the evolutionary development of these host pathogen systems, some mechanism apparently has developed such that a significant proportion of the aphids escape to lay eggs for the overwintering generation.

Very few detailed epizootiological studies have been made of Entomophthorales and aphid populations. In France, researchers (Dedryver, 1978; Dedryver and Robert, 1975; Rabasse and Robert, 1975; Robert et al., 1973) investigated the relationship of _E. neoaphidis_, _T. fresenii_, and _E. planchoniana_ to their aphid host _Aphis fabae_ Scopoli. They concluded that three factors were essential in the control of epizootics: a favorable climate, a well-distributed inoculum, and a widespread spatial distribution of the host.

A quantitative five-year study of _C. obscurus_, _E. neoaphidis_, and _E. planchoniana_ on _Acyrthosiphon pisum_ (Harris) was conducted by Wilding (1975). The only climatic factor that he could relate to incidence of infection was rainfall. Voronina (1971) also thought that rainfall duration was important to entomophthoran epizootics in aphid populations.

Dean and Wilding (1971) studied the spatial and temporal occurrence of _E. neoaphidis_, _E. planchoniana_, and _C. obscurus_ in the cereal aphids _Acyrthosiphon dirhodum_ (Walker) and _Macrosiphum avenae_ (F.) in England. Early season mycosis was caused by _E. planchoniana_, which was replaced by _E. neoaphidis_ and _C. obscurus_ in late July. Although _C. obscurus_ was the most prevalent pathogen, its spatial distribution was adversely influenced by lower relative humidity in open versus sheltered fields. Further studies (Dean and Wilding, 1973) showed an alteration of this progression. Although _E. planchoniana_ occurred first, it was most common following the peak occurrence of _E. neoaphidis_ and _C. obscurus_ before mid-July. A progression of pathogens in the spruce budworm, _C. fumiferana_, was reported by Vandenberg and Soper (1979b). The entomophthoran species were synchronized with instars. _Zoophthora radicans_ was the first to occur and was most abundant in the fifth instar. This was followed by _Entomophthora egressa_ MacLeod et Tyrrell (nom. inval., = _Entomophaga aulicae_ (Reich.) comb. inedit. = _Entomophthora aulicae_ (Reich.) Sorok) in sixth instar and _Conidiobolus_ sp. in prepupae and pupae.

In a study of _C. obscurus_ on the potato aphids

(Shands et al., 1963), fungal infection seemed to be regulated by host density-independent factors. Shands and coworkers concluded that weather was of major importance in the initiation and subsequent spread of Entomophthorales. On the other hand, Missonier et al. (1970) placed a greater importance on an evenly and densely distributed inoculum as being necessary for an epizootic. They found that disease outbreaks followed three days with 90 percent relative humidity for ten hours per day and rain for five hours per day, followed by a mean temperature of 20°C. They thought that epizootics were independent of host density. Wilding (1975) also proposed that host density is of little importance for controlling the spread of *C. obscurus* on pea aphids, whereas he reported a high correlation between infection and the number of airborne entomophthoran conidia. All of these studies involved aphids on annual crops.

In contrast, Soper and MacLeod (in press) conducted a five-year study on *Z. canadensis*, which attacks the forest aphid *S. piniradiatae*. This aphid completes its entire cycle on red pine (*Pinus resinosa* R. T.) and is relatively uncomplicated by the migration of aphids from other crops. Also a red pine plantation is a reasonably stable ecosystem. During this period of study, the disease passed through preepizootic, epizootic, and postepizootic stages. Soper and MacLeod concluded that disease incidence was not limited by climatic factors but, as the Dedryver and Robert (1975) study determined, that host population density and inoculum levels were the major factors controlling epizootics. The preepizootic phase was characterized by a greater distribution of the susceptible population than the inoculum, whereas the postepizootic phase was characterized by greater distribution of the pathogen than the host. It was also found that a population threshold for disease initiation become lower following the epizootic. For purposes of predicting epizootics, the average host density by itself was a poor indicator of population levels; the degree of host aggregation was more important. Given equal average population densities, a less-aggregated population had the greater probability of sustaining an epizootic. Although population density controlled initiation of epizootics (MacLeod et al., 1966), the speed of infection transmission became greater with increasing inoculum density, evidenced by the more rapid kill during the postepizootic phase.

The role of moisture as a limiting ecological factor for entomophthoran epizootics is frequently overstressed. Kenneth and Olmert (1975) reported *T. fresenii* from aphids at Elisha Springs, an oasis in the Israeli desert. Microclimate is surely the more important consideration.

Although it has been widely believed that airborne conidia are responsible for long-distance distribution of the disease, this is apparently not the case. In the recent study (Wilding et al., 1977), epizootics were initiated in the pea aphid on broad beans by the introduction of living diseased aphids. Transmission from the inoculated sites to check plots did not occur despite their close proximity. It has been long known that winged aphids are more susceptible to entomophthoran infection than are the wingless forms (Rockwood, 1950). Probably migrating infected winged aphids enable the long-distance dispersal of Entomophthorales to aphid populations. A study on the green peach aphid on potatoes in Florida found that E. neoaphidis occurred in the alate aphids one week prior to its occurrence in the apterous forms (Figure 2). The migrating aphids sustained 50 percent infection during the early stages of the aphid population buildup. During the period of this study, infection in the wingless aphids did not reach 10 percent. The study of Z. canadensis on S. piniradiatae provided additional evidence for the importance of long-distance dispersal of entomophthorales in aphid populations (Soper, unpubl.). An isolated red pine plantation on St. Josephs Island, Lake Huron, Ontario, was hand infested with the aphids and the population monitored for several years. In the fourth year, Z. canadensis was found at high prevalence within the woolly pine needle aphid population. This fungus is believed specific for this aphid species

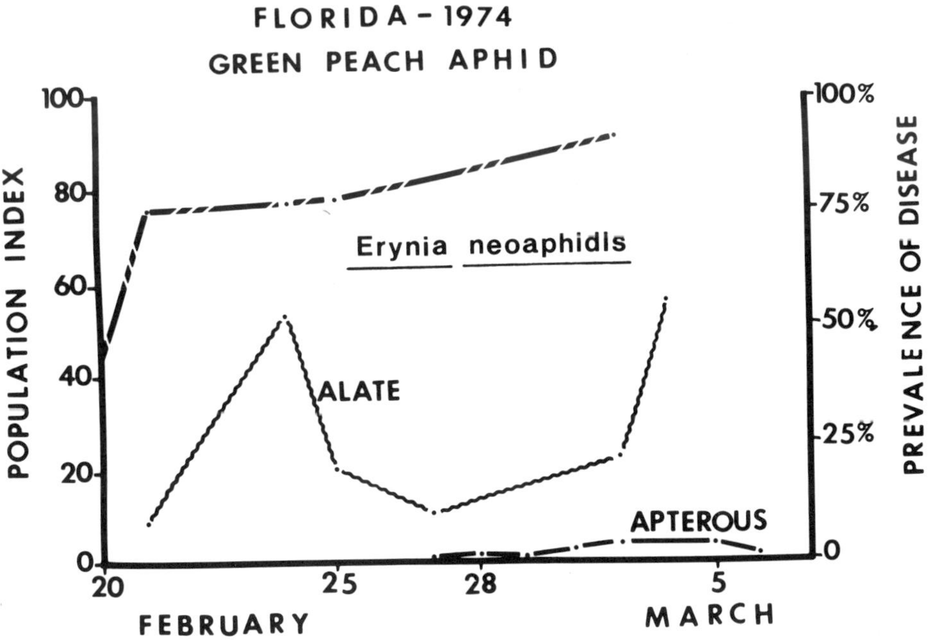

Figure 2
Natural dispersal of Erynia neoaphidis by migrating winged aphids.

(MacLeod et al., 1979). A survey of the general area disclosed an infected population approximately 2 kilometers upwind. It is highly likely that this served as the reservoir infection and that the fungus was transmitted by migrating aphids.

How can IPM Managers use this kind of background information to exploit the entomophthoran complex on potato aphids for their control? There are two basic approaches. One is through the enhancement of infections, and the second is by the development of Entomophthorales as a mycoinsecticide. The strategy is to induce an epizootic early in the season and prevent population buildup beyond the economic threshold. This can be accomplished either through culture practices or by the application of resting spores preconditioned to germinate. The success of either approach depends upon an aphid monitoring system. It is very important to predict the early stages of an epizootic in order that chemical pesticide applications can be discontinued. Monitoring of aphid populations for the occurrence of mycosis is being used extensively in the Soviet Union (Yevlakhova, 1978) and has resulted in a significant reduction in the use of chemical pesticides.

ENHANCEMENT OF INFECTIONS

It would be difficult, if not impossible, to produce consistently a potato crop without the use of fungicides to control early blight, late blight, and the other fungal potato pathogens. Since the Entomophthorales are fungi, it is reasonable to assume that the presence of these fungicides could interfere with the transmission of these pathogens to the aphids. Early field evidence in support of this proposition was presented by Jaques and Patterson (1962) working with the European apple sucker, *Psylla mali* Schmidberger, in the Annapolis Valley, Nova Scotia, Canada. They have shown that in orchards treated with fungicides, not only was the onset of the disease used by *Z. radicans* delayed, but also in some instances epizootics were prevented (Figure 3). Nanne and Radcliffe (1971) demonstrated that green peach aphid populations on Russet Burbank potatoes were significantly greater when treated with diafolatan, dithane M-45, or Bordeaux mixture (Table 2). By late August, the untreated plots had 22 percent diseased aphids as compared to 4 to 5 percent in those treated with fungicides. Although an epizootic finally occurred in all populations, it had been delayed in the fungicide-treated plots.

Mellado et al. (1976) concluded that application of Disyston reduced control of cereal aphids by entomophthorales. In contrast, Furadan had no effect.[1]

[1]Trade names are provided in the appendix to this chapter.

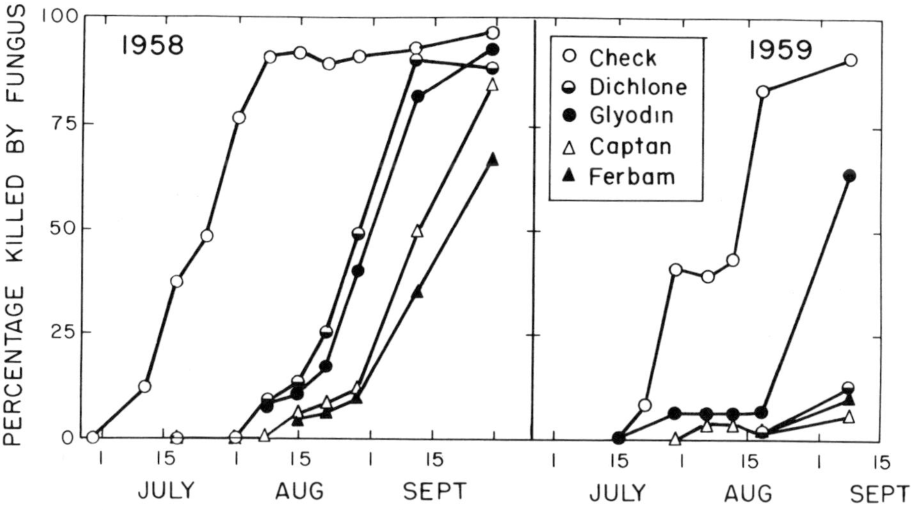

Figure 3
Adverse effect of fungicides on Zoophthora radicans infection in apple suckers (after Jaques and Patterson, 1962).

Under laboratory conditions, Soper et al. (1974) showed differential susceptibility between entomophthorales and the potato pathogen Alternaria solani (Ellis et Martin) Jones et Grant to various fungicides. Although under laboratory conditions all fungicides had some effect on the insect pathogens tested, chlorothalonil had the least effect, whereas benomyl and maneb severely limited their growth. When insecticides were tested, it was found that carbaryl, endosulfan, oxydemeton-methyl, and azinphosmethyl were all detrimental to entomophthorales, whereas demeton had little effect. Fritz (1976) found that chlorothalonil, ethirimol, copper oxychloride, and sulphur had little or no action on Entomophthorales; other fungicides such as benodanil, quinomethionate, chloroneb, and triarimol were highly toxic. In general,

Table 2
Prevalence of Entomophthoran Fungi in Myzus Persicae on Russet Burbank Potatoes

	lb AI/acre/wk kilo AI/ha/wk	Aug. 24 % Disease	Aug. 26 X̄ Aphids per plant
Difolatan	1 ½	5.2 a	5453 a
Dithane M-45	1 ¼	4.0 a	4347 a
Bordeaux	4-4-50	4.5 a	3521 ab
Control	- - -	22.4 b	2063 b

Source: Nanne and Radcliffe, 1971.
Note: Differences indicated at 5 percent significance level.

he found systemic fungicides more effective on mycelial growth than nonsystemics. In similar tests with E. neoaphidis, C. obscurus, and C. thromboides, Zimmerman (1976) found that spore germination was inhibited by Calinix, Imugan, Plantvax, and Saprol. Benomyl and Milstem affected only E. neoaphidis conidia. Of the chemicals tested, only Cercobin-M had little or no effect. Majchrowicz and Mietkiewski (1976), working in Poland, also reported severe in vitro inhibition of entomophthorales by fungicides and insecticides.

Two options are open to the integrated pest managers to circumvent the detrimental effect of agrichemicals on insect fungi. The first is obvious and immediately available: to choose those fungicides, insecticides, and herbicides that have the least effect on entomophthoran species yet will do the job for which they are intended. The second is to introduce pesticide-resistant strains of the aphid pathogens. This alternative is technically feasible. Selection of fungicide-resistant isolates of entomophthorales is currently underway in our laboratory (McCabe, pers. comm.). We have found that Z. radicans can be habituated to chloroneb with little difficulty.

Hall and Dunn (1957) introduced C. obscurus and C. thromboides into advancing populations of the spotted alfalfa aphid, Therioaphis maculata (Buckton), during its initial spread in the San Joaquin Valley, California. More recently, Soper and Milner (unpubl.) introduced Z. radicans into populations of the same aphid in Australia. The fungus was grown in liquid media, consisting of 6 percent sunflower oil, 2 percent dextrose, and 2 percent yeast extract, modified after Latgé et al. (1978b). When the fungus was actively growing in the log phase, it was filtered on Whatman 3 paper and incubated approximately eighteen hours at 25°C. Following incubation, the mycelium produced conidiophores and conidia. In the field, about ten alfalfa plants were enclosed in a large plastic container and the mycelia on the filter paper inverted over an opening in the top. Exposure was from approximately sundown to sunrise the next morning, sufficient time to cause high levels of infection. Within several weeks, the disease had spread sixty meters from the points of the infection epicenters. Successful introductions were made both on irrigated fields at Moonbi, New South Wales, and under dry land conditions at Arding, New South Wales. This same technique easily could be adapted to the introduction of fungicide-resistant isolates in potato fields.

The level of inoculum when the disease first occurs controls the explosiveness of an epizootic. One way to provide high levels of entomophthoran inoculum in potato fields earlier than they usually occur might be through the establishment of reservoirs in intercrops. For example, strips of grain might be planted at intervals between potato plots. It would be expected that grain aphid populations would peak a number of weeks prior to potato aphids. Not only would there be a possibility of the entomopathogen's carrying over from the grain aphids to the potato aphids, but also aphid parasites and predators should increase. Perrin (1975), for example, advocated cutting nettles in mid-June in England to maximize the effect of natural enemies, including Entomophthorales, on nearby aphid infestations on agricultural crops. Although these enhancement techniques should prove beneficial in reducing aphid populations, the ultimate management ploy would be the use of these pathogens as a mycoinsecticide. This usage would give the integrated pest managers the maximum flexibility in utilizing entomophthoran species for aphid control.

DEVELOPMENT OF ENTOMOPHTHORALES AS A MYCOINSECTICIDE

Although in the Soviet Union *C. obscurus* has been used for aphid control on a limited scale (Tsinovski and Egina, 1972), no entomophthoran species currently are registered in the United States for insect control. Nevertheless there have been some important recent developments toward this goal. A program was initiated (Soper et al., 1975; Soper, 1978) to develop Entomophthorales for potato aphid control. Initially this program was centered around *C. thromboides* (Humber et al., 1977) but has since changed its emphasis to *C. obscurus*. The ultimate objective of this research is to develop information to satisfy the U.S. Environmental Protection Agency's requirements for registration. Another important goal is to demonstrate to industry that entomophthorales can be produced economically and will be as effective as chemical aphicides. The program is divided into three major lines of research: safety and monitoring, efficacy, and production.

Safety and Monitoring

It is expected that the entomophthoran species that occur on aphids pose no safety threats to mammals, however, it is still prudent to conduct infectivity tests because some species in the Entomophthorales are known to infect mammals. Rhinophycomycosis entomophthorae is caused by *Conidiobolus coronatus* (Costantin) Batko (= *Entomophthora coronata* (Costantin) Kevorkian) and has been reported attacking horses (Bridges et al., 1962) and chimpanzees (Roy and Cammeron, 1972). Eckert et al. (1972) reported a deep entomophthoran phycomycotic infection from the lungs of a fifteen-month-old boy in the United States. The etiological agent in this mycosis was recently identified by King and Jong (1976) as *Conidiobolus incongruus* Drechsler.

The aphid pathogen *C. thromboides* has been tested

for infectivity to mice (Soper and Bryan, 1974). When mice were subjected to an acute oral dose of *C. thromboides* resting spores, no adverse effects were observed, and it was concluded that large oral doses were harmless to mammals. Similar studies were conducted by Parafiniuk et al. (1976) on *C. obscurus* against guinea pigs. In this case, the animals were subjected to both inhalation and subcutaneous injection of conidia and resting spores; no pathological signs were observed in the guinea pigs during the fifty-five days of the experiment. It is expected that no harmful side effects to humans or domestic animals will occur in the use of the entomophthoran species from aphids, but their effect on beneficial insects is a legitimate concern.

Although the difference in pathogenicity and host range among entomophthoran species might be expected, there are also demonstrable differences in pathogenicity and host range among isolates of the same species. In a recent study by Milner and Soper (unpubl.) *Erynia nourvi* Remaudière et Hennebert (= *Entomophthora exitialis* Hall et Dunn), *C. obscurus*, *Z. radicans*, and *C. thromboides* (a total of twelve strains) were comparatively bioassayed against the spotted alfalfa aphid. It was found that the LC_{50} varied from several hundred conidia per square millimeter to less than one per square millimeter. In this instance, two strains of *Z. radicans* proved to be most pathogenic. In a series of experiments to determine infectivity, various *Z. radicans* isolates were tested against several insect species (Table 3). The isolates

Table 3
Strain selection of *Zoophthora radicans*

Source of *Zoophthora radicans* Isolates	Insects Tested		
	Choristoneura fumiferana	*Musca domestica*	*Myzus persicae*
Choristoneura fumiferana (RS-88), Maine	Susceptible	Susceptible	Resistant
Neodiprion tsugae (RS-13), Alaska	Susceptible	Not tested	Resistant
Empoasca (RS-62), Israel	Resistant	Susceptible	Resistant
Brachycaudus amygdalinus (RS-75), Israel	Resistant	Resistant	Susceptible
Therioaphis maculata (RS-114), Israel	Susceptible	Susceptible	Susceptible

Source: Kenneth, pers. comm.

were obtained from the spruce budworm (*C. fumiferana*) from Maine, the hemlock sawfly (*Neodiprion tsugae* Middleton) from Alaska, a leafhopper (*Empoasca* sp.), and two different aphids from Israel (Brachycaudus amygdalinus (Schouteden) and T. maculata). The isolates of Z. radicans were tested for their ability to infect spruce budworm, the housefly (Musca domestica L.), and the green peach aphid. Responses varied from the spotted alfalfa aphid isolate, which infected all three insects, to the leafhopper isolate, which attacked only the housefly. In addition, it was found that the temperature optimum varied from 20°C from the spruce budworm isolate to 27°C for the Israeli aphid isolate. From this evidence, one would also expect that it might be possible to select isolates with differences in humidity requirements for spore germination. Papierok and Wilding (1979) showed a difference in infectivity between two clones of pea aphids to two strains of C. obscurus. Convincing evidence was presented to show that this was caused by host resistance.

Efficacy

Essential to the development of Entomophthorales as aphicides is the use of a standardized bioassay system. This allows a quantitative comparison of pathogenicity of various isolates and preparations. Wilding (1976) was the first to devise a reliable bioassay for an entomophthoran in which an LD_{50} could be calculated. He tested E. neoaphidis and C. obscurus against the pea aphid, Acyrthosphon pisum (Harris) by placing pea aphids in cavities within a plastic block; the cover contains a hole of known diameter positioned over each aphid. The immobilized aphids are then exposed to conidial showers. The number of discharged spores per unit area can be determined under a microscope and a dosage per aphid calibrated for each exposure. This method, however, is very labor intensive. Latteur (1977b) attempted to improve on this technique by immobilizing aphids on filter paper previously covered with conidia.

A more satisfactory method has been devised by Vandenberg and Soper (1979a) for Z. radicans against the spruce budworm. In this technique, dosage is calibrated from conidial showers in terms of conidia per square millimeter. It has been used successfully with minor variation by Milner and Soper (unpubl.) for Z. radicans against the spotted alfalfa aphid. It has also been used by Holbrook and Soper (unpubl.) for C. thomboides against green peach aphid. In these assays, the pathogen is grown in a suitable liquid media. When the fungus is growing in log phase, the medium is filtered through sterile filter paper and the remaining mycelia rinsed several times with sterile distilled water. Excess moisture is drained off and the fungus incubated in closed dishes for eighteen to twenty-four hours at optimum temperatures under continuous light. For C. thromboides, the fungus was grown on Sabouraud agar containing both maltose and dextrose. Previous experiments showed that the combination of these sugars would stimulate conidial formation (Soper, unpubl.). At the end of this incubation period, the fungus is generally sporulating vigorously. In the spruce budworm assay, water agar plates were exposed to spore showers both before and after groups of sixth-instar larvae. The dosage was then calculated from the conidial densities on these plates. For the spotted alfalfa aphid bioassays, dosage was determined by placement of a coverslip within the infection chamber, which contained the aphids. Following exposure, the coverslip was mounted on a slide and the spore density counted utilizing a microscope.

Bioassays utilizing conidia are extremely useful during the strain selection process and for determination of qualitative changes that might occur to the fungus because of cultural conditions or mutations. Ultimately, however, a bioassay must be utilized that is based on resting spores because this is the stage that would be produced commercially. Such an assay was developed for _C. thromboides_ against _Myzus persicae_ (Holbrook and Soper, unpubl.). It was determined that the unit of infection is not the resting spore but rather the germ conidia that it produces. To obtain quantities of germ conidia, petri dishes containing water agar were sprayed with varying concentrations of resting spores. These were then incubated at 25°C for thirty hours in a sixteen-hour day length. Germ conidia production is maximum under this temperature photoperiod regime at thirty hours (Figure 4). The bioassays were then conducted as previously described except for the use of showers of germ conidia as the inoculum.

Production

The principal objectives of production research are to develop an inexpensive process with high yields of entomophthoran resting spores and to develop suitable formulations for field application. Many studies have been made to develop media for resting spore production (for example, Sawyer, 1929; Egina et al., 1972; Gröner, 1975). Most media have not been suitable for commercialization. With commercial-grade nutrient sources, it was possible to produce large quantities of _C. thromboides_ resting spores (Latgé et al., 1977). Dextrose and corn syrup were found to be the best carbon sources, and yeast extract, soybean flour, and cottonseed flour were the best nitrogen sources. With a nutrient concentration of 12 percent, 3×10^6 spores per milliliter were obtained in five days with a fourteen liter (1) fermenter. It was found that during the recovery procedure, if the resting spores were subjected to high-speed blending in the presence of a weak alcohol solution, high germination rates could be obtained on

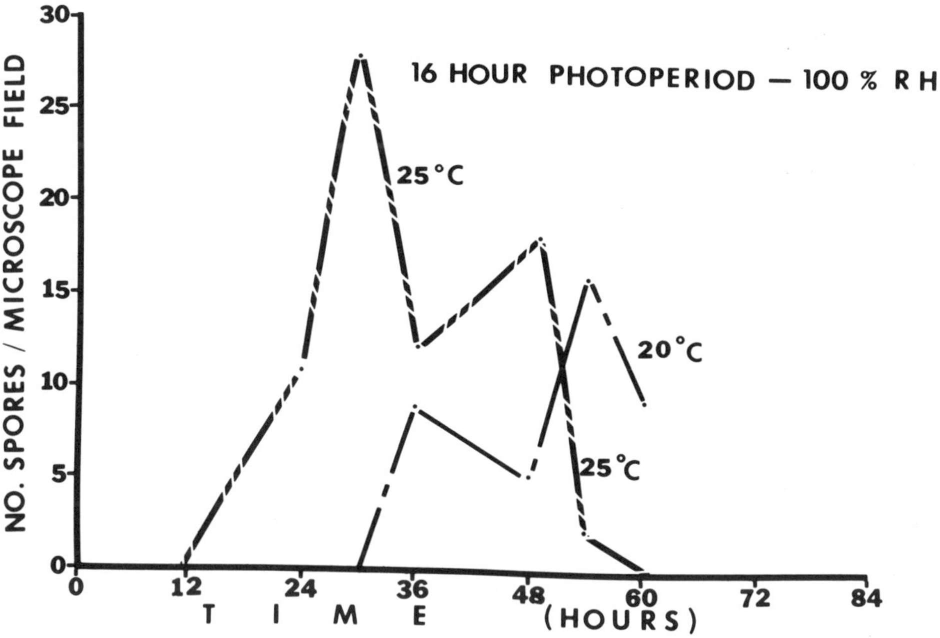

Figure 4
Production of infective units, germ conidia, by *Conidiobolus thromboides. Such production is temperature dependent. Germ conidia may produce secondary spores, which accounts for the bimodal occurrence.*

rehydration of the spores (Soper et al., 1975). Production of C. thromboides resting spores has been successfully scaled up to 250-ℓ pilot plant fermenter by use of industrial-type centrifugation and spray drying for recovery (Soper, 1978). Latgé et al. (1978a) have also been able to produce high yields of C. obscurus resting spores in fermenters by use of commercial grade components. Recently they have developed a media based on corn steep liquor, which has been used successfully to scale up production of C. obscurus to 100-ℓ pilot plant fermenter (Latgé, pers. comm.). To obtain germination of these spores, however, it is necessary to subject them to vernalization process. When the spores are stored at 7°C for two months in the presence of soil, high rates of germination are obtained (Latgé et al., 1979).

The next important step in the development of a mycoinsecticide is formulation. Having the ability to produce large quantities of resting spores with a high level of germination is not sufficient. The spores must also be stabilized to remain viable at ambient temperatures and when applied to the field to cover the foliage adequately. Experiments conducted by Mietkiewski and Soper (1979) were designed to produce a 25 percent wettable powder formulation of C. thromboides.

Accordingly the adjuvants chosen represented those required for this type of formulation. The major portion of any wettable powder is an inert diluent. Those tested in these experiments were Barden and Geopon. Also tested were the dispersing and emulsifying agents Nakal BX 78 and Tamol and wetting agents Selogen W and Renex 35. Spores were mixed with these various adjuvants and held for twenty-four weeks at room temperature. In general, all adjuvants appeared compatible with C. thromboides resting spore viability. An important finding, however, was that two products significantly increased the percentage of resting spore germination. Use of diluent Geopon produced significantly higher germination as compared to the control. Also the dispersing agent Tamol increased spore germination when used at the highest rate tested (10 percent) as compared to 1 percent, 5 percent, or the control; unfortunately none of these compounds stabilized the spores for long-term storage since germination was significantly lower when tested at the termination of the experiment in twenty-four weeks.

Although it is now possible to produce a wettable powder for field testing, the problem of long-term storage remains. It is important that a stabilized resting spore formulation be developed with a shelf life of at least twelve months to allow accumulation of spores during the manufacturing process and provide time for their subsequent distribution. Majchrowicz and Soper (1978) showed that storage of C. thromboides resting spores was possible for several years at 4°C; however, refrigeration is not economically feasible for a commercial preparation. To be practical, a mycoinsecticide must be storable at ambient temperature for at least one year.

CONCLUSIONS

The complex of entomophthoran species that occurs on potato aphids is an extremely important component of the natural enemies complex. Until now, however, their use in integrated pest management for potato aphid control has been only passive. With current technology, much could be done to enhance the activity of these pathogens. Particular attention should be paid to pesticide compatibility. Where it becomes necessary, the introduction of pesticide-resistant entomophthoran strains could be made into potato growing districts. In addition, the natural prevalence of these pathogens should be monitored with the intention of ceasing chemical pesticide applications when entomophthoran epizootics are imminent. The buildup of reservoirs of Entomophthorales and aphids in adjacent noncrop plants should be investigated. This might prove a useful means of transmitting disease from aphids that do not attack potatoes to those that do. Ultimately it is hoped that a mycoinsecticide based on an entomophthoran species from aphids can be made available to IPM managers, providing the ultimate flexibility in manipulating these entomopathogens for insect control.

REFERENCES

Bridges, C. H., W. M. Romans, and C. W. Emmons, 1962, Phycomycosis of Horses Caused by Entomophthora coronata, J. Am. Vet. Med. Assoc. 140:673-677.

Brobyn, P. J., and N. Wilding, 1977, Invasive and Developmental Processes of Entomophthora Species Infecting Aphids, Trans. Br. Mycol. Soc. 69:349-366.

Byford, W. J., and G. J. Reeve, 1969, Entomophthora Species Attacking Aphids in England 1962-66, Trans. Br. Mycol. Soc. 52:342-346.

Dean, G. J., and N. Wilding, 1971, Entomophthora Infecting the Cereal Aphids Metapolophium dirhodum and Sitobion avenae, J. Invertebr. Pathol. 18:169-176.

Dean, G. J., and N. Wilding, 1973, Infection of Cereal Aphids by the Fungus Entomophthora, Annu. Appl. Biol. 74:133-138.

Dedryver, C. A., 1978, Facteurs de limitation des populations d'Aphis fabae dans l'Ouest de la France III. Répartition et incidence des différentes espèces d'Entomophthora dans les populations, Entomophaga 23:137-151.

Dedryver, C. A., and Y. Robert, 1975, Influence du taux d'inoculum de sa répartition et de sa variation sur l'expression et la progression d'une mycose d'insecte, Cas des Entomophthorales sur pucerons. Ann. Phytopathol. 7:346.

Eckert, H. L., G. H. Khoury, R. S. Pore, E. F. Gilbert, and J. R. Gaskell, 1972, Deep Entomophthora Phycomycotic Infection Reported for the First Time in the United States, Chest 61:293-294.

Egina, K. Ya., Ya. P. Tsinovskii, and Z. P. Chaudare, 1972, Growth and Development of the Fungus Entomophthora thaxteriana Petch on Various media, Latvian. Pathol. Nasek. Klesh, (Riga): 95-110.

Engler, R., and M. H. Rogoff, 1976, Entomopathogens: Ecological Manipulation of Natural Associations, Environ. Health Perspect. 14:153-159.

Franz, J. M., 1973, The role of Biological Control in Pest Management, Bull. Lab. Entomol. Agraria (Napoli) 30:235-243.

Fritz, R., 1976, Action de quelques fongicides sur la croissance mycélienne de trois espèces d'Entomophthorales, Entomophaga 21:239-249.

Gröner, A., 1975, Production of Resting Spores of Entomophthora thaxteriana, J. Invertebr. Pathol. 26:393-394.

Hall, I. M., and P. H. Dunn, 1957, Entomophthorous Fungi Parasitic on the Spotted Alfalfa Aphid, Hilgardia 27:159-181.

Hammond, D., and F. R. Holbrook, 1979, Maine Finds Methods to Control Leaf Roll: Results of Pilot Research Program, Spudman 17:18-20.

Humber, R. A., 1978, Entomophthora ignobilis, the Correct Name for a common Fungal Pathogen of Aphids, Mycologia 70:208-210.

Humber, R. A., R. S. Soper, N. Wilding, and G. Remaudière, 1977, The Identification of Certain Widely Studied Strains of Entomophthora Pathogenic for Aphids, Mycotaxon 5:207-210.

Jaques, R. P., and N. A. Patterson, 1962, Control of the Apple Sucker, Psylla mali Schmidb, by the Fungus Entomophthora sphaerosperma (Fresenius), Can. Entomol. 94:818-825.

Keller, S., 1975, Histologische untersuchengen an parasitierten, Entomophthora--infizierten Erbsenblattläusen, Acyrthosiphon pisum, Bull. Soc. Entomol. Suisse 48:247-252.

Kenneth, R., and I. Olmert, 1975, Entomopathogenic Fungi and Their Insect Hosts in Israel: Additions, Irs. J. Entomol.

Kenneth, R., G. Wallis, U. Gerson, and H. N. Plaut, 1972, Observations and Experiments on Triplosporium floridanum (Entomophthorales) Attacking Spider Mites in Israel, J. Invertebr Pathol. 19:366-369.

King, D. S., and S. C. Jong, 1976, Identity of the Etiological Agent of the First Deep Entomophthoraceous Infection of Man in the United States, Mycologia 68:181-183.

Latgé, J.-P., D. Perry, O. Reisinger, B. Papierok, and G. Remaudière, 1979, Induction de la formation des spores de résistance d'Entomophthora obscura Hall et Dunn, C. R. Acad. Sci. Ser. D 288:599-601.

Latgé, J.-P., G. Remaudière, and M. Diaquin, 1978a Un nouveau milieu pour la croissance et la sporulation d'Entomophthorales pathogènes d'Aphides, Ann. Microbiol. Paris 129B:463-476.

Latgé, J.-P., G. Remaudière, and B. Papierok, 1978b, Un example de recherche en lutte biologique: les champignons Entomophthora pathogènes de pucerons, Bull. Soc. Pathol. Exot. 71:196-203.

Latgé, J.-P., R. S. Soper, and C. D. Madore, 1977, Media Suitable for Industrial Production of Entomophthora virulenta Zygospores, Biotechnol. Bioeng, 19:1269-1284.

Latteur, G., 1977a, Sur la possibilité d'infection directe d'Aphides par Entomophthora à partir de sols hébergeant un inoculum naturel, C. R. Acad. Sci. Ser. D 284:2253-2256.

Latteur, G., 1977b, Méthode indirecte d'infection des pucerons par Entomophthora spp., Parasitica 33:80-84.

MacLeod, D. M., J. M. Cammeron, and R. S. Soper, 1966, The Influence of Environmental Conditions on Epizootics Caused by Entomogenous Fungi, Rev. Roum. Biol. Ser. Bot. 2:125-134.

MacLeod, D. M., D. Tyrrell, and R. S. Soper, 1979, *Entomophthora canadensis* N. sp., a Fungus Pathogenic on the Woolly Pine Needle Aphid, *Schizolachnus piniradiatae*, *Can. J. Bot.* 57:2663-2672.

Majchrowicz, I., and R. Mietkiewski, 1976, Wplyu insektycydow i fungicydow stosowanych w ochronie ziemniakow i burakow na rozwoj grzbyow owadobojczczych *in vitro*, *Zesz. Mauk, Akad. Roln. Szczecinie* 53:133-142.

Majchrowicz, I., and R. S. Soper, 1978, Badania nad kielkowaniem i przechowywaniem zarodnikow przetrwalnikowych *Entomphthora virulenta* Hall et Dunn, *Zesz. Nauk, Akad. Rol. Szczecinie* 72:159-166.

Mellado, M., A. M. Flores, and R. Carillo, 1976, Acción de disulfoton y carbofuran en formulación granular en el combate de áfidos en Trigo, *Agro. Sur.* 4:111-118.

Mietkiewski, R., and R. S. Soper, 1979, Progress towards Commercial Formulation of *Entomophthora virulenta* Resting Spores, *Proc. 2nd Intern, Colloq. Invert. Pathol., Prague*, pp. 137-138.

Missonier, J., Y. Robert, And G. Thoizon, 1970, Circumstances épidémiolgiques semblant favoriser le développement des mycoses a Entomophthorales chex trois aphides, *Aphis fabae* Scop., *Capitophorus horni* Borner et *Myzus persicae* Sulz., *Entomophaga* 15:169-190.

Nanne, H. W., and E. B. Radcliffe, 1971, Green Peach Aphid Populations on Potatoes Enhanced by Fungicides, *J. Econ. Entomol.* 64:1569-1570.

Papierok, B., and N. Wilding, 1979, Mise en evidence d'une différence de sensibilité entre 2 clones du puceron du pois, *Acyrthosiphon pisum* Harr. (Homoptères: Aphididae), exposés a 2 souches du champignon Phycomycète: *Entomophthora obscura* Hall and Dunn, *C.R. Acad. Sci. Ser. D* 288:93-95.

Parafiniuk, W., I. Majchrowicz, and J. Januszewski, 1976, Badania wlasnosci patogennych *Entomophthora thaxteriana* (Petch) dla świnek morskich, *Wiadom. Parazytol.* 22:245-251.

Patch, E. M., 1907, The Potato Plant Louse, *Maine Agric. Exp. Stn. Bull.* 147:235-257.

Perrin, R. M., 1975, The Role of the Perennial Stinging nettle, *Urtica dioica*, as a Reservoir of Beneficial Natural Enemies, *Ann. Appl. Biol.* 81:289-297.

Rabasse, J. M., and Y. Robert, 1975, Facteurs de limitation des populations d'*Aphis fabae* dans l'Ouest de la France. II. Incidence des mycosis à *Entomophthora* sur les populations des hôtes primaires et de la féverole de printemps, *Entomophaga* 20:49-63.

Remaudière, G., J.-P. Latgé, and B. Papierok, 1979, Reconsidération taxonomique de *Entomophthora obscura* Hall et Dunn, *Ann. Microbiol. Paris* 130A:151-162.

Remaudière, G., and M. F. Michel, 1971, Première expérimentation écologique sur les Entomophthorales (Phycomycètes) parasites de pucerons en vergers de pêchers, *Entomophaga* 16:75-94.

Robert, Y., J. M. Rabasse, and P. Schetes, 1973, Facteurs de limitation des populations d'Aphis fabae dans l'Ouest de la France. I. Epizootiologie des maladies à Entomophthorales sur féverole de printemps, Entomophaga 17:61-75.

Rockwood, L. P., 1950, Entomogenous Fungi of the Family Entomophthoraceae in the Pacific Northwest, J. Econ. Entomol. 43:704-707.

Roy, A. D., and H. M. Cameron, 1972, Rhinophycomycosis Entomophthoreae Occurring in a Chimpanzee in the Wild in East Africa, Am. J. Trop. Med. Hgy. 21:234-237.

Sawyer, W. H., 1929, Observations on Some Entomogenous Members of the Entomophthoraceae in the Artificial Culture, Am. J. Bot. 16:87-121.

Shands, W. A., I. M. Hall, and G. W. Simpson, 1962, Entomophthoraceous Fungi Attacking the Potato Aphid in Northeastern Maine in 1960, J. Econ. Entomol. 55:174-179.

Shands, W. A., G. W. Simpson, and I. M. Hall, 1963, Importance of Entomogenous Fungi in Controlling Aphids on Potatoes in Northeastern Maine, Maine Agric. Expt. Stn. Bull. T-6., 42 p.

Shands, W. A., G. W. Simpson, and H. E. Wave, 1972, Seasonal Population Trends and Productiveness of the Potato Aphid on Swamp Rose in Northeastern Maine, Maine Life Sci. Agric. Exp. Stn. Tech. Bull. 52, 35 p.

Soper, R. S., 1962, The Woolly Pine Needle Aphid Schizolachnus piniradiate (Davidson), Attacked by the Fungus Entomophthora aphidis (Hoffmann), Can. Dep. For. Entomol. Path. Branch Bi-Mon. Prog. Rep. No. 18.

Soper, R. S., 1978, Development of Entomophthora Species as Possible Microbial Insecticides, Proc. 1st US/USSR Conf. Prod. Select. Stand. Entomopath. Fungi, Riga, May 20-21, pp. 270-282.

Soper, R. S., and T. A. Bryan, 1974, Mammalian Safety of the Aphid-Attacking Fungus, Entomophthora nr. thaxteriana, Environ. Entomol. 3:346-347.

Soper, R. S., F. R. Holbrook, and C. C. Gordon, 1974, Comparative Pesticide Effects on Entomophthora and the Phytopathogen Alternaria solani, Environ. Entomol. 3:560-562.

Soper, R. S., F. R. Holbrook, I. Majchrowicz, and C. C. Gordon, 1975, Production of Entomophthora Resting Spores for Biological Control for Aphids, Maine Life Sci. Agric. Exp. Stn. Tech. Bull. 76, 15 p.

Soper, R. S., and D. M. MacLeod, 1963, Spore Morphology of Entomophthora fresenii Nowakowski, J. Insect. Pathol. 5:478-482.

Soper, R. S., and D. M. MacLeod, 1980, Descriptive Epizootiology of an Aphid Mycosis, USDA Tech. Bull. (in press).

Tingey, W. M., and G. van de Klashorst, 1976, Green Peach Aphid: Magnification of Field Populations on Potatoes, J. Econ. Entomol. 69:363-364.

Tsinovski, Ya., P., and K. Ya. Egina, 1972, Entomophthora Fungi Used in the Control of Aphids. Latvian. Pathol. Nasek. Klesh. (Riga):73-94.

Vandenberg, J. D., and R. S. Soper, 1979a, A Bioassay Technique for Entomophthora sphaerosperma on the Spruce Budworm, Choristoneura fumiferana, J. Invert. Pathol. 33:148-154.

Vandenberg, J. D., and R. S. Soper, 1979b, Prevalence of Entomophthorales Mycoses in Populations of Spruce Budworm, Choristoneura fumiferana, Environ. Entomol. 7:847-853.

Voronina, E. G., 1971, Epizootiae of Entomophthorsis of Acyrthosiphon pisum Harris (Homoptera: Aphidoidea), Rev. Entomol. USSR 4:780-799 (in Russian).

Wallace, D. R., D. M. MacLeod, and A. J. DeLyzer, 1978, Endogenous Light Action in Germination of Entomophthora aphidis resting spores in vitro, Can. For. Serv. Bi-Mon. Res. Notes. No. 34, pp. 24-25.

Wallace, D. R., D. M. MacLeod, C. R. Sullivan, D. Tyrrell, and A. J. DeLyzer, 1976, Induction of Resting Spore Germination in Entomophthora aphidis by Long-Day Light Conditions, Can. J. Bot. 54:1410-1418.

Wilding, N., 1969, Effect of Humidity on the Sporulation of Entomophthora aphidis and E. thaxteriana, Trans. Br. Mycol. Soc. 53:127-130.

Wilding, N., 1970, Entomophthora Conidia in the Air-Spora, J. Gen. Microbiol. 62:149-157.

Wilding, N., 1971, Discharge of Conidia of Entomophthora thaxteriana Petch from the Pea Aphid, Acyrthosiphon pisum Harris, J. Gen. Microbiol. 69:417-422.

Wilding, N., 1973, The Survival of Entomophthora spp. in Mummified Aphids at Different Temperatures and Humidities, J. Invertebr. Pathol. 21:309-311.

Wilding, N., 1975, Entomophthora Species Infecting Pea Aphids, Trans. R. Entomol. Soc. 127:171-183.

Wilding, N., 1976, Determination of the Infectivity of Entomophthora spp., Proc. 1st Internat. Colloq. Invert. Pathol. (Kingston): 296-300.

Wilding, N., and P. J. Brobyn, 1973, Resting Spore Formation by Entomophthora fresenii, Annu. Rep. Rothamsted Exp. Stn. 1973, 198 p.

Wilding, N., P. J. Brobyn, and Best, 1977, Entomophthora Species Controlling Bean Aphids: The Effect of Fungicides on Entomophthora, Annu. Rep. Rothamsted Exp. Stn. 1977 (part 1):102-103.

Yevlakhova, A. A., 1978, Basic Trends in the Use of Entomopathogenic Fungi in the Soviet Union, Proc. 1st US/USSR Conf. Prod. Select. Stand. Entomopath. Fungi, Riga, May 20-21, pp. 36-50.

Zimmermann, G., 1976, Über die wirkung systemischer fungizide auf Aphiden befallenden Entomophthoraceen (Zygomycetes) in vitro, Z. Pflanzenkr. Pflanzenchutz. 83:261-269.

APPENDIX

Common, Trade, and Chemical Names of Some Insecticides and Fungicides

Common Name	Trade Name	Chemical Name
Azinphos-Methyl	Guthion®	O,O-dimethyl S-[(4-oxo-1,2,3-benzotriazin-3(4H)-yl)methyl] phosphorodithioate.
Carbaryl	Sevin	1-naphthyl methylcarbamate.
Carbofuran	Furadan®	2,3-dihydro-2,2-dimethyl-7-benzofuranyl methylcarbamate.
Demeton	Systox	O,O-diethyl O (and S)-[2-(ethylthio)ethyl] phosphorothioates.
Disulfoton	Di-Syston®	O,O-diethyl S-[2-(ethylthio)ethyl] phosphorodithioate.
Endosulfan	Thiodan®	6,7,8,9,10,10-hexachloro-1,5,5a,6,9,9a-hexahydro-6,9-methano-2,4,3-benzodioxathiepin-3-oxide.
Oxydemeton-methyl	Metasystox-R®	S-[2-(ethylsulfinyl)ethyl]O,O-dimethyl phosphorothioate.
Oxythioquinox	Morestan	cyclic S,S-(6-methyl-2,3-quinoxalinediyl) dithioacarbonate.
Benodanil		2-iodo-N-phenylbenzamide (or 2-iodobenzanilide).
Benomyl	Benlate	methyl 1-(butylcarbamoyl)-2-benzimidazole-carbamate.
Bordeaux	Not available	Copper sulfate/hydrated lime mixture.
Captafol	Difolatan	cis-N-[(1,1,2,2-tetrachloroethyl)thio]-4-cyclohexene-1,2-dicarboximide.
Chloraniformethane	Imugan	N-[2,2,2-trichloro-1-(3,4-dichloroanilino) ethyl]=formamide.
Chloroneb	Demosan Tersan SP	1,4-dichloro-2,5-dimethoxybenzene.
Chorothalonil	Bravo	tetrachloroisophthalonitrile.
Ethirimol	Milgo Milstem	5-butyl-2-(ethylamino)-6-methyl-4-pyrimidinol.
Mancozeb	Dithane®M-45	manganous ethylenebis[dithiocarbamate] + (zinc ion).
Maneb	Dithane®M-22 Manzate	manganous ethylenebis[dithocarbamate].
Oxycarboxin	Plantvax	5,6-dihydro-2-methyl-1,4-oxathiin-3-carboxanilide 4,4-dixoide.
Triarimol	Elanco Products-273	α(2,4-dichlorophenyl)-α-phenyl-5-pyrimidine-methanol.
Triforine	Saprol	N,N'-[1,4-piperazinediylbis(2,2,2-trichloro=ethylidene)]bis-[formamide].
Tridemorph	Calixin	2,6-dimethyl-4-tridecylmorpholine.
Thiophanate-Methyl	Cercobin-M	dimethyl [(1,2-phenylene)bis(iminocarbonothioyl)] bis[carbamate].

® = Registered trademark

NOTE ADDED IN PROOF

The following scientific names have been recently changed. The synonyms are noted below:

1. *Erynia canadensis* (MacLeod, Tyrrell et Soper) Humber et Ben Ze'ev (syn. *Zoophthora canadensis*).

2. *Erynia radicans* (Brefeld) Humber et Ben Ze'ev (syn. *Zoophthora radicans*).

Biological Control of Potato Pests

George Tamaki

QUANTITATIVE APPROACHES TO BIOLOGICAL CONTROL

Two past research projects in biological control at the USDA Yakima Agricultural Research Laboratory emphasize the laboratory's direction toward quantitative biological control.

The first begins with this example: A pest management scout reports to the potato grower that he has found an average of twenty green peach aphids (GPA) and two lygus bugs per ten plants. The beneficial insects present are two coccinellids (ladybeetles), one large chrysopid larva (green lacewing), and three geocorids (big-eyed bug). In the present state of the art, the scout can only guess what the effect these predators on the pests will be.

To overcome this problem, our objective was to develop a predictive model based on the reductive impact of the predator complex on the pests. With this model, the scout will place the predator species found into one of three classes, based on their consumption rates as determined by Tamaki and Olsen (1977) and such other workers as Goodarzy and Davis (1958) and Simpson and Burkhardt (1960). (See Table 1.) The three geocorids are placed in class 1, the general feeders. The large chrysopid larva is placed in class 2, the aphidophagous predators that are estimated to consume four times more than class 1 predators. The two ladybird adults are placed in class 3, the most voracious group, which consumes about eight times more than class 1. The feeding differences permitted a concept called predator power. Predator power (K_n) is defined by the formula $K_n = \Sigma N_i C_i$ where N_i is the number of predators in class i and C_i is the relative feeding capacity of that class. In other words the predator power is the potential voracity of the predators rather than simply the number of predators present.

The rate of increase per day of the GPA (q) at different months of the growing season is based on four

Table 1

Average Rate of Predation of Green Peach Aphid by Selected Adult Predators

Species	Number of Predators	Mean (± SE) Predation per Day per Predator
Coccinella transversoguttata Falderman	70	52.70 ± 3.43
Nabis alternatus Parshley	100	10.37 ± 0.62
Anthocoris melanocerus Reuter	50	8.46 ± 0.74
Geocoris bullatus (Say)	80	8.30 ± 0.57
Scymnus marginicollis Mann	90	7.99 ± 0.43
Bembidion spp.	40	6.66 ± 0.67
Geocoris pallens Stål	80	6.47 ± 0.52
Orius tristicolor (White)	90	5.31 ± 0.43

Table 2
Predator Efficacy (E) and Rate of Aphid Increase (q)

	E (Aphids Consumed per Day)	q (Increase per Day)
May	12.69 ± 1.15	1.03 ± 0.04
June	15.80 ± 1.47	1.08 ± 0.01
July	13.17 ± 0.57	1.15 ± 0.01
August	11.98 ± 0.55	1.17 ± 0.00
September	6.39 ± 0.65	1.07 ± 0.00
October	4.90 ± 0.00	1.05 ± 0.00

Note: Based on data collected 1972-1975.

years of field cage studies, shown in Table 2. The predator efficacy (E) was also calculated for each month. These values are dependent on temperature, prey density, the biomass of the prey, the nature and species of the host plant, and other seasonal variation. With q and E available, the scout will need only to calculate the predator power (K_n) and estimate the average number of pests (A_n) per plant. Then the scout could estimate the number of pests likely to be present the next day (A_{n+1}) by the formula

$$A_{n+1} = (A_n - EK_n)\, q, \qquad (1)$$

using the values of E and q corresponding to the month the sample was taken.

Currently we are expanding this model to include the effects of temperature and predator searching ability as influenced by pest and plant densities. Our goal is to develop a simple, practical model for predicting pest population growth trends based on the reductive impact of the natural enemies on the key pests.

Our second approach to quantitative biological control was to evaluate the effect of the natural enemy on a key pest species and determine the effect on crop protection, including the estimate of the prevention of yield loss through biological control (Tamaki and Butt, 1978).

First, in the laboratory, we studied the consumption rate of each life stage of *Perillus bioculatus*, a pentatomid predator, on the eggs and the third and fourth larval stages of the Colorado potato beetle (CPB) (Table 3). Next, we determined the consumption rate of each life stage of the CPB on potato leaves (Table 4). The adult CPB leaf consumption rate was determined for each day of life after emergence. In the field, we determined the size of the CPB population on different sampling dates based on the different life stages (Table 5).

Knowing the size and duration of the different life stages of the CPB in the field, we can calculate their

Table 3
Consumption Rates of Nymphal Stages of *Perillus bioculatus* Fed CPB Eggs and Larvae

Nymphal Stage	Number of CPB Eggs Consumed per Day	Number of CPB Larvae Consumed per Day
1	0.12	0.002
2	3	0.028
3	7	0.099
4	20	0.251
5	39	0.634

Table 4
Developmental and Consumption Rates of *Leptinotarsa decemlineata* Reared on Potato Leaves at 24° ± 7°C

Instar	Number	Average Number of Days to Develop (± SE)	Average Leaf Area Consumed (mm^2) Per Instar	Per Day per Instar
1	70	2.89 ± 0.05	65	23
2	65	2.52 ± 0.09	161	64
3	63	2.84 ± 0.08	517	182
4	62	3.66 ± 0.11	1967	538

Table 5
Field Populations of *Leptinotarsa decemlineata* on Potatoes

Sampling Date	Number of Plants	Average Number per Plant					
		Eggs	Larval Instar 1	2	3	4	Adults
June 7	100	58.6	1.6	0	0	0	0.0
June 16	100	88.9	1.8	0.8	0.5	0	1.4
June 21	50	98.4	20.4	5.4	2.8	2.7	0.5
June 28	50	87.2	7.3	8.1	5.1	9.0	0.6
July 8	50	37.3	2.5	5.3	6.1	15.5	0.7
July 14	50	9.4	0.2	1.2	3.6	9.4	12.0

potential leaf consumption (Table 6). We also measured the total leaf area of the plants on different dates with a photoplanimeter (Table 7) to determine the percentage defoliation of the plants.

With the predator-prey-plant data accumulated, we were able to estimate the total impact of the predators. For instance, in the absence of the predator (Table 8), the field population of the CPB on June 16 is estimated to defoliate 3.7 percent of the plant. If one predator per plant is added to the system, the percentage of defoliation could be decreased to 0.39 percent. At the higher population level of July 14, the addition of two predators per plant had little effect on decreasing the defoliation.

Table 6
Consumption Potential of Field Populations of Leptinotarsa decemlineata on Potato Leaves

Sampling Date	Leaf Area Consumed by Each Life Stage (mm^2) Larval Instar 1	2	3	4	Adult
June 7	36	0	0	0	618
June 16	41	51	91	0	962
June 21	463	345	510	1453	344
June 28	166	517	929	4842	412
July 8	57	338	1111	8339	481
July 14	5	77	656	5057	8244

Note: Based on feeding capacities of each life stage developed in laboratory studies

Table 7
Leaf Area of Potato Plants Taken from Field

Sampling Date	Number of Plants	Average Area (cm^2) (± SE)
June 16	20	1549 ± 142
July 1	10	3208 ± 310
July 14	20	3793 ± 256
August 2	20	6265 ± 402

Table 8
Percentage of Leaf Area Defoliated in Five Days in Absence from and Presence of Predators from Plants

Number of Fifth-Instar Perillus/Plant	Leaf Area Defoliated June 16	June 28	July 14
0.0	3.7	10.7	18.5
0.5	2.04	9.9	17.8
1.0	0.39	9.1	17.2
2.0	--	7.5	15.8

Finally, to estimate the effect of defoliation on plant yield, we used the data collected by Sparks and Woodbury (1967) (Figure 1) and by Takatori et al. (1952), which gave the date corresponding to the height of the potatoes. The loss in yield of U.S. No. 1 potatoes can be estimated by using Figure 2 and the percentage of defoliation given in Table 8. For example, at 50 percent bloom, 10 and 20 percent defoliation will cause about 8 and 14 percent loss in yield of U.S. No. 1, respectively.

Thus, the emphasis is placed not only on the degree of predation on the population of the CPB by the predator P. bioculatus but, more importantly, the degree of prevention of plant injury and yield loss resulting from the biological control of the pest.

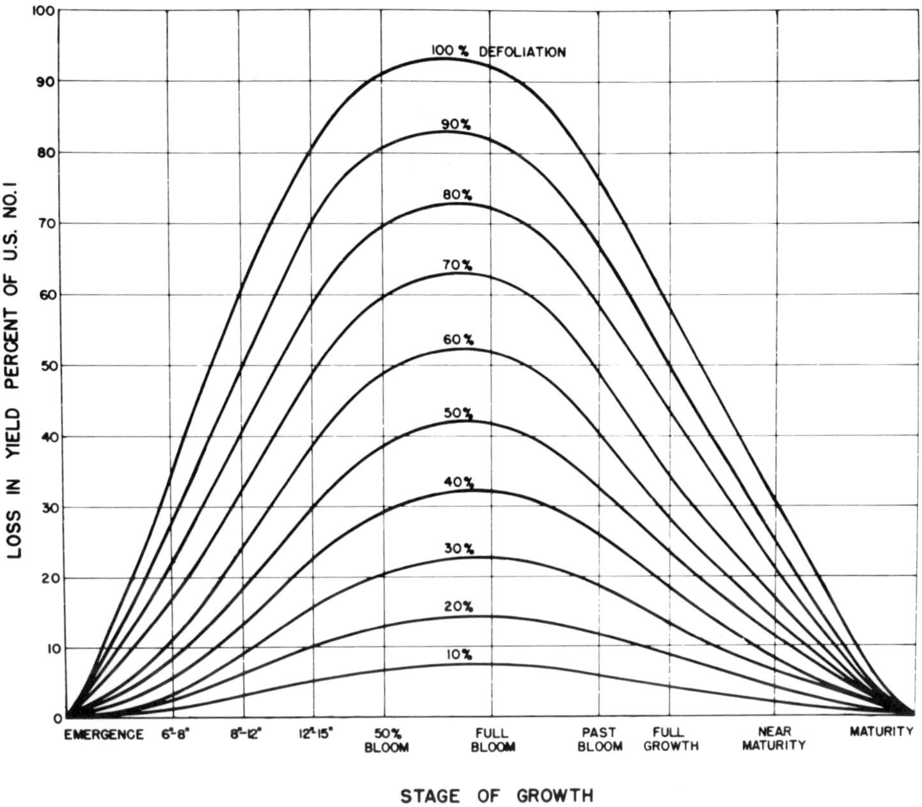

Figure 1
Method of estimating yield loss when percentage of defoliation and stage of growth are known (after Sparks and Woodbury, 1967).

USING THE PREDATOR COMPLEX

The term predator complex is frequently used in discussions concerning biological control, but it has never been fully developed as a concept and its potential or actual role in the control of key insect pests has never been defined. Of course, before the predator complex concept can be explored, the single species approach, in which the biology and ecology of individual species are studied, must be utilized. In many earlier studies, the main criterion used to evaluate a predator species for its potential effectiveness was the number of prey it consumed per day. Usually predators with lower predation rates were judged to be less effective agents when compared to the more voracious predators. But clearly a more quantitative approach is needed to evaluate the reductive effect of natural enemies against

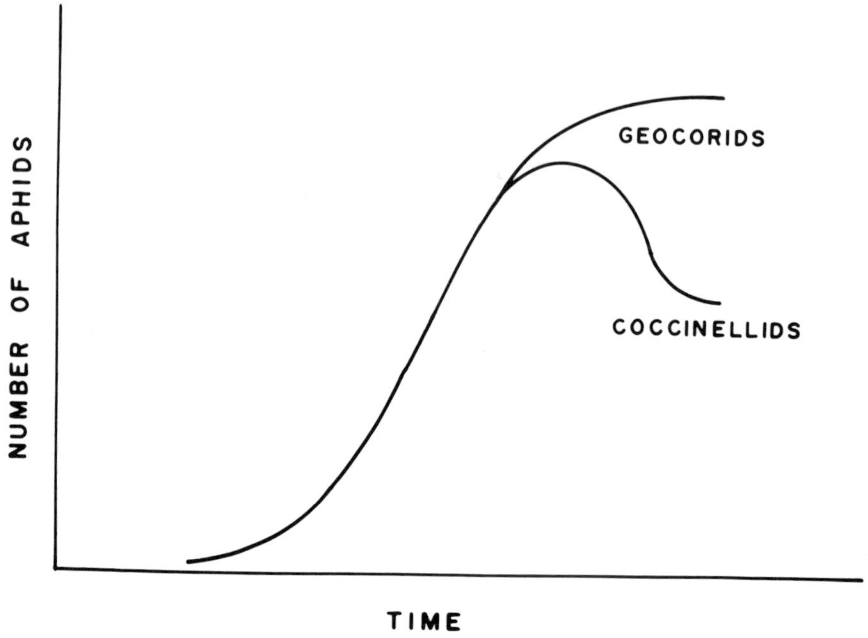

Figure 2
Schematic representation of effect of Coccinella transversoguttata and Geocoris spp. on aphids.

key pests in terms of either the single species or predator complex approach. In single species studies, I use the life table parameter, r_m (the intrinsic rate of increase of the pest), and the consumption rate of the predator for a predator-prey quantitative interaction. This is especially useful in comparing different predator species and pest species with different r_m.

For instance, Geocoris pallens Stål consumes about 6.5 ± 0.5 SE green peach aphids per day (Tamaki and Olsen, 1977). (For argument, assume G. pallens prefers young nymphs.) The intrinsic rate of increase, r_m of the GPA is 0.40 on radish (Tamaki and Olsen, 1979). Then we calculate R = the number of reproductive GPA kept at a zero rate of growth by one G. pallens by using the formula:

$$R = \frac{C_p}{(e^{r_m} - 1)}, \qquad (2)$$

where C_p is the predator consumption rate and r_m the instrinsic rate of increase per female per day. Thus one Geocoris adult would limit 13.2 reproductive GPA^s to zero population growth.

The more voracious predator, Coccinella transversoguttata Falderman, consumes 52.7 ± 3.43 GPA per day (Tamaki and Olsen, 1977). From Equation 2 and the r_m of the GPA on radish, one C. transversoguttata would limit 107.2 reproductive GPA^s to zero population growth.

In evaluating these two predators, we would tend automatically to consider C. transversoguttata a more effective predator than G. pallens. This is especially evident during the asymptotic phase of the schematic growth curve of the GPA shown in Figure 2; C. transversoguttata is a more effective predator than G. pallens. In field studies, Tamaki and Weeks (1973) demonstrated that released coccinellids decreased the populations of the GPA, while other GPA populations were still increasing in plots treated by releases of geocorids, nabids, or chrysopids.

Pest Densities and Predation

Although the consumption rates of different predator species are a means of evaluating their effectiveness, we should not accept this criterion as a generalization. Another equally important criterion is the relationship of the predator species to the population growth phase of the pest. To illustrate, Figure 3 shows

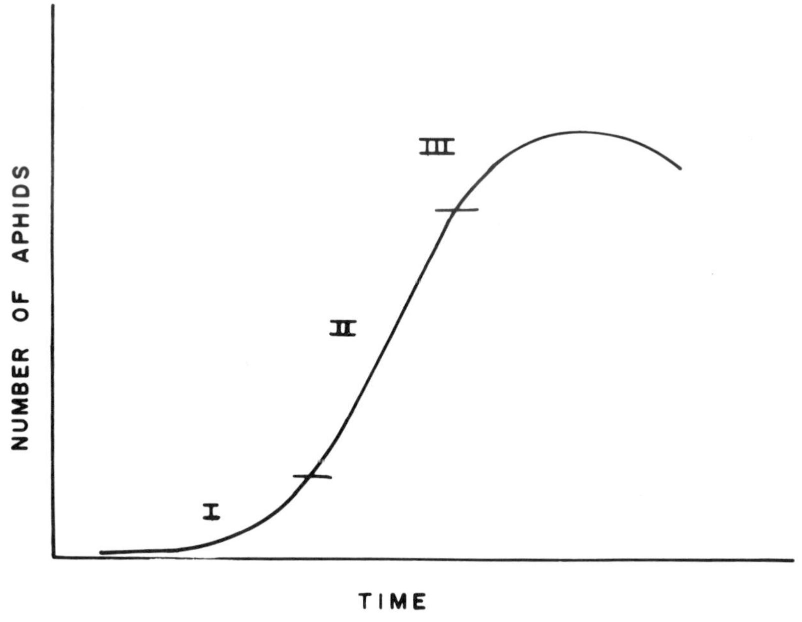

Figure 3
Generalized growth curve of the pest population. Phase I: Incipient growth; Phase II: Accelerated growth; Phase III: Asymptotic growth

a generalized growth curve of the pest population divided into three phases. Phase I is incipient growth, phase II accelerated growth, and phase III asymptotic growth.

As Table 9 shows, the predator species responded differently to low and high densities (Tamaki and Long, 1978). The numerical responses of geocorids and nabids to high densities of GPA were up to 1.2 times that of low densities. We arbitrarily considered a difference of less than 1.5 times as essentially no substantial response to increasing aphid density. The large coccinellids responded to aphid densities by ovipositing an average of six times more eggs on plants with high than with low densities of aphids. Syrphid predators also showed a differential response to high aphid densities, with ratios of 2.8 for the larvae and 5.18 for eggs during both years.

Adult green lacewings (Chrysopidae) laid 6.9 times more eggs on plants with high than with low aphid densities.

The classification of the predator species into phase I, II, or III predators is based on presence and effect during each phase. For instance, the adult coccinellids oviposit during the latter part of phase II, so in phase III the predation rate increases because the effect of the adults is increased by the presence of predatory larval progeny. Many biological control successes have occurred during phase III when a once rapidly increasing pest population has declined rapidly.

Geocorids, on the other hand, are present on many of the vegetable and field crops before the pest populations actually become established. For that reason and because they feed on only a few insects per day, have a wide prey range, and can survive on plant food in the absence of insect food, geocorids are best classified as phase I predators (Tamaki and Weeks, 1972). Geocorids generally are the most abundant predators on vegetable crops in the Yakima Valley, but they have been considered a general predator or an ineffective predator because they feed on so few prey per day. Of course, a predator should be evaluated on the basis of prey consumed, its prey preference, and functional or numerical responses, but it also should be evaluated relative to the growth phase of the pest population.

As a phase I predator, geocorids contribute to biological control by delaying the establishment of the phase II population growth of the pest. In this way, the phase I predators contribute to pest control strategy. For example, they may interfere with an insect-vector-virus interaction by delaying arrival of the vector into fields, preventing early infection of young seedlings, and thereby reducing yield loss (Figure 4). Also, with key pest species, the delay in occurrence of phase II may eliminate the need for early insecticide protection.

Table 9
Numerical Response Ratios of Different Stages of Predator Species to Low and High Densities of Green Peach Aphid

Species	Density, 1974			Density, 1975			1974-1975
	Low	High	Ratio(H:L)	Low	High	Ratio(H:L)	Ratio(H:L)
Coccinella							
Adult	23.2	63.0	2.7:1	32.2	53.0	1.6:1	2.1:1
Larva	0.8	8.4	10.5:1	4.0	5.4	1.4:1	2.9:1
Egg	27.8	266.8	9.6:1	26.0	54.0	2.1:1	6.0:1
Geocoris							
Adult	79.4	98.0	1.2:1	42.8	43.2	1.0:1	1.2:1
Larva	36.2	40.6	1.1:1	1.0	5.4	5.4:1	1.2:1
Nabis							
Adult	3.2	4.2	1.3:1	2.2	2.4	1.2:1	1.2:1
Larva	--	--	--	2.4	1.8	0.8:1	0.8:1
Scymnus							
Adult	2.2	3.8	1.7:1	7.8	6.8	0.9:1	1.1:1
Nymph	--	0.4	--	3.8	3.0	0.8:1	0.9:1
Syrphid							
Larva	0.6	0.8	1.3:1	0.2	1.4	7.0:1	2.8:1
Egg	1.2	19.0	15.8:1	5.2	14.2	2.7:1	5.2:1
Chrysopa							
Larva	0.2	0.2	1.0:1	0.4	1.2	3.0:1	2.3:1
Egg	3.2	15.6	4.9:1	--	6.4	--	6.9:1
Spiders	3.0	3.0	1.0:1	8.6	5.0	0.6:1	0.7:1
Orius sp.	1.6	4.0	2.5:1	1.8	--	--	1.2:1
Bembidion sp.	4.0	8.8	2.2:1	0.4	1.4	3.5:1	2.3:1
Others[a]	0.3	0.2	0.7:1	1.8	1.4	0.7:1	0.8:1
Aphids	125	588	4.7:1	145	535	3.7:1	4.2:1

[a] Stethorus, Deraeocoris, Anthocoris, and Staphalinids.

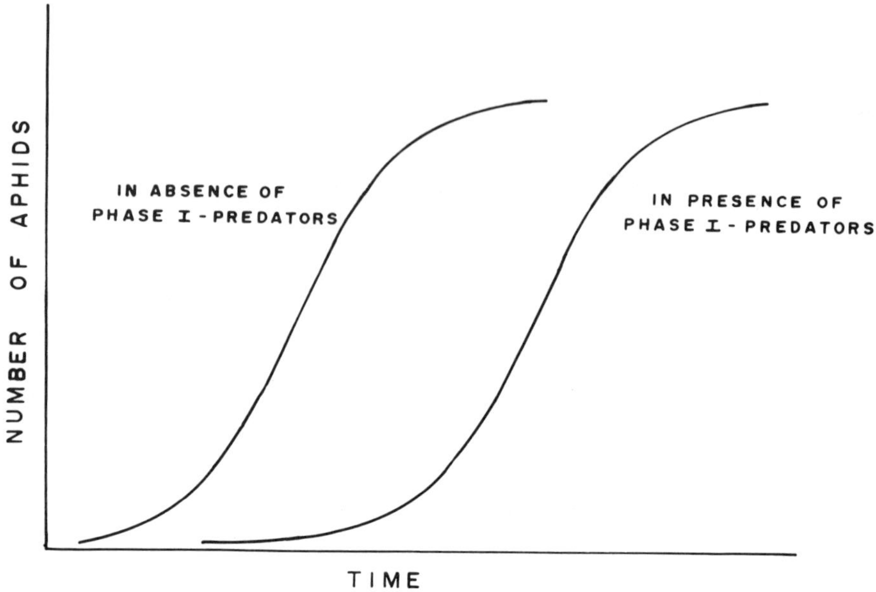

Figure 4
Schematic illustration of effect of Phase I predators such as geocorids in delaying population growth of a pest.

Predator Complex as a Single Species Concept

In the single species approach, voraciousness, density-dependent action (Nicholson, 1958), and numerical or functional responses (Holling, 1966) are used in evaluating predator effectiveness. In the predator complex approach, all species of predators attacking the pest at a given time are considered as a single unit and the unit is evaluated. Thus, in the predator complex approach, coccinellids actually may not be any more important than geocorids in overall effect on the population growth curve of the GPA.

Biological Control of GPA

The predator complex reacts as an idealized predator species in the control of GPA. During phase I of the population growth of GPA on potatoes and sugarbeets in the Yakima Valley, geocorids are the primary predators. As the population of GPA increases and moves toward phase II, *Nabis* spp. and *Scymnus marginocollis* Mannerheim appear. In the latter part of phase II, the more voracious predators such as syrphids and chrysopids appear in response to the increase in pest density, as do the coccinellid adults (Tamaki and Long, 1978). The coccinellid species, along with their progeny, then have a major part in causing a decline in the population growth rate as it moves into phase III.

Each predacious species has a role and is associated with one or more of the growth phases of the GPA population. Thus a species may contribute to the delay of the accelerated growth, to decelerated growth, or to a decline in the population of GPA. Because predators with low prey consumption rates are usually abundant at low pest densities, the few GPA they consume may have as great (or greater) an effect on the population of GPA as a more voracious predator feeding on the high populations of GPA. Figure 5 illustrates this point. It shows the

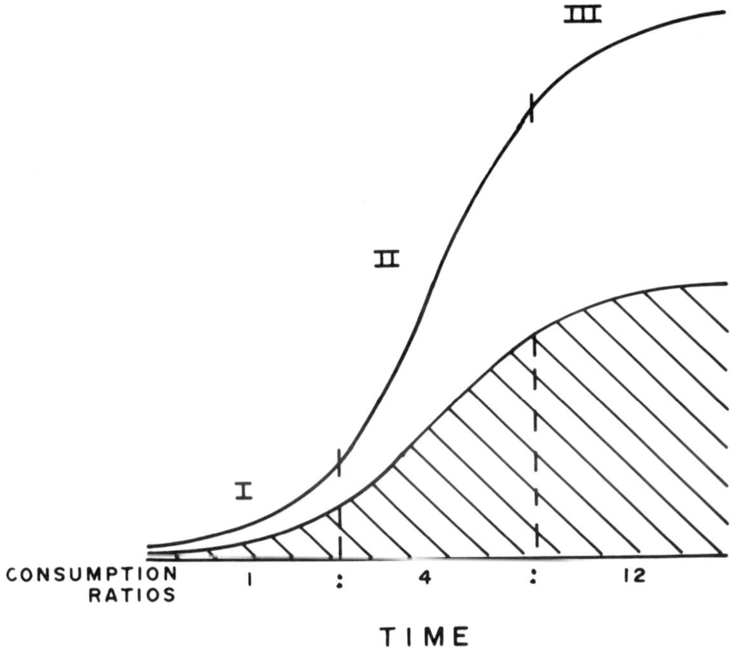

Figure 5
Schematic population growth curve of aphids. The shaded area represents the amount of predation required to reduce the population by half.

schematic growth curve of the GPA and the amount of predation required (shaded area) to reduce the population by half. Thus for a predator to produce the same percentage of reduction in phases I, II, and III, its estimated consumption rates would have to be one, four, and twelve times as great, respectively, in each of these phases.

In fact natural biological control of the GPA in the Yakima Valley is reasonably effective because the predator complex contains a number of different species that attack the GPA in all three phases of its population growth. However, the GPA is a vector of potato leafroll virus (PLRV) and this affects its management strategy on potatoes. The emphasis is on phase I of the growth curve because the economic threshold for the pest is so low. As a result, current biological research at Yakima emphasizes geocorids and nabids that are phase I or early phase II predators.

Still another method of favoring the predators is to manipulate the pest population by suppressing it before it arrives in the potato fields. This approach is being investigated in a pilot study underway in the Yakima Valley. An attempt is being made to suppress GPA populations in their early spring habitats where millions are generated before they migrate into potato fields in early summer. Peach trees are one generating site; another is cultivated orchard floors where broadleaf weeds grow profusely between cultivations and generate 70 million migrants per hectare (Tamaki and Olsen, 1979); the third is the banks of drainage ditches fed by warm springs (Wallis and Turner, 1969; Tamaki et al., 1979) where summer (parthenogenic) forms are able to survive when the winter is mild. The broadleaf weeds in the orchards therefore will be replaced, and ditch banks will be provided with permanent grass cover since grass is not a host of the GPA.

In attacking the vector-GPA-virus interaction, the strategy is to reduce the reservoir of spring migrants of the GPA to potato and other field crops. This should allow the phase I predators to delay the move of the population into phase II.

Colorado Potato Beetle

The situation of the Colorado potato beetle (CPB), *Leptinotarsa decemlineata* (Say), is quite different. The only two natural enemies in the Yakima Valley are the pentatomid predator, *Perillus bioculatus* (F.), and the tachinid parasite, *Doryphorophaga doryphorae* (Riley). A one-species predator complex, even when it is supported by a tachinid fly, has little effect on the CPB. Harcourt (1971) reported that natural control of CPB occurs only when the CPB eats its food supply so fast that it starves to death.

Currently *D. doryphorae* is being reared and studied to determine how the flies overwinter and why they appear late in the season after CPB has done its damage.

Also the ecology and biology of *P. bioculatus* (Tamaki and Butt, 1978) are being studied. The predator has been found to be limited in its effectiveness because it is never abundant, it never wastes food, feeding upon the prey until it is sucked dry, and it prefers not to eat alone (commonly two predators will feed on the same CPB larva while hundreds of other larvae surround them).

Thus, because of its feeding behavior, *P. bioculatus* cannot function as an effective predator.

SUMMARY

In summary, pest control strategy must be based on the evaluation of the predator complex and not solely on the individual predator species. The goal in biological control should be to utilize the entire complex of parasite, predators, and pathogens as a single reductive force.

REFERENCES

Goodarzy, K., and D. W. Davis, 1958, Natural Enemies of the Spotted Alfalfa Aphid in Utah, *J. Econ. Entomol*. 51:612-616.

Harcourt, D. G., 1971, Population Dynamics of *Leptinotarsa decemlineata* (Say) in Eastern Ontario. III. Major Population Process, *Can. Entomol*. 103:1049-1061.

Holling, C. S., 1966, The Functional Response of Invertebrate Predators to Prey Density, *Mem. Entomol. Soc. Can. No. 48*, 86 p.

Nicholson, A. J., 1958, Dynamics of Insect Populations, *Ann. Rev. Entomol*. 3:107-136.

Simpson, R. G., and C. C. Burkhardt, 1960, Biology and Evaluation of Certain Predators of *Therioaphis maculata* (Buckton), *J. Econ. Entomol*. 53:89-94.

Sparks, W. C., and G. W. Woodbury, 1967, Stages of Potato Plant Growth, *Idaho Agric. Exp. Stn. Bull. 309*, 22 p.

Takatori, F. H., W. C. Sparks, and G. W. Woodbury, 1952, A Study of Simulated Hail Injury on Potatoes, *Univ. Idaho Res. Bull. 22*, 15 p.

Tamaki, G., and B. A. Butt, 1978, Impact of *Perillus bioculatus* on the Colorado Potato Beetle and Plant Damage, *USDA Tech. Bull. 1581*, 11 p.

Tamaki, G., and G. E. Long, 1978, Predator Complex of the Green Peach Aphid on Sugarbeets: Expansion of the Predator Power and Efficacy Model, *Environ. Entomol*. 7:835-842.

Tamaki, G., and D. Olsen, 1977, Feeding Potential of Predators of *Myzus persicae*, *J. Entomol. Soc. B.C*. 74:23-26.

Tamaki, G., and D. Olsen, 1979, Evaluation of Orchard Weed Hosts of Green Peach Aphid and the Production of Winged Migrants, *Environ. Entomol*. 8:314-317.

Tamaki, G., L. Fox, and B. A. Butt, 1979, Ecology of the Green Peach Aphid as a Vector of Beet Western Yellows Virus of Sugarbeets, *USDA Tech. Bull. 1599*, 16 p.

Tamaki, G., and R. E. Weeks, 1972, Biology and Ecology of Two Predators *Geocoris pallens* Stål and *G. bullatus* (Say), *USDA Tech. Bull. 1446*, 46 p.

Tamaki, G., and R. E. Weeks, 1973, Impact of Predators on Populations of Green Peach Aphids on Field-Grown Sugarbeets, Environ. Entomol. 2:345-349.

Wallis, R. L., and J. E. Turner, 1969, Burning Weeds in Drainage Ditches to Suppress Populations of Green Peach Aphids and the Incidence of Beet Western Yellows Disease in Sugarbeets, J. Econ. Entomol. 62:307-310.

GPA-CAST: A Computerized Model for Green Peach Aphid Management on Potatoes

Zane Smilowitz

In Pennsylvania many insect species damage potatoes. The most economically important of these are the green peach aphid (Myzus persicae), the Colorado potato beetle (Leptinotarsa decemlineata), and the potato leafhopper (Empoasca fabae). Approximately $60-80 per acre is spent annually in controlling these species (USDA, 1977). In 1975 an integrated pest management (IPM) program for potato insects was initiated at the Pennsylvania State University, concentrating initially on the green peach aphid (GPA). In 1978 a program called GPA-CAST was put into operation, designed to maximize cost-benefits of chemical control by recommending insecticide applications only when pests attain damaging population levels, to reduce environmental contamination, and to delay or prevent the pests from developing insecticide resistance.

The management strategy employed in the program involves reduced use of systemic insecticide at planting. GPA populations are then estimated weekly through forecasting and monitoring techniques during the growing season, and if populations are predicted to reach economic status, a selective aphicide is recommended.

The rationale behind this strategy may be understood by comparing it to conventional control. Conventional control is a unilateral approach, and insecticides are considered to be the only important mortality factor for the pest population. Insects are not tolerated on the crop, and pest populations are considered capable of causing economic damage each year. In an IPM program subeconomic pest populations are tolerated, yearly population levels are considered to be dynamic, and there are many sources of pest mortality, including natural enemies. Thus, in the potato IPM program the mortality expected from natural enemies is viewed as compensation for the reduced rate of insecticide applied. If populations are not contained within subeconomic levels by natural factors, then a selective aphicide is applied.

DEVELOPMENT OF IPM TECHNOLOGY FOR GREEN PEACH APHID

An IPM program requires three categories of information: pest component, economic component, and implementation and delivery component. The pest component includes factors affecting pest density, among them rates of birth, development, and major biotic and abiotic mortality factors. The economic component requires knowledge of plant growth and yields, crop price, and probable yield loss caused by the pest when making pest control recommendations. The implementation and delivery component deals with methods by which the biological and economic data are analyzed and a recommendation is developed and presented to the grower in an acceptable format for information utilization.

Pest Component

The GPA model was developed through laboratory and field research. Field experimentation permitted understanding population growth characteristics of aphids in a field environment where the impact of predators, parasites, and *Entomophthora* fungi was reduced. Laboratory experiments were conducted to determine the developmental rates of aphids under constant temperatures. An analysis of the laboratory results gave 4± 1°C as the low temperature threshold for GPA development and 30± 1°C as the upper limit (Whalon and Smilowitz, 1977).

These two developmental thresholds were used to adapt a modified sign wave algorithm (Allen, 1976) to calculate accumulated GPA degree days (DD). Actual field DD were calculated by recording maximum and minimum temperatures with a hygrothermograph in the canopy of a potato field. Initiation of DD accumulation in the spring was based on aphid immigration data. GPA movement was monitored using yellow plastic utility tubs filled with water. These water traps were placed on the northwest and southeast sides of potato plots and were visited periodically during the week. Accumulation of DD was initiated when peak winged GPA numbers were detected in the pan traps.

Mathematical equations for predicting field populations of GPA were derived from research conducted at the Pennsylvania State University Rock Springs Research Farm from 1975 through 1978 (Whalon and Smilowitz, 1979). Twenty plots, measuring 7.2 meters x 12 meters (m), were treated with aldicarb 15G, a soil systemic insecticide, at rates of 0.00, 0.57, 1.68, 2.79 or 3.93 kilograms (kg) per hectare (ha) in a randomized complete block design. Each treatment was replicated four times. Fertilizer, herbicide, fungicide, and vine killer were applied according to standard production procedures. Aphid predators and other competing insects were eliminated or held at very low population levels with carbaryl (0.57 kg active ingredient ha) sprayed at five to seven day intervals. Carbaryl acts as an insecticide check because at this application rate it is relatively nontoxic to the GPA. GPA populations were estimated weekly by sampling 120 leaves, 3 on each of 40 plants in each treatment block.

Data from these plots were fit to the exponential growth equation

$$N_{DD} = N_A e^{r(DD)} \tag{1}$$

where N is green peach aphid density at time (DD), N_A is the original density at $DD = 0$, e is the constant 2.718 (base of the natural logarithm system), and r is the intrinsic rate of GPA population increase per DD. Data were collected from peak immigration of GPA to late August. The exponential growth equation was derived from each aldicarb treatment in every year by regressing the $\log_e$ of GPA densities on accumulated DD. Parameters N_A and r of Equation 1 were compared statistically using homogeneity of coefficients analysis. Pooled equations were obtained by combining mean densities over years within aldicarb treatment (Table 1).

The exponential prediction equations (Table 1), DD calculating algorithm, economic injury level index, and a plotting algorithm were combined into a Fortran computer model for predicting GPA field populations. Predictions were made as follows. A given equation was chosen based on the amount of aldicarb applied at planting. Predictions were initiated when peach aphid flights were detected in each grower's field. DD used in the prediction equation were obtained from maximum and minimum temperatures recorded by hyrgrothermographs placed in each field. When a five-to seven-day forecast was desired, maximum and minimum temperatures from weather forecasts were used.

Natural Enemy Population

In developing the potato pest management program, much consideration was given to natural enemies. Thus experiments were performed to evaluate the potential of using natural enemies to regulate GPA densities. In these tests natural enemies were eliminated from a field plot with carbaryl applied on a five-to seven-day schedule, and GPA densities were compared with an untreated adjacent field. This test was conducted for three seasons beginning in 1975. GPA populations were estimated weekly in both 0.4 ha plots by sampling 40 plants or 120 leaves in each plot. GPA densities in field plots where natural enemies were eliminated ranged from fifteen-to a thousand-fold greater than those plots where natural enemies were conserved (Figure 1). On this

Table 1

Regression Equations and Statistics from $\log_e$ of Pooled Mean Green Peach Aphid Densities (N) Regressed Against Accumulated Degree Days (DD)

Year	Aldicarb (kg AI/ha)	Equation: $N_{DD} = \ell n N_A + r^{DD}$	R^2	Standard Error ($\ell n N_A$)	Standard Error r
1975–1977	0.0	$N_{DD} = -1.40 + 0.0008\text{-}DD$	40	0.147	0.00017
1975–1977	0.57	$N_{DD} = -3.09 + 0.0025\text{-}DD$	63	0.416	0.00047
1975–1977	1.68				
	2.79	$N_{DD} = -4.57 + 0.0026\text{-}DD$	58	1.010	0.00073
	3.93				

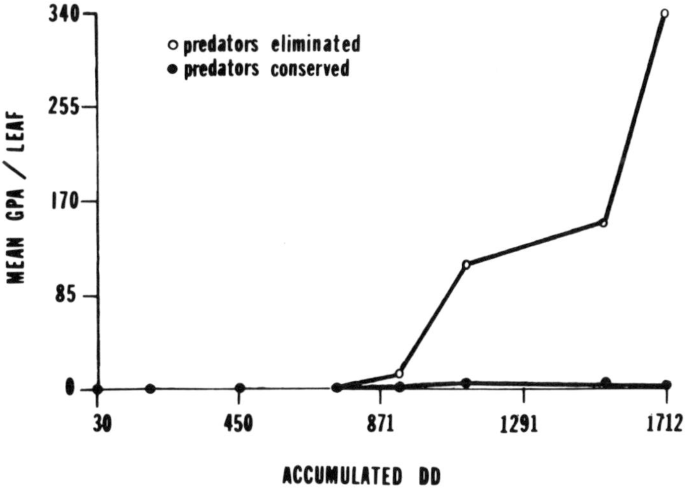

Figure 1
Influence of eliminating predators on GPA population density.

basis the effect of natural enemies on GPA population growth was considered sufficient to warrant their incorporation into the IPM program.

To quantify and predict the effect of predators, a temperature-mediated functional response model for predators of GPA is being developed through our program. Five variables are being incorporated into the model for each predator species: prey size, predator size, prey density, predator density, and temperature. The predator size, prey size, and density are being determined in order to estimate the number of pests consumed by an individual of a specific size. These variables are being studied at different temperatures. The resulting data will increase the precision in forecasting GPA population densities by projecting natural enemy populations and their associated effect on GPA populations. In this manner insecticides and natural enemies could be managed to optimize their combined effect on GPA populations.

Economic Considerations

Information on the economic consequences of GPA damage is essential for an IPM program on GPA. To fulfill this need, data are required on plant growth, crop yields, values, control costs, and cost-benefits associated with control recommendations. Economic injury levels generally are employed as a single fixed value estimate for these variables. Potato plant growth, however, is dynamic and can be thought of as having two growth phases: foliar production and tuberization.

Optimum production of foliage will, within limits, result in maximum tuber yield. Thus high GPA populations that occur following maximum foliage production will have less effect on tuber production than will stresses on the plant prior to maximum foliage production. To understand better the relationship of insect numbers to yield reduction, more information is needed about potato phenology.

Temperature-dependent models for potato foliage production, as well as models for tuber-foliar relationships, have been developed at our laboratory. Foliage production was used as the criterion for evaluating damage because the GPA attacks foliage, and a high correlation was found between seasonal foliage production and tuber yield ($R^2 = 0.86$). Our models, which estimate foliage and tuber production in the absence and presence of insect pests, were employed in developing crop loss estimates.

The experimental methods used were based on varying GPA populations by applying different levels of aldicarb at planting and comparing foliage and tuber production relative to the GPA population densities. The aldicarb rates and sampling procedures were the same as previously described. First, natural enemies and competing insects were reduced or eliminated with carbaryl, and GPA population densities were estimated weekly. Then foliage and tubers were sampled from each treatment both on alternate weeks or weekly during the growing season and from a final tuber harvest. To estimate plant growth, foliage and tubers were removed and weighed from one row of each treatment. Dry weight estimates of foliage were obtained from subsamples. Foliar weights and accumulated GPA numbers, obtained from each treatment, were analysed by regression for the specific harvest period to estimate the effect of GPA on production (Figure 2). Percentage loss was determined from differences in foliar production relative to maximum, the latter being estimated from the Y intercept where GPA = 0 in Figure 2. A least-squares method was used to compare percentage loss and GPA numbers to obtain the percentage loss per GPA for each harvest period. This was referred to as the relative damage index (RDI). The accumulated DD for plant growth from green row corresponding to each harvest period were then regressed against the RDI, giving the relationship of expected percentage loss per GPA per leaf for foliar production throughout the season. This was termed the seasonal damage index (SDI) (Figure 3). Percentage loss per GPA for tubers was determined from the foliage-to-tuber ratios. The values obtained from these data estimated the dynamic effect of GPA on crop loss throughout the season. By employing the plant growth model to determine the stage of plant growth and using the SDI, a prediction of the expected crop loss resulting from GPA can be estimated for any period during the growing season.

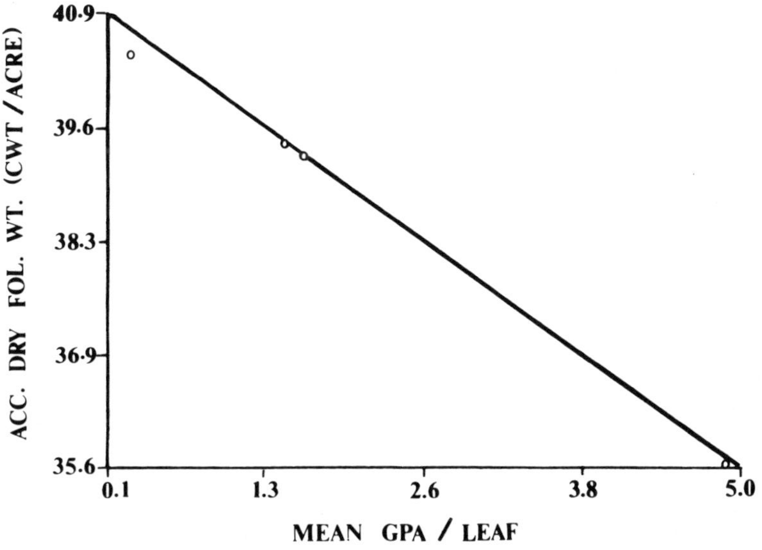

Figure 2
Influence of GPA density on foliage production.

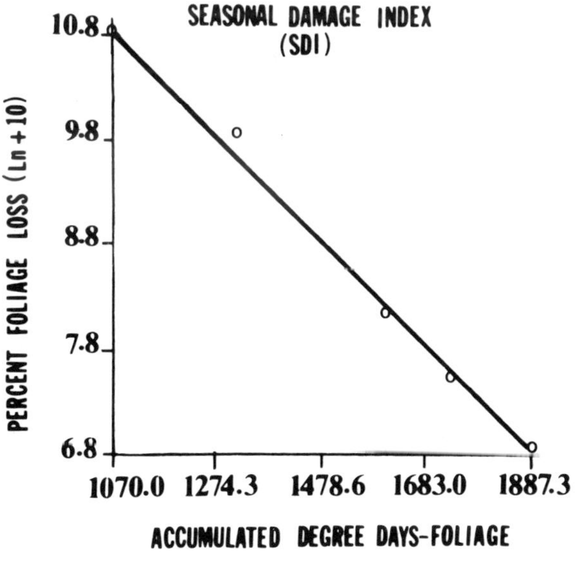

Figure 3
Changes in percentage loss of foliage per GPA in relationship to accumulated DD plant growth.

The next step, assigning an economic value to insect damage, was accomplished by converting the SDI to dollar loss. A market model was constructed to fulfill this requirement using the estimated prices and potential yields. Flexibility was built into the model to allow for changing prices and yields. Five prices were assigned: $88, $94, $99, $105, and $110 dollars per metric ton (MT) and yields were 27, 29, 31, 34, and 36 metric tons per hectare (MT/ha). A matrix of crop values, constructed from these yields and prices, allowed decisions to be tied to changing crop value estimates. The dollar loss established using this model and the SDI was called the monetary damage index (MDI). Like the SDI, it is a dynamic parameter that changes during the season. The MDI converts the percentage loss per GPA per leaf (SDI) to dollar loss per GPA per leaf.

The procedure was simple for converting the SDI into MDI once a market model was established. The SDI constitutes the percentage foliage loss, and it must be converted to tuber loss before estimating MDI. A constant ratio of 6.02 to 1.00 exists between total foliage and tuber production. The conversion is made by dividing SDI by 6.02 to obtain the TSDI (tuber seasonal damage index). Crop values are then established from yields and commodity prices. In the market model utilized in this program MT/ha crop values ranged from $2376 to $3960 per ha. The MDI was obtained from the equation MDI = TSDI x crop value/100, which gives the dollar loss per GPA per leaf found in the field. The MDI does not indicate whether a treatment would be needed, only the economic loss caused by GPA.

Next, economic injury levels (EIL) had to be determined. Before making a spray recommendation, cost-benefits must be considered, and the grower should realize a reduction in loss and a gain in crop value equivalent to or greater than the management costs. Because MDI is the dollar loss per GPA, the resulting answer is the number of GPA needed to equal cost of control. In addition, employment of a crop value matrix model allows the grower to obtain a number of EILs, which could reflect changing market values. High market crop value would give lowest EILs and low values high EILs.

Implementation and Delivery

Implementation of the potato pest management program was accomplished through integration of the biological and economic components as shown in Figure 4.

GPA-CAST utilized weather forecasts and present population numbers in estimating future population densities. The predicted maximum and minimum temperatures for the coming three-day period were used to determine the accumulated DD, and increases in GPA densities over the next three days were determined. The

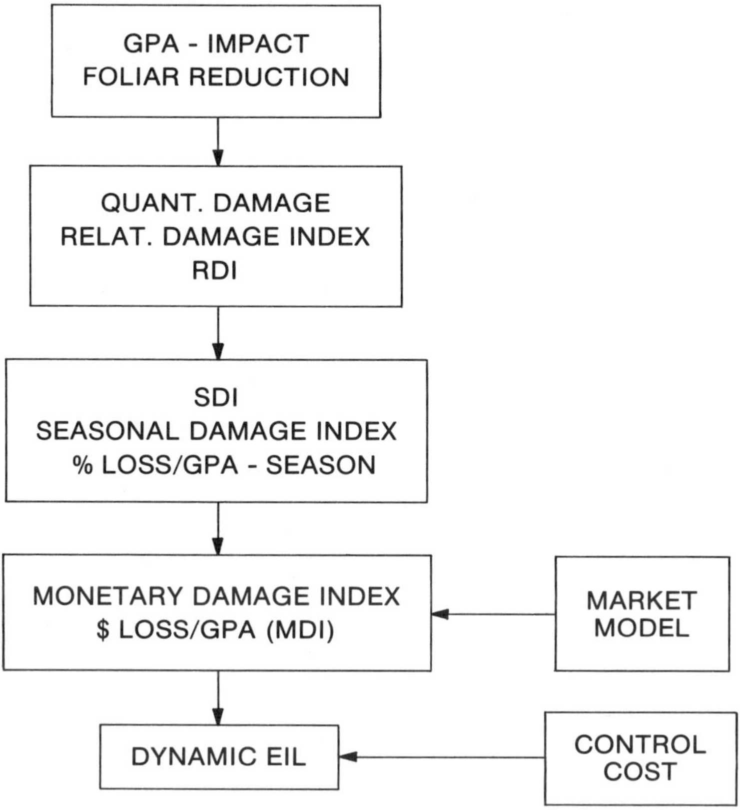

Figure 4
Dynamic economic injury levels for green peach aphid.

predicted temperatures were also used to estimate plant growth stage, MDI, and ultimately EIL. If the predicted GPA density equals or exceeds the EIL, spray is recommended. Some advantages of the dynamic system over the static EIL are that the yield and price (as well as management costs) are incorporated into making a control recommendation. In addition, the system improves the precision of the estimates by incorporating plant susceptibility to damage into the recommendation.

Research initiated in 1975 on the development of an IPM system for GPA has produced considerable technology and improved our capabilities to make control decisions that result in benefits for the grower. Unless grower-oriented delivery systems are available, however, much of this technology will go unused. Systems that are simple, accurate, and practical for growers must be developed. With this in mind, the advantages and disadvantages of several monitoring and delivery systems were examined. Four delivery systems are available, and these are scheduled for further testing: a hand-held programmable calculator, a dedicated microcomputer, a Radio Shack TRS-80 microcomputer, and a pencil and paper technique. I shall discuss only one system, the microcomputer TRS-80.

The level II unit with 16K capacity, numeric keyboard, and tape drive can be purchased for approximately $850. The unit could be set up at the farm or in grower cooperative, consultant, or county extension offices. Operation of the TRS-80 at the farm would differ from other areas because the grower would not have to contact someone else to record data and obtain a control recommendation. The present system is designed to produce recommendations concerning late blight (*Phytophthora infestans*) and GPA.

To use the system, the grower calls the operation at the green row stage of plant development to establish a file and initiate the process. Environmental data are given to the operator weekly to determine plant growth and obtain blight recommendations. The example that will be used to illustrate the method for GPA-CAST will start at 600 accumulated DD plant growth (approximately three to four weeks past green row) and 0 accumulated DD GPA. First, the grower's file is inserted into the disc drive. The accumulated plant growth DD, GPA DD, and GPA per leaf are entered into the program. Because it is the beginning of the season for GPA-CAST, 0 will be the input for GPA density and GPA DD. The GPA season begins when the grower finds a minimum of five GPA per fifty leaves. A series of questions is now asked of the grower.

Q. "Is this the beginning of the GPA season?" Enter "yes" if five or more GPA found per fifty leaves.

Q. "How many GPA found per fifty leaves?" Enter "12" (or whatever number is found in the sample).

Q. Do you prefer using a fixed price and yield for estimating crop value or a matrix model?" (The question being asked is how crop value should be determined for estimating EILs. If the grower has a price contract and can estimate yields, a fixed price may be used. If price is variable or unknown, a matrix model would be more appropriate. For this example the matrix model will be employed because it best demonstrates the range and flexibility of the program.) Enter "M."(The program designates five expected yields, 27-36 MT/ha, and five potential prices, $88-$110/MT.)

Q. "How many days have you observed maximum and minimum temperatures?" Enter "7."

Q. "What is the cost of control per hectare?" Enter $30.00 (or whatever material and application costs are estimated to be). Output displays: accumulated DD plant growth, accumulated DD GPA, and predicted GPA numbers. (It is assumed that the GPA sample was taken on day 1 at the beginning of the current environmental data period.)

Q. "Do you desire a change in the prices of yields previously entered?" (This question enables the grower to benefit from changing market prices and growing conditions.) Enter "N" for No. Output: twenty-five EILs are output corresponding to each price and yield and a predicted GPA number.

Decisions: It is from this output that a control decision is made. If the predicted GPA population exceeds the EIL at the selected crop value, a spray recommendation is made. If the GPA population is less than the EIL, a prediction of whether it will exceed the EIL within the next three days is made, based on local weather forecasts.

Q. "What are the predicted maximum and minimum temperatures for the next three days?" Enter (expected temperatures).

Output: Predicted accumulated DD plant growth for each of the three days. Predicted accumulated DD GPA, predicted GPA number for each of the three days.

Matrix of EILs and a predicted GPA population are estimated for each of the three days.

If the predicted GPA number exceeds the EIL for a selected crop value during this three-day period, a spray is recommended. If the populations are not economical during this time frame, the grower is advised to contact the agent the following week.

REFERENCES

Allen, J. C., 1976, A Modified Sine Wave Method for Calculating Degree Days, Environ. Entomol. 5:387-396.

U.S. Department of Agriculture, 1977, Cooperative Plant Pest Report, 1:49-52.

Whalon, M. E., and Z. Smilowitz, 1977, Determination of Constant Temperature Developmental Thresholds for *Myzus persicae* (Sulz.), N.Y. Entomol. Soc. 85:206.

Whalon, M. E., and Z. Smilowitz, 1979, Temperature-Dependent Model for Predicting Field Populations of Green Peach Aphid, *Myzus persicae* (Homoptera: Aphididae), Can. Ent. 111:1025-1032.

Considerations for the Management of the European Corn Borer on Potatoes

George G. Kennedy and *T. E. Anderson*

North Carolina annually produces about 17,000 acres of potatoes with a value of $16 million (Anon., 1979). Approximately 75 percent of the total harvest is processed for chips, with the remainder being sold for fresh consumption. Potatoes, grown primarily as a spring crop in the eastern coastal counties, are planted in late February and March and harvested in June. The most commonly grown varieties are Pungo, Superior, Norchip, and Atlantic. The North Carolina potato crop fits into a marketing sequence of successively maturing crops that stretches up the Atlantic seaboard from Florida to Maine. Virtually the entire crop is marketed at harvest; hence there is little or no long-term storage.

Commercial potato yields in North Carolina typically are rather low, with the averages from 1976, 1977, and 1978 being 145, 165, and 150 hundredweight per acre, respectively (Anon., 1979). A number of factors contribute to the low yields, including marketing pressure for early harvest, poorly drained fields, low soil pH, hard pans, lack of crop rotation, air pollution, lesion nematode, insects, weeds, and soil-diseases (Averre and Sorensen, 1979). Consistently higher commercial yields are clearly possible, and some growers have achieved yields of almost 400 hundredweight per acre. Average yields for a properly managed crop,

---

We thank K. A. Sorensen and J. W. Van Duyn, Department of Entomology, North Carolina State University, and D. N. Ferro, Department of Entomology, University of Massachusetts, for sharing their results and ideas on the management of potato insect problems and V. B. Covington for his technical assistance. We also thank C. G. Eckenrode, R. L. Rabb, R. E. Stinner, K. A. Sorensen, and G. T. Weekman for their constructive reviews of this manuscript.

however, should probably approach 300 hundredweight per acre (Averre and Sorensen, 1979).

INSECT PESTS

In general, the spring potato crop in North Carolina is attacked by relatively few serious insect pests. Of those that are commonly encountered, all except the European corn borer (ECB) and perhaps the seed-corn maggot are readily controlled. The more important insect pests are listed in Table 1.

Table 1
Insects Pests Commonly Attacking Spring Potatoes in North Carolina

Pest	Importance
Seed-corn maggot (*Hylemya platura*)	-Attacks seed piece; sporadic occurrence; losses rarely serious, associated with blackleg.
Flea beetles (*Epitrix* spp.)	-Early season defoliator; readily controlled by systemic insecticides.
Green peach aphid (*Myzus persicae*)	-Rarely a problem; if large populations develop, they are typically late and very close to harvest.
Colorado potato beetle (*Leptinotarsa decemlineata*)	-Large populations are observed every year, although their distribution is patchy; readily controlled.
European corn borer (*Ostrinia nubilalis*)	-Considered the most serious insect pest; difficult to control; present every year.

Although the seed-corn maggot, *Hylemya platura* (Meigen), occurs only sporadically and rarely causes serious yield reductions directly, it may be of considerable indirect importance as a vector of the blackleg pathogen *Erwinia carotovora* (Bonde, 1930, 1939; Carter, 1973), especially by introducing it into fields previously free from the disease.

Flea beetles, *Epitrix spp.*, have rarely been a problem in recent years because of the widespread use of granular systemic insecticides at planting; however, the use of these materials has been declining. Currently only 50 percent or less of the spring potato acreage is treated with a systemic insecticide at planting. If this trend continues and alternative procedures for managing flea beetles are not adopted, damaging flea beetle populations are likely to be encountered more frequently.

The Colorado potato beetle, *Leptinotarsa decemlineata* (Say), is present every year, and damaging populations commonly develop if it is not controlled. In North Carolina, however, this insect is readily controlled with a number of insecticides, including carbaryl and carbofuran. Typically insecticide applications applied for control of the ECB help keep the Colorado potato beetle populations from reaching damaging levels.

The ECB, Ostrinia nubilalis (Hbn), is currently perceived by North Carolina potato growers as the most serious insect pest attacking their crops. This insect was introduced into North America (New York and Massachusetts), probably in shipments of broom corn, from central Europe in the early 1900s (Caffery and Worthley, 1927). It has subsequently spread throughout most of the eastern and central United States (Showers, 1979). Three ecotypes of the ECB can be distinguished on the basis of the photoperiod x temperature interaction resulting in the initiation of diapause (Showers et al., 1975; Showers 1979). The southern ecotype, adapted to regions of moderate climate, typically has three to four generations per year (Jones et al., 1939; DuRant, 1969; Sparks and Showers, 1975, Showers, 1979). In North Carolina, there are typically three generations per year, but it is the first generation that figures so prominently.

Although the ECB has a very wide host range (Barber, 1926; Hodgson, 1928) that includes feral species as well as crops (such as sweet corn, field corn, snap beans, lima beans, beets, celery, southern peas, pepper, and potatoes), potatoes are the principal host of the moth's first generation in North Carolina and on the Eastern Shore of Virginia (Hofmaster, 1975).

The ECB overwinters as a diapausing fifth instar larva in above-ground crop debris, such as corn stalks and numerous wild hosts (Hodgson, 1928). Light trap records and field-cage studies indicate that emergence of adults of the overwintering generation lasts approximately fifty to fifty-five days beginning in early April and peaking in late May (Figure 1). Corn is still being planted in April, but potato plants are well developed and are an attractive host for oviposition. The eggs are laid on the foliage and typically hatch in seven to ten days. Newly eclosed larvae feed on the leaf surface or in leaf veins for several days prior to boring into the stalk, where they complete their development and pupate (Jones et al., 1939; Caffery and Worthley, 1927). Larval and pupal development requires approximately six weeks at the temperatures typically prevailing during the spring in eastern North Carolina (Matteson and Decker, 1965; Anderson and Kennedy, unpubl). Adults of the spring generation begin to emerge in very late May and early June.

Field survey data (Anderson and Kennedy, unpubl.) indicate that most of the eggs deposited by moths of the spring generation in the potato growing region of North Carolina are laid on the maturing potato crop, although a smaller but significant portion is deposited on the

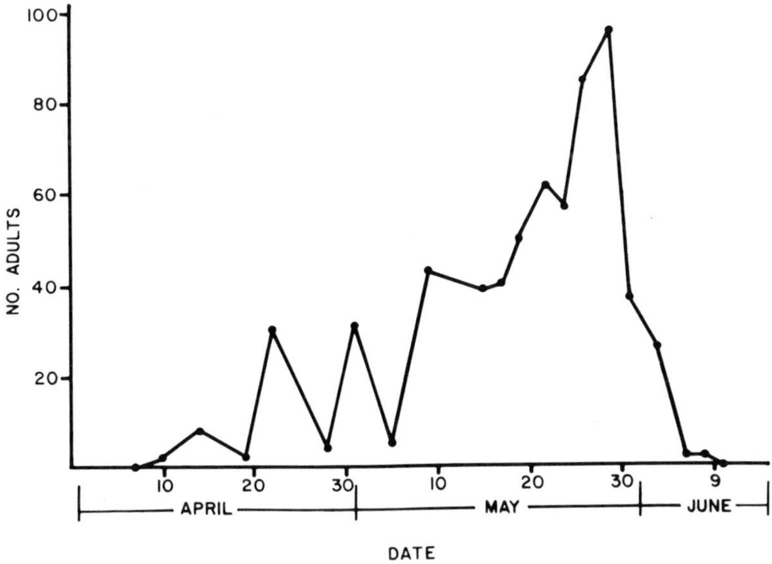

Figure 1
Emergence of ECB moths of overwintering generation, recorded in field cages at Clayton, N. C., 1978.

emerging corn. Those eggs deposited on potatoes are ill fated. The June harvest of the potato crop cuts short the development of resulting larvae, as well as those remaining first-generation larvae and pupae.

Larval damage to the plant is concentrated in the lower portion of the main stems but can be found throughout the plant (Table 2). The larvae feed primarily in the pith but may also damage the xylem and occasionally the phloem. Tunneled stems are weakened and often break when whipped by the wind; the foliage of damaged stems often wilts during periods of high temperature. In addition, there is some evidence that the feeding injury predisposes the plant to blackleg disease caused by Erwinia spp. (Nielsen, 1963).

Table 2
Distribution of European Corn Borer Injury on Irish Potato Plants

	Portion of Plant		
Plant Part	Lower Third	Middle Third	Upper Third
Main stem	32 a	18 b	6 c
Lateral stem	14 b	15 b	14 b
Petiole	0 d	1 d	0 d

Note: Values are percentages based on examination of 253 damage sites. Means not followed by the same letter are significantly different at $p \leqslant 0.05$.; mean separation for all values.

CONTROL AND MANAGEMENT OF THE EUROPEAN CORN BORER

For many years the ECB was controlled on North Carolina potatoes through the prophylactic use of insecticides. Weekly sprays throughout April and May were common. The strategy was to maintain a lethal dose of insecticide on the plant surface to kill the newly eclosed larvae before they tunneled into the stem. This was important since neither the eggs nor the larger larvae within the stems were controlled by commonly used insecticides.

In recent years, the number of insecticide applications necessary to control the ECB on potatoes has been reduced from an average of six to eight to an average of three to four, accomplished through the use of light traps to monitor the emergence of moths of the overwintering generation. Under this program insecticide applications are initiated when thirty or more ECB moths are trapped over a period of five consecutive days.[1] This does not generally occur until the last week in April or early May. Once insecticide treatments are initiated, repeated applications are made every five to seven days until about two weeks before harvest (Figure 2). Once again the basic strategy is to maintain a lethal dose of insecticide on the plant surface during the period of maximum egg hatch to prevent larval establishment. The reduction in the number of insecticide applications necessary to control the ECB is the result of eliminating those treatments that were applied in April before the ECB was active.

The use of light traps to monitor ECB activity poses certain difficulties, including the requirement for a power source and the need to sort the ECBs from the remainder of the often-considerable moth catch. In a quest for ways to obviate these difficulties, we evaluated the potential for using sex-pheromone-baited traps to monitor ECB moth activity.

The situation with regard to the sex pheromone of the ECB is quite complex. Investigations have demonstrated the existence of two distinct population types. One, the Z strain, produces and is attracted to a 97:3 blend of (Z)-11-and (E)-11-tetradecenyl acetate (tda); the other, the E strain, produces and is attracted primarily to a mixture of the same two compounds having a Z:E ratio of 4:96 (Carde et al., 1975; Kochansky et al., 1975; Klun et al., 1975). The Z strain is widely

---

[1] The program was developed and instituted by K. A. Sorensen, Department of Entomology, North Carolina State University, and is part of a larger pest management program, Pest Alert, developed by Dr. Sorensen and C. W. Averre, Department of Plant Pathology, NCSU.

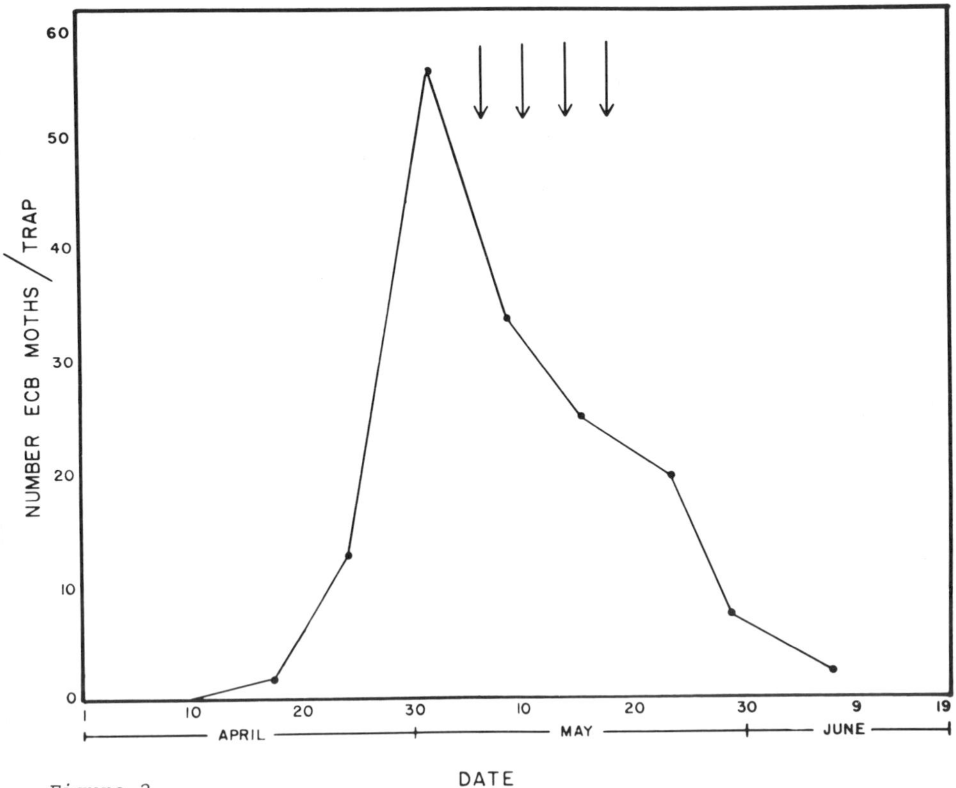

Figure 2
ECB moth activity of overwintering generation as indicated by light trap in Carteret County, N. C., 1978. Ordinate is number of ECB moths per trap every five days. Arrows indicate when insecticide treatments would be applied. Harvest would be during first week of June.

distributed throughout the United States and Europe, while the E strain appears to be restricted to a relatively few localities (Klun et al., 1975). In the United States, the E strain has been reported only in New York where it occurs allopatrically and Pennsylvania and New Jersey where it occurs sympatrically with the Z strain (Kochansky et al., 1975; Klun et al., 1975; Carde et al., 1975, 1978). Available data indicate that where the two strains occur sympatrically, hybridization occurs. F_1 hybrid females produce a sex pheromone with the Z-to-E ratio of 35-to-65, while the F_1 hybrid males respond more intensely to the 35-to-65 mixture than to either the 97-to-3 or the 3-to-97 mixtures (Klun and Maini, 1979).

Because the pheromone response pattern of the ECB in North Carolina was not known and this information was essential to the development of a sex pheromone-based monitoring system, a study was conducted to determine which Z-to-E isomeric blend of 11-tda was the most attractive. We found that both the Z and the E strains of the ECB, and very likely the hybrid as well, are present in North Carolina (see Kennedy and Anderson, 1980; Klun and Maini, 1979; Carde et al., 1978). The greatest number of males were captured in traps baited with a blend of 96Z:4E-11-tda, the least in traps baited with a blend of 4Z:96E-11-tda, and an intermediate number in those baited with blends of 50Z:50E or 35Z:65E (Table 3). The numbers of males captured by the various blends throughout the season were not well correlated (Table 4), indicating that none of the blends tested could be used alone to monitor activity of the entire ECB moth population accurately.

Further, even though both traps and pheromone sources were changed frequently (see McLeod and Starratt, 1978), the number of moths captured in the pheromone-baited traps was very small compared to the

Table 3
Number of Male *O. nubilalis* Caught in Traps Baited with Different Isomeric Blends (Z:E) of 11-Tetradecenyl Acetate, Wake County, N.C.

Proportion of Isomer Z	E	$\bar{X}$ Total Males per Location per Season 1977	1978	1979
96	4	16a	11a	12a
50	50	7b	8a	6ab
35	65	-	-	7ab
4	96	3b	4b	6b
Unbaited		0c	0c	0c

Note: Means from nine locations in 1977, seven in 1978, and fifteen in 1979. The 35:65 blend was included only in 1979.
Mean separation vertical by Duncan's multiple range test; $p \leqslant 0.05$.

Table 4
Correlation Coefficients for Number of *O. nubilalis* Males Captured by Traps Baited with Various Blends (Z:E) of 11-Tetradecenyl Acetate

Correlation (blend vs. blend)	Year 1978	1979
96Z: 4E vs. 4Z:96E	- 0.1937 (63)	+ 0.3973 (59)**
96Z: 4E vs. 50Z:50E	+ 0.1978 (72)	+ 0.2994 (63)*
96Z: 4E vs. 35Z:65E	--	+ 0.2100 (65)
4Z:96E vs. 50Z:50E	- 0.3033 (69)**	+ 0.1259 (60)
50Z:50E vs. 35Z:65E	--	+ 0.1648 (61)

Note: Values in parentheses are numbers of observations. Observations in which no moths were caught in both blends were not included in the calculations of correlation coefficients.
*$p \leqslant 0.05$
**$p \leqslant 0.01$

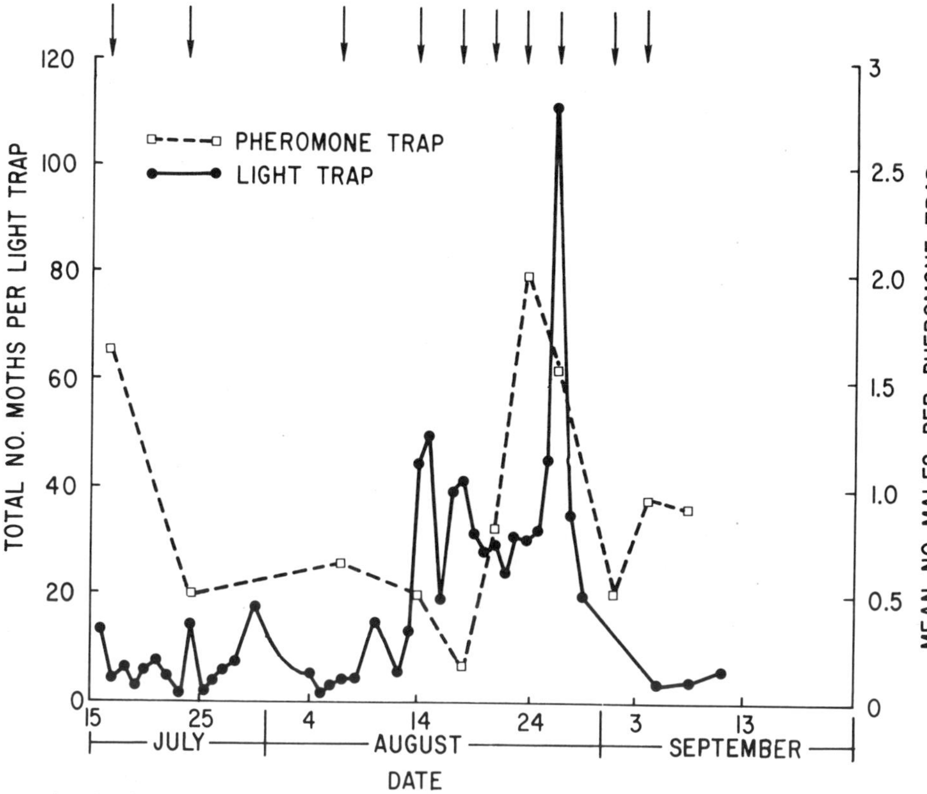

Figure 3
Comparison of number of ECB moths captured in light (males and females) and pheromone (males only) traps. Values for pheromone traps represent $\overline{X}$ total moth catch from all traps baited with various blends tested. Arrows indicate when pheromone traps were changed. Wake County, N. C. 1978.

number captured in the light trap (Figure 3). To date, pheromone-baited traps have failed to provide a reliable indication of ECB moth activity in spring-planted potatoes. Even when ECB moths could be readily observed in potato fields, sex-pheromone-baited traps placed both in and around the field failed to capture a significant number of moths.

Overall these data indicate that sex-pheromone-baited traps are not adequate for monitoring ECB moth activity for the purpose of timing insecticide treatments. Additional research on ECB pheromone constituents and blends, on pheromone dispensing systems, and on trap design and placement may improve the effectiveness of this technique.

SEVERITY OF THE EUROPEAN CORN BORER

The ECB is perceived by North Carolina potato growers as their most serious insect pest problem. Even under the present program of initiating insecticide treatments based upon moth activity, North Carolina potato growers spend about $200,000 annually to control the ECB on some 17,000 acres. As in the past, the present management program assumes that significant reductions in potato yields will result if the ECB populations are allowed to develop unchecked. Yet the literature on this subject is vague and, at times, contradictory, with little data available.

In Connecticut, Turner and Zappe (1937) measured yields of individual infested and uninfested plants and found no differences in yield, although their data were highly variable. Pepper (1938) expressed doubt that the ECB would ever be serious on early potatoes in New Jersey but considered it to be a potentially serious threat to the late crop. Working on the Eastern Shore of Virginia, Jones et al. (1939) concluded that the level of damage to potatoes caused by a given ECB population varied greatly depending upon the general growing conditions. During at least one year of their study, they observed that plants having an average of three to three and a half borers per plant at harvest "produced lower yields of prime potatoes and had a tendency to produce more cull potatoes than did plants infested with an average of 0.73 borers per plant at harvest". However, during two other years, they were unable to show significant differences in yield between infested and uninfested plants. On the contrary, Beard (1940), working in Connecticut, reported "very good yields of potatoes . . . from fields with infestations which would be considered severe." Indeed he reported larger tuber size when the ECB infestation was high. Bray (1961) found that in Delaware tuber yields were not affected by heavy natural populations of the ECB; unfortunately no data on population levels were provided, although the population was high enough to destroy completely some vines in the more heavily infested plots.

In addition to the direct effects of ECB feeding, yields may be affected by soft rots (blackleg), which secondarily attack the ECB-injured vines (Merrill, 1957). Nielsen (1963), working in North Carolina, showed a highly significant correlation ($r = 0.627$) between ECB infestation and the recorded incidences of blackleg, but was unable to show a reduction in yield attributable to the ECB-blackleg complex. He believed, however, that failure to control the ECB might result in yield reductions due to infestion of damaged vines by the blackleg bacteria.

Theoretically a number of factors can interact to influence the effect of ECB injury on potato yields, including blackleg (Merrill, 1957; Nielsen, 1963), phenological stage of the plant at the time the damage occurs (Sparks, 1977), plant variety, and the occurrence of other insects and diseases, as well as a number of abiotic factors such as temperature, water stress, and fertility. The possibilities for interactions among these and other factors make it difficult to compare directly the results of these different studies and makes their application to the North Carolina spring potato crop tenuous at best.

Because an understanding of the relationship between pest-induced plant damage and yield is basic to the development of a pest management program and because this information is lacking for the ECB on spring potatoes in North Carolina, over the last several years we have been engaged in an effort to obtain this information.

In these studies we established a wide range of ECB infestation levels in a large number of small plots (twenty-eight plants per plot) through a combination of artificially infesting plots with varying numbers of ECB egg masses and reducing the natural infestations of others with insecticide treatments (carbaryl, which our data indicate has no direct effect on yields other than that attributable to insect control). The egg masses are placed in the plots when natural ECB oviposition was at its peak. Numerous parameters considered likely to influence yield were measured or calculated; these included number of stems with ECB tunnels, percentage of stems with ECB tunnels, number of dead plants, $\overline{X}$ number of tunnels per plant, number of stems with blackleg, and percentage of stems with blackleg. The results obtained during 1979 are illustrative of the relationships found. They involve the variety Pungo planted on March 16 and harvested on June 27. All damage measurements were taken one week prior to harvest. Data were subjected to both simple and multiple-regression procedures. The stepwise regression procedure for maximum R^2 improvement was used to obtain those models giving the maximum R^2 value (Helwig and Council, 1979).

The single-variable regression models expressing the yields of U.S. No. 1 potatoes as a function of the various plant injury parameters are presented in Table 5. From these it is apparent that ECB injury alone, whether expressed as $\overline{X}$ number of tunnels per plant, number of stems with tunnels, or percentage of stems with tunnels, was not significantly related to yield of U.S. No. 1 potatoes. The scatter diagram in Figure 4 illustrates the lack of association between $\overline{X}$ number of tunnels per plant and yield of U.S. No. 1 potatoes. Blackleg incidence had a significant negative relationship to potato yield, although this explained only slightly less than 15 percent of the variation in yield (Figure 5). The addition of other independent variables to the regression model resulted in only slight increases in R^2 and actually decreased the significance level of the regression.

Table 5
Relationship of Yield of U.S. No. 1 Pungo Potatoes to Various Measurements of Plant Injury Associated with Infestations of European Corn Borer, 1979

Independent Variable	Regression Coefficient (B)[a]	Intercept	R^2	F[b]	Prob>F
$\overline{X}$ Number tunnels/plant	- 0.0293	50.99	0.0019	0.10	0.749
Percentage stems with tunnels	+ 0.0350	47.39	0.0087	0.47	0.497
Number stems with tunnels	+ 0.0362	46.07	0.0224	1.22	0.275
Number dead plants[c]	- 1.6155	51.46	0.0466	2.59	0.114
Percentage stems with blackleg	- 0.2417	53.74	0.1473	9.16	0.004
Number stems with blackleg	- 0.1794	53.83	0.1478	9.19	0.004

[a] B values are for single-variable models.
[b] Error degrees of freedom = 53.
[c] Attributable to a variety of causes; few plots contained dead plants and never more than three.

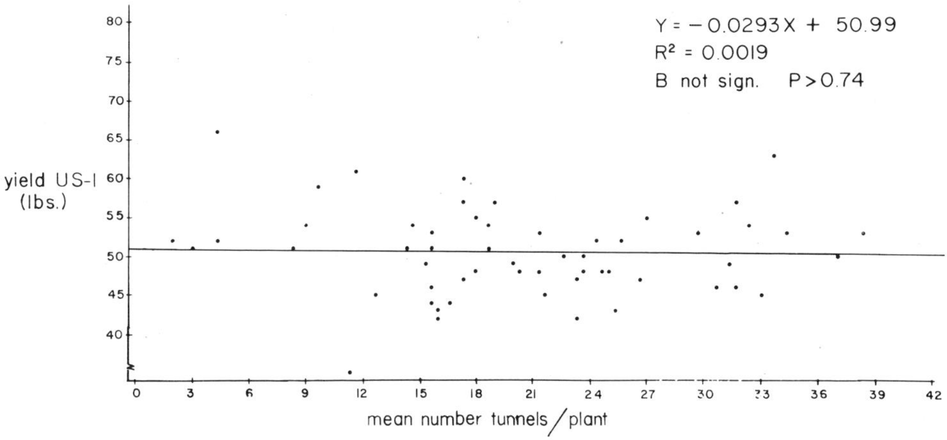

Figure 4
Relationship between $\overline{X}$ number of ECB tunnels per plant and yield of U.S. No. 1 potatoes per plot.

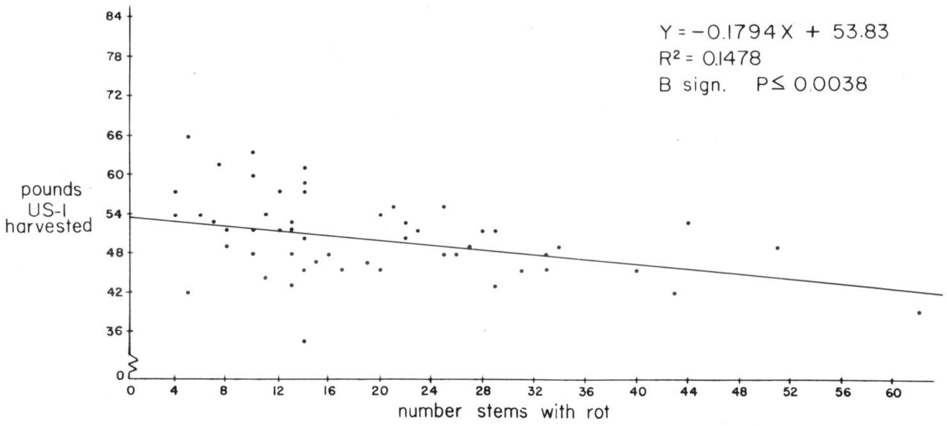

Figure 5
Relationship between number of stems with blackleg and yield of U.S. No. 1 potatoes per plot.

In this study even the highest ECB infestations (up to thirty-eight tunnels per plant), which were far in excess of those normally encountered under natural conditions, had no direct adverse effect on yield of U.S. No. 1 potatoes. In most years in North Carolina (and in this study), peak ECB oviposition occurs about the time the plants are in the 50 percent bloom stage, but feeding damage typically does not become apparent until the plants are full grown and approaching maturity, when the plant can withstand considerable loss of foliage with little reduction in yield (Sparks, 1977). Although the ECB is not a defoliator, it often produces foliage loss through the destruction of branches and stems. The relationship between ECB damage and potato yields undoubtedly is influenced by the phenological stage of the plant when the injury occurs. For the spring crop in North Carolina, this seems to vary little from year to year, suggesting that the ECB by itself is not the serious pest of potatoes it is perceived to be. In other areas or in late season crops where the ECB injury occurs when the plants are in a different phenological stage, ECB infestations may have a significant direct adverse effect on yield, however this information remains to be developed.

The most significant factor affecting yield in this study was blackleg disease (Table 5), which previously had been associated with ECB injury (Merrill, 1957; Nielsen, 1963). Our data also revealed a significant positive relationship between blackleg incidence and the $\overline{X}$ number of tunnels per plant (Figure 6); the regression explains about 13 percent of the variance in blackleg incidence. Addition of the number of stems with tunnels to the model did not improve the regression.

Obviously factors other than the ECB also influence the incidence and severity of blackleg. Such factors include but are not restricted to inoculum density, inoculum source (soil, seed piece, or insect), moisture, temperature, and the availability of entrance sites other than those caused by ECB (such as hail, wind, and mechanical injury) (Aleck and Harrison, 1978; Burr et al., 1977; Kloepper et al., 1979; Bonde, 1950; Nielsen, 1963). The potential of blackleg for reducing potato yields and the relationship between ECB damage and blackleg incidence necessitates the development of a

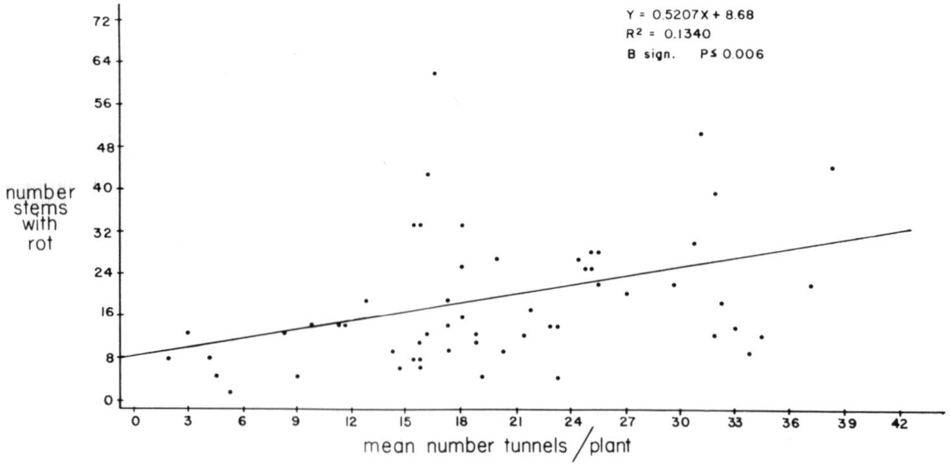

Figure 6
Relationship between $\overline{X}$ number of ECB tunnels per plant and number of stems with blackleg symptoms.

more thorough understanding of the epidemiology of blackleg and the role played by the ECB before the decision is made to cease or relax our attempts to manage the ECB. It may be possible to manage blackleg independently of the ECB through the use of *Erwinia*-free stocks (Graham and Hardie, 1971) or antibiotic treatment of seed pieces (Bonde, 1955). If this is not possible, we may be able to determine the intensity with which to manage the ECB in a given location and season by considering the incidence of blackleg inoculum on the seed pieces and either the level of blackleg inoculum in the soil or the history of blackleg in the field.

ONGOING RESEARCH AND FUTURE PROSPECTS

Our present research efforts are directed toward enhancing our understanding of the ecology of the ECB as it relates to the spring potato crop and the incidence of blackleg. To date, we have preliminary evidence that the role of ECB in the epidemiology of blackleg may involve more than merely providing a site of entry for the pathogen (Anderson, Echandi, and Kennedy, unpubl.). In a laboratory study, viable bacteria (*Erwinia carotovora* subsp. *atroseptica*) have been recovered from the cuticle, the gut, and the frass of ECB larvae fed diseased snap beans (Table 6). Because ECB larvae often tunnel into

Table 6
Occurrence of *Erwinia carotovora* subsp. *atroseptica* on Cuticle, in Gut, and in Frass of ECB Larvae Fed on Infected Green Beans

	Percent Infested Larvae	
Location	Larvae Fed Healthy Beans	Larvae Fed Infested Beans
Cuticle	26.7	100.0
Gut	6.7	76.5
Frass	----	40.0

more than one potato stem during the course of their development, they may play a role in the spread of the blackleg pathogen among plants; however, additional research is needed to determine if such spread actually occurs and what significance it has under field conditions. Further research is needed to determine if ECB adults play any role in the spread of the blackleg pathogen as has been reported for certain Diptera (Kloepper et al., 1979).

In order to enhance our understanding of ECB ecology and to improve the efficiency with which we can manage the ECB on potatoes, we are attempting utilization of temperature-dependent models to predict the spring emergence of overwintering ECB moths and the development of the first generation. We hope to be able to use these models ultimately to predict when adults of the overwintering generation will emerge in the spring and when development of the first spring generation will be complete. It is hoped that these models will greatly enhance the efficiency of the existing Pest Alert program.

According to Beck (1967), once the ECB has completed diapause development, temperature is the only factor influencing diapause termination and postdiapause morphogenesis. Thus, by placing larvae that had completed diapause development (but had not yet begun postdiapause morphogenesis) at a variety of temperatures over the normal temperature range for ECB development, we were able to generate the curves shown in Figure 7. They

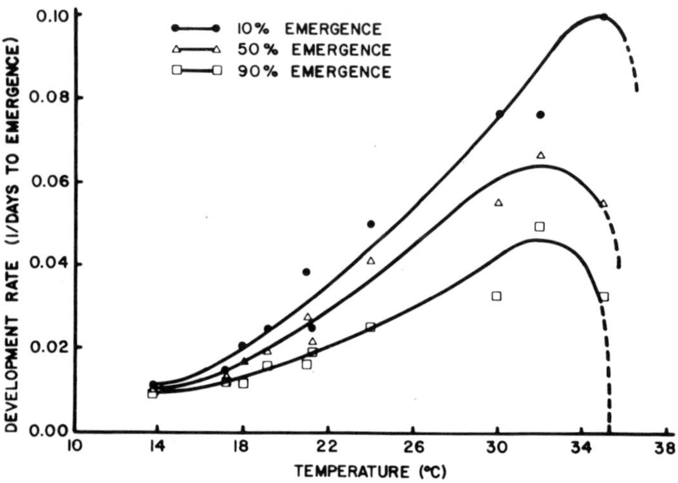

Figure 7
Development rate for postdiapause morphogenesis of ECB larvae at constant temperatures. The three curves represent rates for 10, 50, and 90 percent of the ECB population under study. Completion of development was marked by adult emergence.

represent the rates of postdiapause morphogenesis as measured by adult emergence for 10, 50, and 90 percent of the ECB population under study. These development rate data can be utilized with temperature data from the field to predict adult emergence in the spring, using previously published models (Stinner et al., 1975; Logan et al., 1976).

Developmental parameters of the nondiapausing first generation of ECBs can be determined in much the same way. The curve for egg development from oviposition to eclosion is shown in Figure 8, and those for larval and pupal development are shown in Figure 9. Only a single curve for egg development was constructed because egg development is relatively uniform, with over 90 percent of the eggs at a given temperature hatching on the same day. Larval and pupal development rates are more

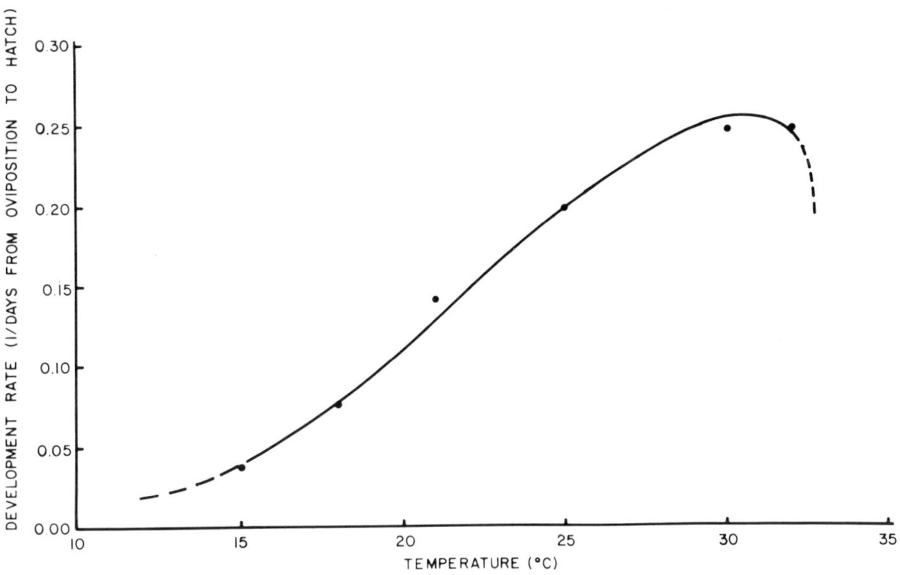

Figure 8
Development rate from oviposition to eclosion for ECB eggs at constant temperatures.

variable, and the three curves in Figure 9 represent the developmental rates for 10, 50, and 90 percent of the population under study. These curves too can be utilized with temperature data from the field to predict when emergence of the first spring generation will occur. Similar curves for larval development can be used in predicting when the maximum larval feeding injury (that caused by fourth and fifth instars) is likely to occur. Such information, coupled with a model of plant phenological development and an understanding of the relationship of plant phenology, ECB injury, and yields, may prove of considerable value for predicting yield losses likely to result from a given ECB infestation. Ultimately it would be desirable to use the ECB models with a model of the epidemiology of blackleg (when and if such a model is developed) to provide a comprehensive pest management model for the ECB and *Erwinia* on potatoes in North Carolina.

One of the likely consequences of more efficiently managing the ECB-blackleg complex is a further reduction in the quantities of insecticide regularly applied to the potato crop. The result, almost certainly, will be that damaging Colorado potato beetle populations will occur more frequently. At present, such populations are often prevented from developing by measures used to control the ECB. An awareness of this possibility should enable us to utilize the information on Colorado potato beetle management being developed in other states and to apply

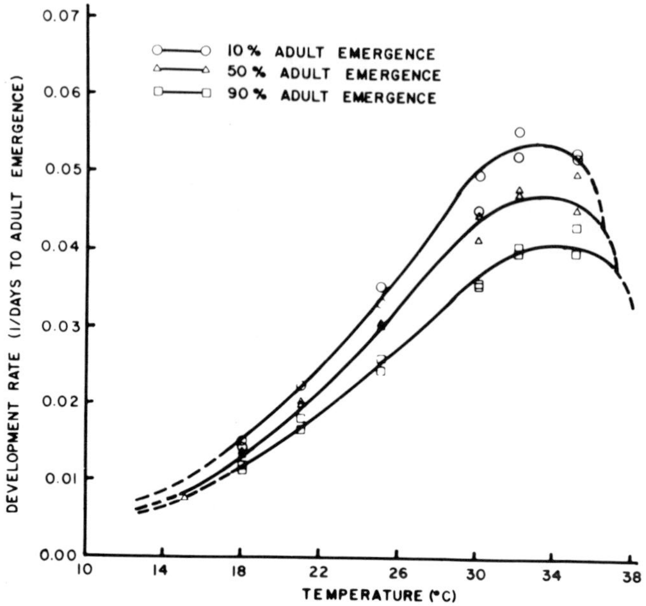

Figure 9
Development rate from egg hatch to adult emergence for nondiapausing ECB at constant temperatures. The three curves represent rates for 10, 50, and 90 percent of the ECB population under study.

procedures for efficiently managing this pest before it becomes a severe problem.

REFERENCES

Anonymous, 1979, North Carolina Agricultural Statistics, North Carolina Crop and Livestock Reporting Service, Raleigh, N.C. 72 p.

Aleck, J. R., and M. D. Harrison, 1978, The Influence of Inoculum Density and Environment on the Development of Potato Blackleg, Am. Pot. J. 55:479-494.

Averre, C. W., and K. A. Sorensen, 1979, Potato Pest Alert, North Carolina Agric. Ext. Service, Raleigh, N.C. 92 p.

Barber, G. W., 1926, A Two Year Study of the Development of the European Corn Borer in New England, J. Agric. Res. 32:1053-1068.

Beard, R. L., 1940, Effect of Corn Borers on Potato Yield, Conn. Agr. Exp. Stn. Bull. 434:275-276.

Beck, S. D., 1967, Water Intake and the Termination of Diapause in the European Corn Borer, Ostrinia nubilalis, J. Insect Physiol. 13:739-750.

Bonde, R., 1930, Some Conditions Determining Seed-Piece Decay and Blackleg Induced by Maggots, *Phytopathology* 20:128.

Bonde, R., 1939, The Role of Insects in the Dissemination of Potato Blackleg and Seed-Piece Decay, *J. Agr. Res.* 59:889-917.

Bonde, R., 1950, Factors Affecting Potato Blackleg and Seed-Piece Decay, *Bull. Maine Agr. Exp. Stn.* 482

Bonde, R., 1955, Antibiotic Treatment of Seed Potatoes in Relation to Seed-Piece Decay, Blackleg, Plant Growth and Yield Rate, *Plant Dis. Rpt.* 39:120-123.

Bray, D. F., 1961, European Corn Borer Control in Potatoes, *J. Econ. Entomol.* 54:782-784.

Burr, T. J., M. N. Schrath, and D. N. Wright, 1977, Survival of Potato Blackleg (*Erwinia carotovora* var. *carotovora*) and Soft Rot Bacteria (*E. carotovora* var. *atroseptica*), *Cal. Agr.* 31:12-13.

Caffery, D. J., and L. H. Worthley, 1927, A Progress Report on the Investigations of the European Corn Borer, *USDA Bull. 1476,* 155 p.

Carde, R. T., J. Kochansky, J. F. Stimmel, A. G. Wheeler, Jr., and W. L. Roelofs, 1975, Sex Pheromone of the European Corn Borer (*Ostrinia nubilalis*): cis and trans - Responding Males in Pennsylvania, *Environ. Entomol.* 4:413-414.

Carde, R. T., W. L. Roelofs, R. G. Harrison, A. T. Vawter, P. F. Brussard, A. Mutuura, and E. Munroe, 1978, European Corn Borer: Pheromone Polymorphism or Sibling Species? *Science* 199:555-556.

Carter, W., 1973, *Insects in Relation to Plant Disease,* 2d ed., Wiley-Interscience, New York, 759 p.

DuRant, J. A., 1969, Seasonal History of the European Corn Borer in Florence, South Carolina, *J. Econ. Entomol.* 62:1071-1075.

Graham, D. C., and J. L. Hardie, 1971, Prospects for the Control of Potato Blackleg Disease by the Use of Stem Cuttings, *Proc. 6th Brit. Insect. Fungic. Conf.* 1:219-224.

Helwig, J. T., and K. A. Council, eds., 1979, *SAS Users Guide, 1979 edition,* SAS Institute, Raleigh, N.C. 494 p.

Hodgson, B. E., 1928, The Host Plants of the European Corn Borer in New England, *USDA Tech. Bull. No. 77.*

Hofmaster, R. N., 1975, The Status of the European Corn Borer and Its Control on Irish Potatoes, *Veg. Gro. News* 29:1.

Jones, D. W., H. G. Walker, and L. D. Anderson, 1939, The European Corn Borer on the Eastern Shore of Virginia, *Virg. Truck. Exp. Stn. Bull.* 102:1621-1648.

Kennedy, G. G. and T. E. Anderson, 1980, European Corn Borer Trapping in North Carolina with Various Sex Pheromone Component Blends, *J. Econ. Entomol.* 73:642-646.

Kloepper, J. W., M. D. Harrison, and J. W. Brewer, 1979, The Association of Erwinia carotovora var. atroseptica and Erwinia carotovora var. carotovora with Insects in Colorado, Am. Pot. J. 56:351-361.

Klun, J. A., and Cooperators, 1975, Insect Sex Pheromones: Intraspecific Pheromonal Variability of Ostrinia nubilalis in North America and Europe, Environ. Entomol. 4:891-894.

Klun, J. A., and S. Maini, 1979, Genetic Basis of an Insect Chemical Communication System: The European Corn Borer, Environ. Entomol. 8:423-426.

Kochansky, J., R. T. Carde, J. Liebherr, and W. L. Roelofs, 1975, Sex Pheromone of the European Corn Borer, Ostrinia nubilalis (Lepidoptera: Pyralidae), in New York, J. Chem. Ecol. 1:225-231.

Logan, J. A., D. J. Wollkind, S. C. Hoyt, and L. K. Tanigoshi, 1976, An Analytic Model for Description of Temperature Dependent Rate Phenomona in Arthropods, Environ. Entomol. 5:1133-1140.

McLeod, D. G. R., and A. N. Starratt, 1978, Some Factors Influencing Pheromone Trap Catches of the European Corn Borer, Ostrinia nubilalis (Lepidoptera: Pyralidae), Can. Entomol. 110-51-55.

Matteson, J. W., and G. C. Decker, 1965, Development of the European Corn Borer at Controlled Constant and Variable Temperatures, J. Econ. Entomol. 58:344-349.

Merrill, L. G. 1957, European Corn Borer: A Problem for Potato Growers, Market Growers J. 86:2, 20-21.

Nielsen, L. W., 1963, Blackleg of Irish Potato Plants Following European Corn Borer Damage, Plant Dis. Rpt. 47:272-275.

Pepper, B. P., 1938, European Corn Borer Infestations on Potatoes, Hints to Potato Growers, 19:1-4.

Showers, W. B., 1979, Effect of Diapause on the Migration of the European Corn Borer into the Southeastern United States, in, R. L. Rabb and G. G. Kennedy, eds., Movement of Highly Mobile Insects: Concepts and Methodology in Research, pp. 420-430, Univ. Graphics, N.C. State Univ., Raleigh, N.C., 456 p.

Showers, W. R., H. C. Chaing, A. J. Keaster, R. E. Hill, G. L. Reed, A. N. Sparks, and G. J. Musick, 1975, Ecotypes of the European Corn Borer in North America, Environ. Entomol. 4:753-760.

Sparks, A. N., and W. B. Showers, 1975, Current Status of European Corn Borer in South Georgia, J. Georgia Entomol. Soc. 10:347-352.

Sparks, W. C., 1977, In L. Chicarappa, ed., Crop Loss Assessment Methods, p. 133, FAO Manual on the Evaluation and Prevention of Losses by Pests, Diseases and Weeds, Suppl. 2.

Stinner, R. E., G. D. Butler, Jr., and J. S. Bacheler, 1975, Simulation of Temperature-Dependent Development in Population Dynamics Models, Can. Entomol. 107:1167-1174.

Turner, N., and M. P. Zappe, 1937, European Corn Borer in Potatoes, Conn. Agr. Exp. Stn. Bull. 396:345-346.

Management of Plant Parasitic Nematodes in Potato Production

George W. Bird

Nematode pathogens play a significant role in the worldwide production of potatoes. The potato cyst nematodes (*Globodera rostochiensis* and *G. pallida*) are by far the most damaging. It has been estimated that 10 percent of the U.S. potato crop is lost annually to plant parasitic nematodes (Society of Nematologists, 1970). But in the United States, *G. pallida* is not known to be present, *G. rostochiensis* is not widely distributed, and *Globodera* spp. have little, if any influence on potato production (Spears, 1968). The major U.S. nematode pathogens of potatoes include the potato rot nematode (*Ditylenchus destructor*), root-knot nematodes (*Meloidogyne* spp.), root-lesion nematodes (*Pratylechus* spp.), and stubby-root nematodes (*Trichodorus* spp.). All of these nematodes are present in Michigan. *Pratylenchus penetrans* and *Meloidogyne hapla* are the most serious nematode pathogens of potatoes in Michigan. Both species are associated with potato production in both mineral and organic soils; however, *P. penetrans* is most frequently associated with losses in mineral soil and *M. hapla* in organic soils. *D. destructor* has been a problem in a few isolated cases, and *Trichodorus* spp. have not been associated with potato losses in Michigan.

Relatively little is known about the influence of *P. penetrans* on the growth and development of *Solanum tuberosum*. My objective here is to review the role of *P. penetrans* in potato production and summarize the

---

I wish to express appreciation to all members of the Michigan State University Nematology Program for their contributions to the potato project. Significant support has also come from the faculty of the Departmet of Crops and Soil Science and the Michigan Potato Industry Commission.

potato nematode research conducted in Michigan State University during the past seven years.

ROOT-LESION NEMATODE

Literature

Hastings and Bosher (1938) reported that P. penetrans retarded the growth of potato seedlings by 59.6 percent. Oostenbrink (1954, 1958) and Dickerson et al. (1964) confirmed the pathogenicity of this host-parasite relationship and reported yield losses between 20 and 50 percent. Oostenbrink (1966) found a linear relationship between initial population densities of P. penetrans and tuber yield, and information about the relationship was expanded by Olthof et al. (1973) and Olthof and Potter (1973). The initial population density tolerance limit for P. penetrans was estimated at 1.0 per gram of soil (Seinhorst, 1950). In a synthesis of earlier work, Oostenbrink (1966) identified a tolerance limit range of 0.4 to 1.0 P. penetrans per gram (g) in sandy soil and 0.7 to 2.0 per g in loam and organic soils. Olthof and Potter (1973) established a damage threshold of 2.0 per g of soil, which Barker and Olthof (1976) used in a report of tolerance limits for a number of crops.

Bernard (1973) and Bernard and Laughlin (1976), investigating the ecology and economics of P. penetrans associated with potato production in Michigan, established that different potato cultivars had different pathogenic and host suitability responses to P. penetrans. Superior was the most susceptible of five cultivars to P. penetrans, and Russet Burbank the most tolerant. Russet Burbank, however, was the best host, as evaluated by nematode reproductive potential. A moderate initial population density significantly increased tuber yield of Katahdin, whereas higher densities resulted in a significant reduction in yields. Symptoms were confined to reductions in tuber weight and root weight. Kimpinski (1979) reported that the amount of water translocated in the stems of S. tuberosum infected with P. penetrans was less than in nematode-free plants.

Distribution

Approximately 2 percent (810 acres) of Michigan's potato acreage was sampled in 1975 for occurrence and population densities of plant parasitic nematodes. The state was divided into sixteen potato growing regions of five different production categories (Table 1). A total of 162 commercial potato production sites were selected in cooperation with county cooperative extension service personnel. Each site represented 5.0 acres and was

Table 1

Root-Lesion Nematode (*Pratylenchus penetrans*) Survey of Commercial Potato Production in Michigan, 1975

Acreage, Region (Soil)	Sites Sampled	*P. penetrans* Problem Sites (%)	Nematicide-Treated Sites (%)
I MORE THAN 10,000 ACRES			
Montcalm (mineral)	39	42	62
II 5,000-10,000 ACRES			
Bay (mineral)	36	25	11
III 2,000-5,000 ACRES			
Allegan (organic)	12	67	92
Antrim (mineral)	10	70	0
Dickinson-Iron (mineral)	10	80	90
Lapeer (organic)	0	--	--
Monroe (mineral)	9	78	78
Presque Isle (mineral)	10	50	20
IV 500-2,000 ACRES			
Delta (mineral)	6	67	0
Emmet (mineral)	3	67	67
Houghton (mineral)	10	90	0
Jackson (organic)	9	67	67
Luce (mineral)	0	--	--
Macomb (mineral)	0	--	--
V Fewer than 500 ACRES			
Manistee (mineral)	3	100	0
Van Buren (mineral)	5	60	60

Note: The survey included about 2 percent of Michigan's potato acreage.

sampled twice during the growing season at two degree-day ranges using a base of 50°F (100-1400 DD_{50} and 1400-2500 DD_{50}). Soil ($100 cm^3$) and root (1.0 g) samples were processed by the centrifugation-flotation and shaker techniques, respectively, and the nematodes identified and counted. Data pertaining to the soil type, potato cultivar, pesticide usage, and cropping history were collected for each location sampled.

There was considerable variation among the sites and geographic locations in the occurrence and population density of root-lesion nematodes associated with potato production in Michigan. *Pratylenchus* spp. were recovered from 63.6 percent of the sites sampled. *P. penetrans* (50.5 percent) was by far the most common species. *P. crenatus* (19.4 percent), *P. neglectus* (19.2 percent), and other *Pratylenchus* spp. (10.9 percent) were also recovered, frequently concomitant with *P. penetrans*. Forty-two percent of the sites sampled were treated with a nematicide in 1975 (Table 1). These had a mean population density of forty-two *P. penetrans* per 100 cm^3 soil and 1.0 g of root tissue. The nontreated sites had a mean population density about four-fold greater (160 *P. penetrans* per 100 cm^3 soil and 1.0 g root tissue). Using the tolerance limit of $P_i = 50$ *P. penetrans* per 100 cm^3 soil, it was estimated that 49 percent of the sites sampled had a root-lesion nematode problem.

Loss Assessment Model

A more comprehensive crop loss assessment system was designed to evaluate the effects of this host-parasite relationship on firm, farm, enterprise, county, district, and state. The assessment system uses five input parameters: name of *S. tuberosum* cultivar, name and rate of nematicide, accumulated degree days at a base of 50°F (DD_{50}) at the time of sampling, *P. penetrans* per 100 cm^3 and g root tissue, and nematode control cost and projected crop value. Initial *P. penetrans* densities (P_i) are estimated for three cultivar groups in the presence and absence of nematode control from *P. penetrans* densities at a specific DD_{50}. Initial population densities are used to estimate tuber yields [yield (tolerant cv) = $0.54 \times \log (P_i + 1) + 3.58$; yield (susceptible cv) = $0.76 \times \log (P_i + 1) + 1.56$; yield (highly susceptible cv) = $0.98 \times \log (P_i + 1) + 3.39$]. A processor-based algorithm is used to compute crop losses for desired levels of interaction (for example, farm, enterprise, county, district, or state) and evaluate alternate agroecosystem design options.

Influence on Fertilizer Uptake and Utilization

Relatively little is known about the interaction of plant-parasitic nematode damage and its influence on nutrient uptake and utilization. Crop yield increases from soil fumigation are associated with the control of various plant diseases, nematodes, insects, and weeds (Wilson, 1969). Certain soil fumigants applied at specific rates control root diseases of potatoes caused by fungi and plant-parasitic nematodes (Cetas and Harrison, 1963; Miller and Hawkins, 1969; Workman et al. 1977). Several organocarbamate pesticides control both plant parasitic nematodes and insects. In addition to the fungicidal and nematicidal properties, many edaphic pesticides affect soil nitrogen transformation (Elliott et al., 1977; Wolcott et al., 1960).

More research emphasis has been placed on the influence of nutrients on nematode reproduction than on the joint influence of nutrients and nematodes on plant ontogeny. Dolliver (1961) found that moderately reduced NPK fertilization of Wando peas resulted in increased population densities of P. penetrans. Nitrogen is reported to have two effects on nematode populations. Its application can result in nematode population density increase, presumably by providing more feeding sites through stimulation of root growth; however, most reports indicate that addition of nitrogen results in lower nematode numbers (Castaner, 1966; Kimpinski et al, 1976; Patterson and Bergeson, 1967; Ross, 1969; Walker, 1971; Oteifa, 1955; Heald and Burton, 1968). It has been suggested that ammonia is responsible for this phenomenon (Eno et al., 1955; Oteifa, 1955; Walker, 1971). Oteifa (1955) speculated that ammonium ions are inhibitory to egg hatching.

Several forms of nitrogen have been studied, and not all are equally effective in their influence on nematode populations. Walker (1971) found that nitrate was least effective in reducing numbers of P. penetrans. Heald and Burton (1968) reported that organic nitrogen in the form of activated sewage sludge reduced nematode populations more than did inorganic nitrogen. Unless phytotoxic levels are used, nematode population densities decrease with increasing amounts of nitrogen. Huber and Watson (1974) reported that the form of nitrogen was very important for the control of many plant diseases; however, Workman et al. (1977) were unable to find any differences in soil ammonium or nitrate and found varying levels of nitrate in potato petioles between fumigated and nonfumigated plots. Evidently other factors are operating directly on the nematode or indirectly through the host. Nematode populations in Manitoba soils generally decrease in clay with increasing nitrogen, but numbers increased in sand, with increased nitrogen indicating that soil factors may be important (Kimpinski and Welch, 1971).

Joint Action of Nitrogen and Nematicides

The joint action of nitrogen fertilizer and nematicides was investigated using three potato varieties (Superior, Onaway, and Russet Burbank), three rates of nitrogen (84, 168, and 336 kg per ha), five edaphic pesticides (disulfoton, aldicarb, 1,3-D + MIC, carbofuran and thiofanox), and an insecticide spray to minimize the role of foliar feeding insects.

In 1977, disulfoton, aldicarb, and disulfoton plus 1,3-D + MIC significantly increased the yield of Superior at all nitrogen rates. The yield increases ranged from 37 to 56 percent for Superior, 15 to 35 percent for Onaway, and 13 to 17 percent for Russet Burbank (Table 2). Nitrogen had no detectable influence on yield. Population densities of *Pratylenchus penetrans* were significantly reduced by disulfoton plus 1,3-D + MIC and aldicarb. Superior was most susceptible to *P. penetrans*, Onaway intermediate, and Russet Burbank moderately tolerant. Neither the potato variety or nitrogen significantly influenced the population densities of *P. penetrans*.

In 1978 aldicarb and 1,3-D + MIC significantly increased yields of Superior to all nitrogen rates (Table 2). Nitrogen also significantly increased yields, but only where aldicarb and 1,3-D + MIC were applied. Total yield was increased 12 to 14 percent and 17 to 23 percent, respectively. Aldicarb and 1,3-D + MIC appeared to increase tuber set, and nitrogen increased tuber size. Although all of the nematicides provided some control of *P. penetrans*, aldicarb had the highest degree of efficacy and gave excellent season-long control.

Joint Action of Phosphorus and Nematicides

The joint action of phosphorus fertilizer and nematicides was investigated in 1979 using two potato varieties (Superior and Russet Burbank), three rates of phosphorus (0, 22, and 65 kg P/ha (0, 50 and 150 lb P_2O_5/A, respectively), two edaphic pesticides--aldicarb and 1,3-D + MIC (Temik 15G at 3.4 kg per ha active ingredient and Vorlex at 94 ℓ per ha, respectively), and an insecticide spray to minimize the role of foliar feeding insects. The experimental design was a 3 x 3 split factorial with five replications. Varieties existed as whole plots, while phosphorus and nematicide treatments were randomized within each replication. The two experiments were conducted on a McBride sandy loam soil (alfic fragiothods) at the Montcalm Research Farm in west-central Michigan. This soil had a Bray P_1 phosphorus test of 240 parts per million (ppm).

Despite the very high phosphorus soil test, significant yield increases were obtained with both varieties at both rates of applied phosphorus fertilizer (Table 3). Aldicarb and 1,3-D + MIC also significantly

Table 3
Influence of Selected Treatments on Tuber Yield

Phosphorus (lb/A)	Treatment	Yield (cwt/acre)[a] Superior	Yield (cwt/acre)[a] Russet Burbank
0	Check	236.7a	231.9a
	Aldicarb 3.0 lb a.i./acre	280.7ab	268.5ab
	1,3-D + MIC 10.0 gal/acre	314.9bc	309.7bc
50	Check	273.3ab	279.9b
	Aldicarb 3.0 lb a.i./acre	316.7bc	301.0bc
	1,3-D + MIC 10.0 gal/acre	348.9c	346.7c
150	Check	313.7bc	322.6bc
	Aldicarb 3.0 lb a.i./acre	364.1c	310.2bc
	1,3-D + MIC 10.0 gal/acre	366.7c	342.4c

[a] Column means followed by the same letter are not significantly different ($p \leqslant 0.05$) according to the Student-Newman-Keuls multiple range test.

increased yields of both varieties, particularly at the lower levels of applied phosphorus. Maximum yields for both varieties were obtained with the highest level of applied phosphorus in combination with either aldicarb or 1,3-D + MIC. Aldicarb and 1,3-D + MIC contributed toward increased size of tubers for Superior; however, phosphorus fertilizer contributed more to size for Russet Burbanks.

Population densities of *Pratylenchus penetrans* in soil and root tissue were significantly reduced by aldicarb, while 1,3-D + MIC had little or no control of the nematode. These data differ from previous studies on this farm where 1,3-D + MIC has given good early season control of *P. penetrans*.

This study demonstrated that the Bray P_1 soil test on sandy soil does not predict phosphorus availability for potatoes as well as it does with other field crops. A combination of nematicide and phosphorus fertilizer was required for optimum potato production. Because 1,3-D + MIC showed little control for *P. penetrans*, it must be exerting some other beneficial effect on the potato.

JOINT ACTION OF ROOT-LESION NEMATODES AND COLORADO POTATO BEETLES

The objective of this study was to examine the effect of *P. penetrans* and *Leptinotarsa decemlineata* on the growth, development, and yield of *S. tuberosum*, including observations on how plant defoliation by *L. decemlineata* affects the population dynamics of

Table 2
Influence of Nitrogen and Three Pesticides on Yield of Three Potato Cultivars.

Treatment		Yield (quintal/ha)				
Nitrogen (kg/ha)	Pesticide	Superior (1977)	Onaway (1977)	Russet Burbank (1977)	Mean (1977)	Superior (1978)
84	Control	133a[a]	157a	224a	171a	289a
	Aldicarb[b]	193bc	195ab	236ab	209bc	340bc
	Disulfoton + 1,3-D + MIC[b,c]	205bc	223bc	287ab	240de	356c
168	Control	118a	195ab	239ab	185ab	314ab
	Aldicarb[b]	220c	236bc	272ab	244de	389d
	Disulfoton + 1,3-D + MIC[b,c]	212bc	227bc	274ab	240de	400d
336	Control	96a	160a	217a	168a	304ab
	Aldicarb[b]	195bc	216bc	279ab	228cde	419d
	Disulfoton + 1,3-D + MIC[b,c]	200bc	240c	302b	249e	416d
Mean	Control	124w	170w	227w	175w	308y
	Aldicarb[b]	202y	216y	259y	227y	382z
	Disulfoton + 1,3-D + MIC[b,c]	205y	231y	288y	241z	390z

[a]Comparable column means followed by the same letter are not significantly different ($p \leqslant 0.05$) according to the Student-Newman-Keuls multiple range test.
[b]3.4 kg/ha a.i., in fertilizer furrow at planting.
[c]93.5 l/ha broadcast injected to 20 cm three weeks before planting.

P. penetrans. Seed pieces (cv Superior) were planted on May 24, 1979, at the Montcalm Potato Research Farm in Entrican, Michigan. Each plot consisted of two rows 1.83 m in length and 0.86 m apart, with 20.5-30.5 cm space between plants. Twenty-seven insect cages, measuring 6 ft by 6 ft by 6 ft ($1.83m^3$), were erected over the plots after planting.

Cages selected at random were assigned one of nine different treatments. A treatment consisted of one of three population levels of L. decemlineata and one of three population levels of P. penetrans. Population levels were achieved by stapling egg masses of L. decemlineata to the leaves of the newly emerged plants within the cages, allowing them to hatch, and then manipulating the plant densities to either zero, ten or twenty beetles per plant. Low initial population levels of P. penetrans were achieved by preplant fumigation with Telone II (20 gal per acre) on May 1, 1979. The medium population levels represented natural field populations. The high population levels of P. penetrans were achieved by complementing the natural field populations with a liquid suspension of nematodes obtained from potato roots cultured in the laboratory. A 10 ml aliquant of the nematode suspension was applied at planting to provide an additional 500 nematodes to the root rhizosphere of each potato plant.

Plant growth and development was monitored at various intervals throughout the season by randomly selecting a plant from one specific row of each plot and then returning each to the laboratory for analysis. In the laboratory, root weight, stem weight, leaf weight, leaf surface area, tuber weight, and number were recorded at each sampling date. Soil and root populations of nematodes were estimated from samples taken at these times. Soil samples (centrifugation-flotation technique) for nematode analysis were taken by removing the soil adjacent to plant roots. Root samples (shaker technique) were derived from plants returned to the laboratory for plant growth analysis. At harvest, the remaining rows within each plot were harvested, graded, and weighed. During the season plants were maintained under normal commercial irrigation and disease control practices. A completely randomized two-factorial design was used in the experiment, with each treatment replicated three times.

Larval feeding by L. decemlineata influenced the population dynamics of P. penetrans and the growth and development of the potato plant. The direct effect of the Colorado potato beetle feeding was a reduction in leaf weight, with increasing beetle density. As beetles reduced the leaf dry weight, there was a corresponding reduction in the plant root dry weight, which was lowest in the twenty beetles per plant treatments. These changes are directly reflected in the reduced plant and tuber weights through time for the beetle-infested treatments. Beetle feeding had no significant influence on soil population densities of *P. penetrans*. The root population density of *P. penetrans* was lower in the beetle-infested treatments than in the beetle-free treatments. By reducing the leaf weight through time, beetle feeding decreased the size of the total root system and the number of *P. penetrans* per gram of root tissue was affected.

NEMATICIDE TRIALS

Chemical nematicides can be used to reduce nematode population densities below a dynamic economic threshold. Both fumigant and nonfumigant nematicides are available for use in potato production. Many of the fumigant nematicides also reduce soil population densities of fungal pathogens of *S. tuberosum* and may provide a limited amount of weed control. Most of the nonfumigant nematicides are also systemic insecticides. The most common fumigant nematicides used in potato production are the dichloropropenes (especially in mixture with methylisothiocyanate). Aldicarb is the most widely used nonfumigant nematicide.

Sixteen field experiments were used by the MSU nematology program during the 1974-1979 growing seasons to evaluate the influence of aldicarb and 1,3-D + MIC on the control of *P. penetrans* and tuber yield of two potato cultivars (Superior and Russet Burbank). Fourteen of the experiments were conducted at the MSU Montcalm Potato Research Farm (Entrican, Michigan) and the other two at the MSU Vegetable Research Farm (Sodus, Michigan). Each treatment was replicated four to twelve times in a randomized block design, depending on the experiment. At the Montcalm Potato Research Farm, each plot consisted of four rows (two rows at Sodus) 50 ft in length and 35 in apart with 8 to 12 in spacing between the seed pieces. 1,3-D + MIC (10 gal per acre broadcast) was injected to a soil depth of 6 to 8 in fourteen to twenty-one days before planting. Aldicarb (2.0 lb a.i. per acre, in row) was applied at planting. Soil and root population densities of *P. penetrans* were monitored throughout the growing season as previously described for the survey. The center two rows of each plot were harvested (all rows at Sodus) and the tubers graded and weighed. Foliar insecticides were used to minimize the influence of insects on tuber production.

Aldicarb (2.0 lb a.i. per acre) provided season-long control of *P. penetrans* associated with *S. tuberosum* (cvs Superior and Russet Burbank) in 1976 and 1977. Nematode population densities at the end of the growing season (P_f) were less than those at the beginning of the growing season (P_i). A multiple-year economic analysis was used to determine if aldicarb controls *P. penetrans* for more than one growing season. Tuber yield increases were greater with 1.3-D + MIC than with aldicarb (Table 4).

Four years of field data indicated that for Superior, the log (\$/ha) = 3.39 - 0.098 x log (P_i + 1), and for Russet Burbank, the log (\$/ha) = 3.58 - 0.054 x log (P_i + 1). A model was developed assuming that P_f + 2.0 P_i in the absence of a nematicide and P_f = 0.3 P_i when treated with aldicarb, and overwintering mortality (P_i for subsequent growing season) = 0.5 P_f. The model was evaluated for both cultivars at P_i = 60 *P. penetrans* per 100 cm^3 soil. The cost of the aldicarb treatment was set at \$96.37 per hectare and the cost of a systemic insecticide (\$24.71 per hectare) was included where aldicarb was not used. The model was run for both cultivars for five sequential growing seasons. Evaluations at the end of the first, second, and third seasons, were done on a net profits basis. This sequence of treatment resulted in the lowest P_f. Four years of continuous treatment with aldicarb continued to increase net profits with Superior; however, net profits were greatest with Russet Burbank when no nematicide application was used during the fourth year. A fifth annual application of aldicarb for control of *P. penetrans* associated with Superior was not economically advantageous. The higher the P_i in this pest-crop ecosystem, the longer it takes for a control strategy such as aldicarb to reduce the P_f so that a subsequent P_i is below the economic threshold. The length of time a population can be maintained below the economic threshold is a function of the relationship between P_i, subsequent P_f, overwintering mortality, and natural enemies. In this case, a multiple-year understanding for both the biology and economics of pest-crop ecosystem is essential for the development of efficient control programs for *P. penetrans*.

PEST-CROP ECOSYSTEM MODEL (POTATOES)

In 1975, the population dynamics of *P. penetrans* was modeled in relation to the influence of soil temperature on nematode reproductive potential. Each nematode in the population was monitored daily to determine its developmental status and reproductive contributions in conjunction with potato pests (Figure 1). A model for growth and development of *S. tuberosum* was constructed with the assimilation rate as a function of light intensity and air temperature (Figure 2). Assimilates were partitioned into leaves, stems, roots, and tubers every day. The models were integrated into a pest-crop ecosystem model designed to evaluate the influence of different initial population densities of *P. penetrans* on the growth, development, and tuber yield of *S. tuberosum* and the response of the nematode population to variation

Table 4
Tuber Yields of Three Potato Cultivars Associated with Fumigant and Nonfumigant Nematicide Application in Eighteen Experiments, 1974-1979

	Yield (ctw/acre)						
Cultivar and Treatment	1979	1978	1977	1976	1976	1974	$\overline{x}$
Superior							
Nontreated control	275	284	129	146	168	70	183
Aldicarb (2.0 lb/acre)	321	349	182	180	293	101	250
1,3-D + MIC (10 gal/acre)	344	358	204	-	300	-	295
Russet Burbank							
Nontreated control	278	-	211	278	301	272	264
Aldicarb (2.0 lb/acre)	293	-	237	399	366	348	321
1,3-D + MIC (10 gal/acre)	333	-	256	390	467	-	362

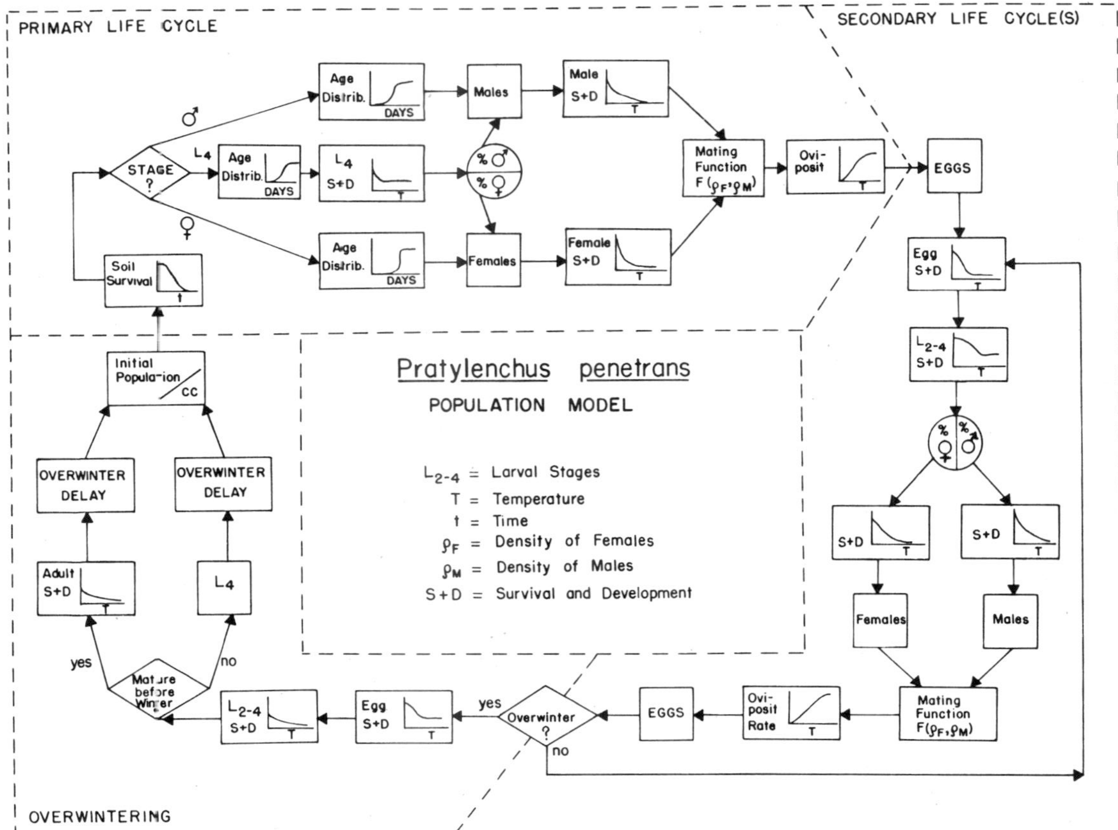

Figure 1
Conceptual diagram of life history of Pratylenchus penetrans (root-lesion nematode) model used in computer simulation (POTATOPEST), developed by M. Sarette (N.Y. Agric. Expt., Geneva) and G. W. Bird (Michigan State University).

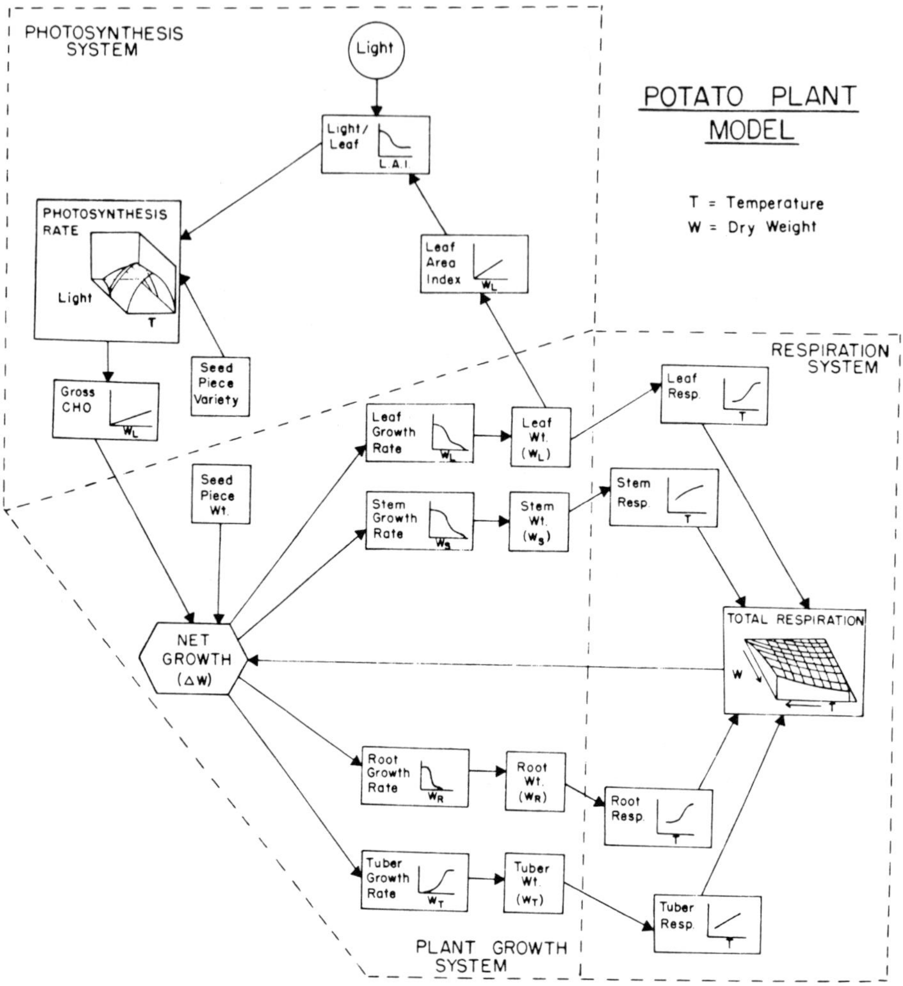

Figure 2
Conceptual diagram of ontogeny of potato plant model used in POTATOPEST.

in the growth and development patterns and cultivars of the host (Figure 3). The most significant contribution of the initial model was to identify specific areas where detailed quantitative biological information is required before a more refined simulation can be developed and validated. The simulation has been used extensively in academic courses for teaching the dynamic economic threshold concepts of pest management studies. It has also been used in extension programs.

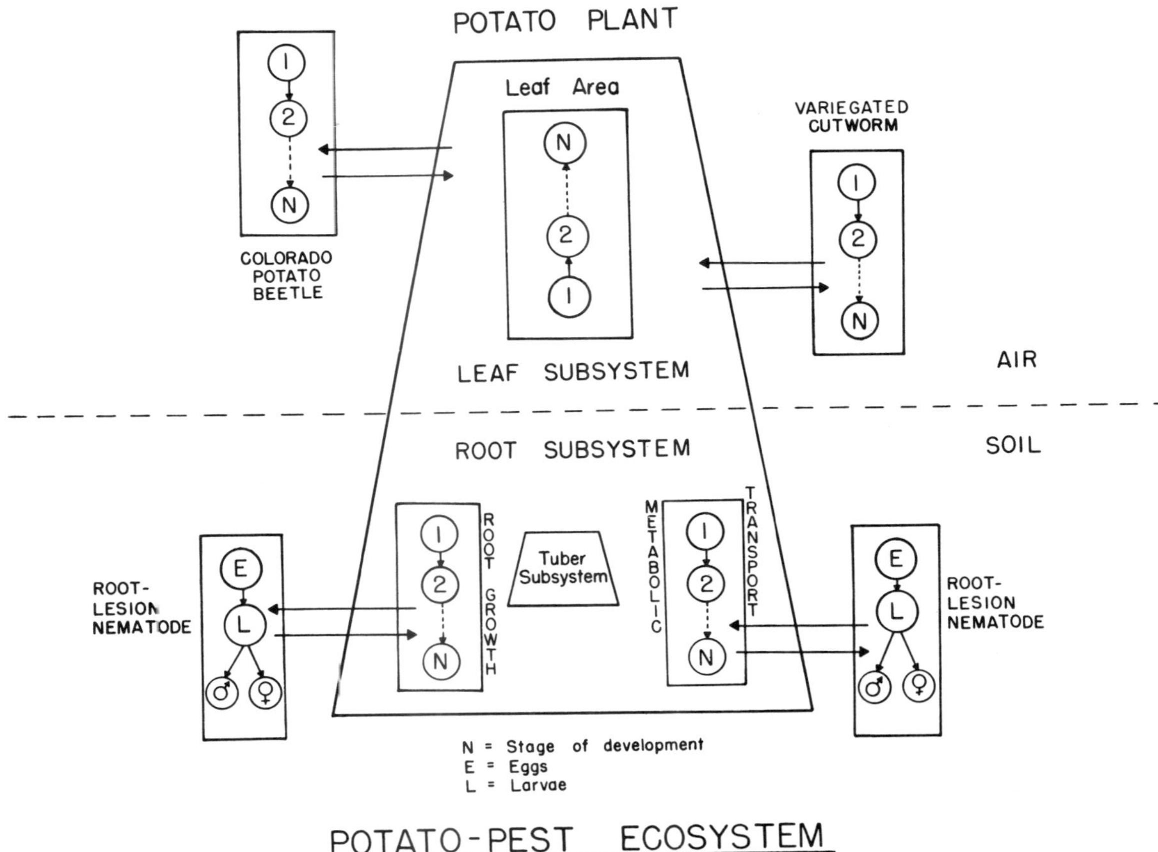

Figure 3
Conceptual diagram of interaction model used in POTATOPEST.

NEMATODE POPULATION MANAGEMENT

Pest Management

Integrated pest management (IPM) is the interdisciplinary activity that deals with the development, evaluation, and implementation of pest control strategies designed to result in favorable economic, ecologic, and sociologic consequences. It has received considerable attention during the past few years, including several major reviews. The components of IPM programs usually include biological monitoring systems, environmental monitoring systems, pest-crop ecosystem models, production system management recommendations, and information delivery systems. Programs should be designed to optimize harmonious synchrony among the components of human societies and the natural environment (Figure 4). This usually requires a transdisciplinary approach in which scientists from different disciplines work as a team with continual conceptual synthesis and intellectual interchange. The process must also include a mutually accepted set of systems objectives developed within the general framework of the principles of systems science (the coordinated study of the structure and interactions among related entities). The majority of scientists engaged in public sector IPM activities have disciplinary backgrounds in entomology, plant pathology, nematology, or weed science. Professional economists, ecologists, sociologists, meteorologists, and systems scientists are also essential for the successful development and implementation of IPM programs.

IPM programs must be economical and operational within the overall objectives of production systems. This requires proper interfacing of public-sector and private-sector activities. Implementation includes: collection and delivery of information required by the management system, employment of appropriate IPM procedures, evaluation of the effect of IPM on the production system and natural environment, and use of evaluation information for the selection of additional implementation procedures and development of future IPM programs.

The production system owner or a representative is usually the manager of pests at the firm-level, farm, or enterprise basis (Figure 5). Production enterprises may employ private consultants or custom applicators for recommendation and implementation of IPM strategies. Private enterprise should be the basis of most IPM implementation production systems. Public-sector involvement in IPM is most effective if the focal point is to provide district IPM information to the private sector and maintain appropriate programs at the county, district, state, and regional levels.

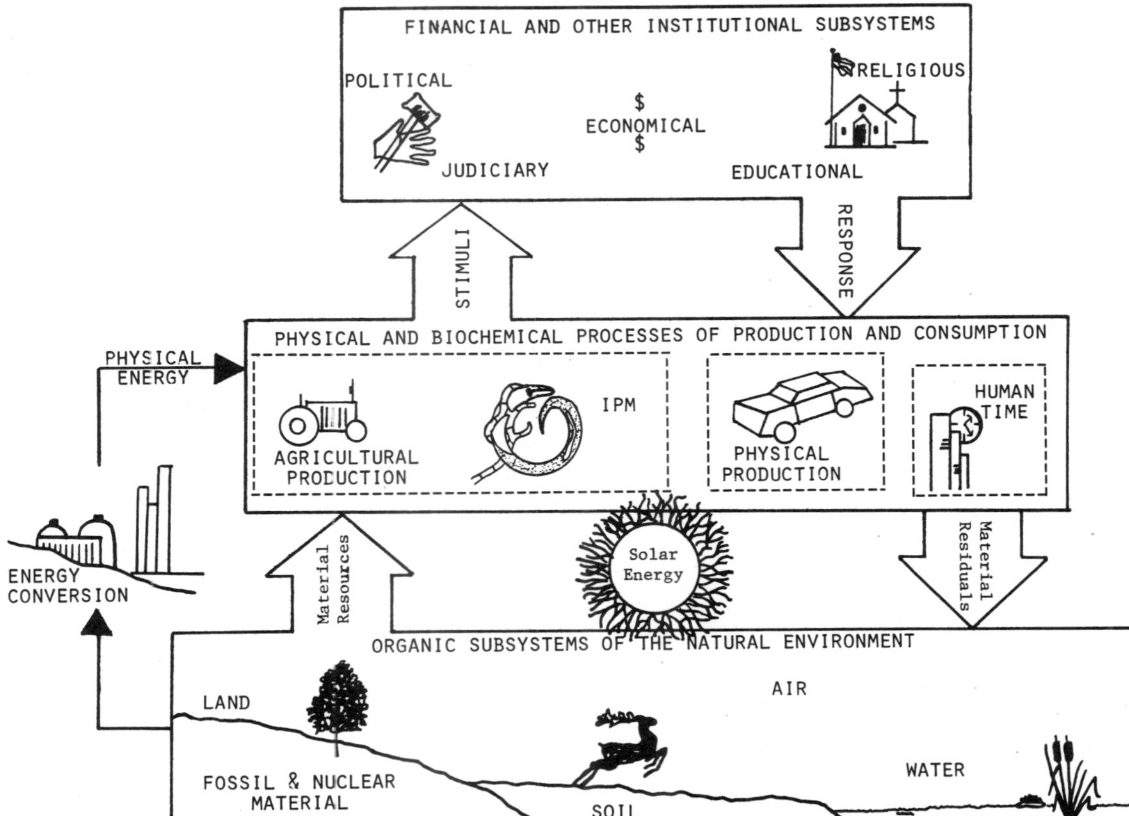

Figure 4
Conceptual elements of relationships among institutions, production systems, and natural resources.

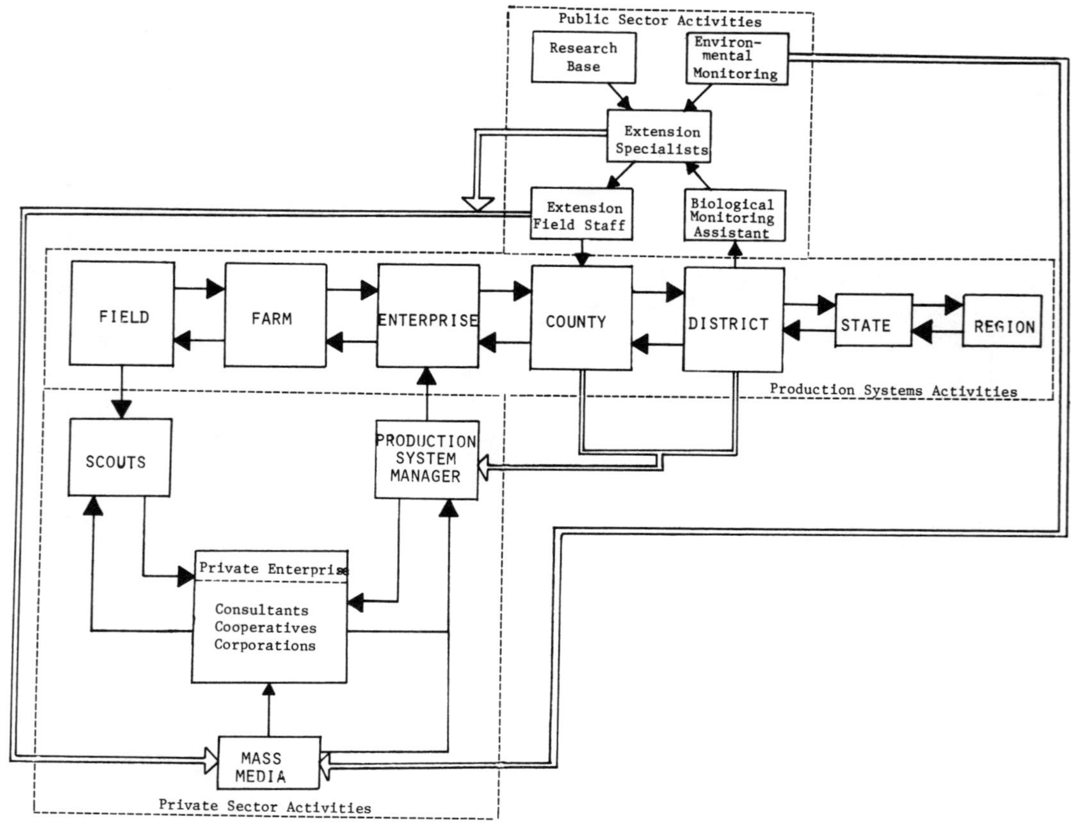

Figure 5
Conceptual diagram of relationships between the public and private sectors. Private and production system activities associated with pest management.

Nematode Management in Potato Production

Nematode management programs for potato production can contain several components, including exclusion, population reduction, and use of resistant cultivars. Regulatory procedures and the use of nematode-free seed pieces have been successful in containing *G. rostochiensis* in North America. On a more limited basis, similar procedures have been used to prevent the spread of *D. destructor* and *Meloidogyne* species. Once a nematode pathogen has become established in a potato production site and its population density has exceeded the dynamic economic threshold, it may be necessary to employ population reduction procedures or to use resistant potato cultivars. A number of potato cultivars are resistant to specific races of *G. rostochiensis* and *G. pallida*. Studies at Michigan State University have shown the cv Superior is highly susceptible to *P. penetrans* and cv Russet Burbank is relatively tolerant. The six other cultivars evaluated were intermediate.

In many situations it may be necessary to reduce the nematode population density below the dynamic economic threshold. This can be achieved through appropriate crop rotation and the use of cover crops. Many crops are suitable for rotation with potatoes in noninfected sites. It is essential, therefore, that a biological monitoring program be used to determine if a nematode problem exists for any given site. Information about the specific nature and distribution of nematode problems is also important.

REFERENCES

Barker, K. R., and T. H. A. Olthof, 1976, Relationships between Nematode Population Densities and Crop Responses, *Ann. Rev. Phytopathol*. 14:327-353.

Bernard, E. C., 1973, the Pathogenicity and Population Management of *Pratylenchus penetrans* on Potato, M. S. thesis, Dept. of Entomology, Michigan State University, 44p.

Bernard, E. C., and C. W. Laughlin, 1976, Relative Susceptibility of Selected Cultivars of Potato to *Pratylenchus penetrans*, *J. Nematol*. 8:239-242.

Castaner, D., 1966, The Relationship of Numbers of *Helicotylenchus microlobus* in Nitrogen Soil Ammendments, *Iowa State J. Sci*. 41:125-135.

Cetas, R. C., and M. B. Harrison, 1963, Evaluation of Soil Fumigants for Control of Early Maturity Wilt on Potatoes on Long Island, *Phytopathology* 53:347-348.

Chase, R. W., N. R. Thompson, R. L. Ledebuhr, C. M. Hanson, and R. B. Kitchen, 1978, Plot Harvester for Potatoes, *Am. Potato J*. 55:235-237.

Dickerson, O. J., H. M. Darling, and G. D. Griffin, 1964, Pathogenicity and Population Trends of Pratylenchus penetrans on Potato and Corn, Phytopathology 54:317-322.

Dolliver, J. S., 1961, Population Levels of Pratylenchus penetrans as Influenced by Treatments Affecting Dry Weight of Wando Pea Plants, Phytopathology 51:364-367.

Elliot, J. M., C. F. Marks, and C. M. Tu, 1977, Effects of Pratylenchus penetrans in Flue Cured Tobacco, Can. J. Plant Sci. 58:143-154.

Eno, C. F., W. G. Blue, and J. M. Good, Jr., 1955, The Effect of Anhydrous Ammonia on Nematodes, Fungi, Bacteria and Nitrification in Some Florida Soils, Proc. Soil Sci, Soc. Amer. 19:55-58.

Hastings, R. J., and J. E. Bosher, 1938, A Study of the Pathogenicity of the Meadow Nematode and Associated Fungus Cylindrocarpon radicicola Wr., Can. J. Res. 16 (sec. C):225-229.

Heald, C. M., and G. W. Burton, 1968, Effect of Organic and Inorganic Nitrogen on Nematode Populations on Turf, Plant Dis. Rep. 52:46-48.

Huber, D. M., and R. D. Watson, 1974, Nitrogen Form and Plant Disease, Ann. Rev. Phytopathology 12:139-165.

Kimpinski, J., 1979, Root-Lesion Nematodes in Potatoes, Am. Potato J. 56:79-86.

Kimpinski, J., H. R. Wallace, and R. B. Cunningham, 1976, Influence of Some Environmental Factors on Populations of Pratylenchus minyus in Wheat, J. Nematol. 8:310-314.

Kimpinski, J., and H. E. Welch, 1971, The Ecology of Nematodes in Manitoba Soils, Nematologica 17:308-318.

Miller, P. M., and A. Hawkins, 1969, Long Term Effects of Preplant Fumigation on Potato Fields, Am. Potato J. 46:387-397.

Olthof, T. H. A., C. F. Marks, and J. M. Elliot, 1973, Relationship between Population Densities of Pratylenchus penetrans and Crop Losses in Flue-Cured Tobacco in Ontario, J. Nematol. 5:158-162.

Olthof, T. H. A., and J. W. Potter, 1973, The Relationship between Population Densities of Pratylenchus penetrans and Crop Losses in Summer-Maturing Vegetables in Ontario, Phytopathology 63:577-582.

Oostenbrink, M., 1954, Over de betekensis van vrijlevende wortelaaljtes in land- en tuinbouw, Versl. Meded. Plantenziektenkundigen Dienst. 124-196-233.

Oostenbrink, M., 1958, An Inoculation Trial with Pratylenchus penetrans in Potatoes, Nematologica 3:30-33.

Oostenbrink, M., 1966, Major Characteristics of the Relationship between Nematodes and Plants, Meded. LandbHoogesch. Wageningen 66:1-46.

Oteifa, B. A., 1955, Nitrogen Source of the Host Nutrition in Relation to Infection by a Root-Knot Nematode, Meloidogyne incognita, Plant Dis. Rep. 39:902-903.

Patterson, Sis M. T., and G. B. Bergeson, 1967, Influence of Temperature, Photoperiod, and Nutrition on Reproduction, Male-Female-Juvenile Ratio, and Root to Soil Migration of *Pratylenchus penetrans*, *Plant Dis. Rep*. 51:78-82.

Ross, J. P., 1959, Nitrogen Fertilization on the Response of Soybeans Infected with *Heterodera glycines*, *Plant Dis. Rep*. 43:1284-1286.

Ross, J. P., 1969, Effect of *Heterodera glycines* on Yields of Nonnodulating Soybeans Grown at Various Nitrogen Levels, *J. Nematol*. 1:40-42.

Seinhorst, J. W., 1950, De betekenis van de toestand van de grond voor optreden van aantasting door het stengelaaljte (*Ditylenchus dipsaci* (Kuhn) Filipjev), *Neth. J. Plant Pathol*. 61:188-190.

Sighn, I. Gajendra, and P. N. Misra, 1978, Effect of Some Insecticides on Aphid Population and Yield of Potato Crop, *Indian J. Agric. Sci*. 48:12-16.

Society of Nematologists, Committee on Crop Losses, 1970, Estimated Crop Losses from Plant-Parasitic Nematodes in the United States, *Spec. Publ. No. 1, J. Nematology*, 4p.

Spears, Joseph F., 1968, *The Golden Nematode Handbook: Survey, Laboratory, Control and Quarantine Procedures*. Agric. Handbook No. 353. USDA, Washington.

Walker, J. T., 1971, Populations of *Pratylenchus penetrans* Relative to Decomposing Nitrogenous Soil Amendments, *J. Nematol*. 3:43-49.

Wilson, J. D., 1969, Soil Fumigation, *Am. Potato J*. 45:414-426.

Wolcott, A. R., F. Maciak, L. N. Shepherd, and R. E. Lucas, 1960, Effects of Telone on Nitrogen Transformations and on Growth of Celery in Organic Soil, *Down to Earth* 16:10-14.

Workman, M., M. D. Harrison, and C. H. Livingstone, 1977, Effects of Soil Fumigation with Telone C on Soil and Plant Nitrogen, Verticillium Wilt and Yield on Norland and Norgold Potatoes, *Am. Potato J*. 54:103-113.

Management of Late Blight

William E. Fry

Potato late blight is capable of tremendous devastation of potato foliage and tubers, and consequently the disease is well known to growers, agricultural scientists, and a small segment of the general populace. It was this disease that destroyed potato foliage and tubers in Ireland in the mid-1840s and led to the Irish potato famine (Large, 1940). Since then, the disease has been a problem in most areas where potatoes have been cultivated. In the United States, the disease is generally most severe in the humid, high-rainfall potato production areas of the East, from Florida to Maine. However, epidemics of potato late blight in California and in the Pacific Northwest during the last five years indicate that even these drier production areas are not completely free of the disease. My objective here is to describe the potential for managing potato late blight by integrating several approaches and methods.

POTATO LATE BLIGHT EPIDEMIOLOGY

There are several probable sources of the late blight pathogen (*Phytophthora infestans*). It can be present in infected seed tubers; it can be blown in from infected cull potatoes or volunteers; or it can be blown in from a neighboring field of infected potatoes. Regardless of the initial source (initial inoculum), the amount of initial disease may be quite small but still very important.

Disease incidence can increase rapidly in amount because of the large reproductive potential of the asexual generations of the pathogen. For example, on a susceptible potato cultivar there may be $1\text{-}5 \times 10^5$ sporangia produced from a single lesion (Bruck et al., 1980). Not all sporangia are effective in initiating a new lesion, but even if only 0.5 percent of them are successful, there could be a twenty-five-hundred-fold increase in lesions each week. Not many weeks at this rate of replication are required before most of the foliage has been destroyed or infected. Sporangia produced from a lesion on the foliage are transported by wind or rain-splashed water droplets to healthy tissue where infection can occur. Sporangia require water to germinate directly and penetrate foliage or to give rise to zoospores, which subsequently germinate and penetrate foliage. Under the proper conditions (susceptible host, high humidity, and daily periods of leaf wetness) the time required for penetration, lesion development, and sporulation (one generation) may be only five days.

Foilage infection is not the only concern. Sporangia can be washed through the soil to tubers and initiate infection. Infections not visible at harvest may develop during storage. Severe infections provide opportunities for soft-rotting bacteria to develop. Thus tuber blight can lead to a soft-rot problem in storage.

TRADITIONAL APPROACH TO POTATO LATE BLIGHT CONTROL

Because late blight can destroy potato foliage rapidly and because of the problems associated with infected tubers, growers in the Northeast traditionally have used regular applications of protectant fungicides to prevent its occurrence. Many factors influence the number of applications made, but this number may range from an average of five to more than ten each season (Fry et al., 1979b; James, 1973). These repeated applications are necessary to replace that which is removed by a variety of forces and to cover newly produced tissue. A large amount of fungicide is applied to control late blight in the Northeast, and about 35 percent of all that applied by all growers in the Northeast in 1971 was applied to potatoes (Andrilenas, 1974) (Table 1). Nearly all the potato acreage received fungicide application in 1971 (Andrilenas, 1975) (Table 2). Thus, this one disease on a single crop accounted for a large proportion of the fungicide applied by growers in the Northeast.

Growers are encouraged to avoid introducing _P. infestans_ into their production areas. The destruction of infected cull potatoes (Bonde and Schultz, 1943; Peterson, 1947) and the use of seed tubers free of late blight are recommended. In fact in New York State, tubers with an incidence of late blight infection greater than 1 percent cannot be sold as certified seed (Anonymous, 1978).

The large amount of fungicide is probably not necessary on all fields of potatoes in all years, for three reasons. First, some cultivars are sufficiently resistant to _P. infestans_ that adequate disease suppression could be achieved with less fungicide than that necessary to suppress disease development in

Table 1
Pesticides Used on Potatoes, 1971

	Proportion of Acreage Treated (%)			Quantities (1000 pounds)			Proportion of Total Pesticide Applied to Potatoes (%)
	Mountain	Northeast	United States	Mountain	Northeast	United States	
Fungicide	23	96	49	189	2463	4124	36
Herbicides	38	93	51	45	1451	2178	18
Insecticides	65	97	77	258	621	2770	11

Source: Andrilenas, 1974, 1975.

Table 2
Pesticides Used on Several Crops in the Northeast

	Fungicide		Herbicide		Insecticide	
	1000 Pounds	Proportion of Acreage	1000 Pounds	Proportion of Acreage	1000 Pounds	Proportion of Acreage
Corn	-	1	5250	86	155	9
Grains	-	1	266	29	15	8
Soybeans	-	-	207	52	27	7
Potatoes	2463	96	1451	93	621	97
Apples	2943	85	128	33	2403	96

Source: Andrilenas, 1974, 1975.

susceptible cultivars. Second, the pathogen apparently is absent from some production areas in some years. Adequate knowledge of the size of the initial pathogen population would help growers to determine the need for fungicide sprays. Third, in some years the macroenvironment is not especially favorable to the growth and development of *P. infestans* in potato foliage (Nugent, 1950). Consequently there might be so few generations during the season that only a very small proportion of the foliage would be affected.

COMPONENTS OF LATE BLIGHT MANAGEMENT

Because potato late blight development is influenced strongly by many factors, an integration of several approaches can enhance the efficiency and efficacy of disease suppression relative to that achieved by sole reliance on a rigid fungicide spray program. The major components of this integrated program have been identified during the last thirty or more years, but their quantification and subsequent integration were initiated less than ten years ago. These efforts will continue, so the integrations developed subsequently are likely to be more efficient than that described here. The components of an integrated late blight management program are host resistance, factors that influence the amount and composition of the initial pathogen population, weather and other environmental factors that can be used to predict disease development, and therapeutant fungicides.

Host Resistance

Some potato clones (resistant) do not support as prolific development of *P. infestans* as do other clones (susceptible), and potato breeders have sought resistant clones to aid in late blight management. An extremely high level of resistance, which nearly totally suppressed *P. infestans* development, was identified early in this century (Ferris, 1955; Van der Plank, 1963). This resistance was conferred by a single gene from *Solanum demissum*. Unfortunately the ability to overcome or avoid the effects of this gene appeared quickly within the *P. infestans* population. Other single genes were also overcome by the pathogen population within a few years of their large-scale use (Malcolmson and Black, 1966; Niederhauser and Cobb, 1959; Thurston, 1971). Thus it appears that the pathogen population is sufficiently variable to render these single genes nearly useless in late blight suppression.

Other potato clones have lower levels of resistance; pathogen development is slowed but not prevented (Guzman-N, 1964), and the effect of these clones relative to more susceptible ones is to suppress the rate of epidemic development (Fry, 1975; Thurston, 1971). The pathogen population has not adapted dramatically to clones with this rate-reducing resistance (Nelson, 1978) so that such clones may be an important long-term component of a disease management program. For example, through years of observation, growers have learned that the cultivar Sebago is more resistant than most other commercially available cultivars and that Sebago may require less fungicide than a more susceptible cultivar to suppress disease adequately. Growers on Prince Edward Island, Canada, made fewer applications to Sebago than to more susceptible cultivars (James et al., 1973).

At Cornell University we have attempted to quantify the levels of rate-reducing resistance in cultivars and to determine how this resistance can be integrated with fungicide use. The quantification has been done in relative rather than absolute terms; the effects of resistance and fungicide on epidemic development were assessed. We have used two statistics to make comparisons: the apparent infection rate (Van der Plank, 1963) and the area under the disease progress curve. Both host resistance and weekly applications of a protectant fungicide reduce the apparent infection rate (Figure 1). By determining the influence of weekly fungicide applications made to a susceptible cultivar, one can calculate the relative resistance of varieties in terms of fungicide equivalents (Fry, 1975). For example,

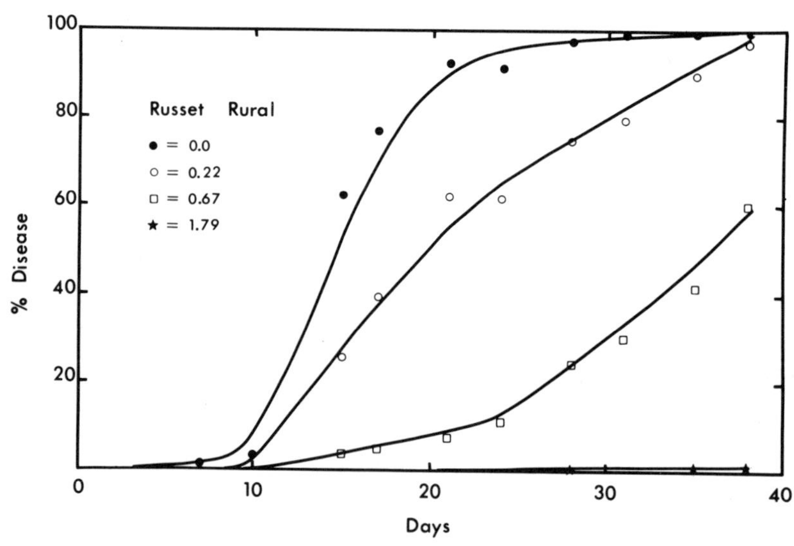

Figure 1
Effect of different weekly dosages of mancozeb on potato late blight epidemics. Mancozeb was applied weekly at 0.22, 0.67, or 1.79 kg/ha in 935 liters. Inoculation occurred on day 0 (from Fry, 1975, by permission of The American Phytopathological Society).

the greater resistance in Sebago relative to that of Russet Rural was equivalent to weekly application of 0.5 kilogram (kg) mancozeb per hectare (ha) made to Russet Rural. For comparison the relative resistances of several potato cultivars are given in Figure 2.

By comparing areas under disease progress curves, we determined that fungicides can be integrated with host resistance by limited adjustment of the interval between applications, as well as by adjustment of the dosage of regular applications (Fry, 1978). For example, when we adjusted fungicide dosage by altering the interval between fungicide sprays from two to nineteen days, the effect on disease development (area under the disease

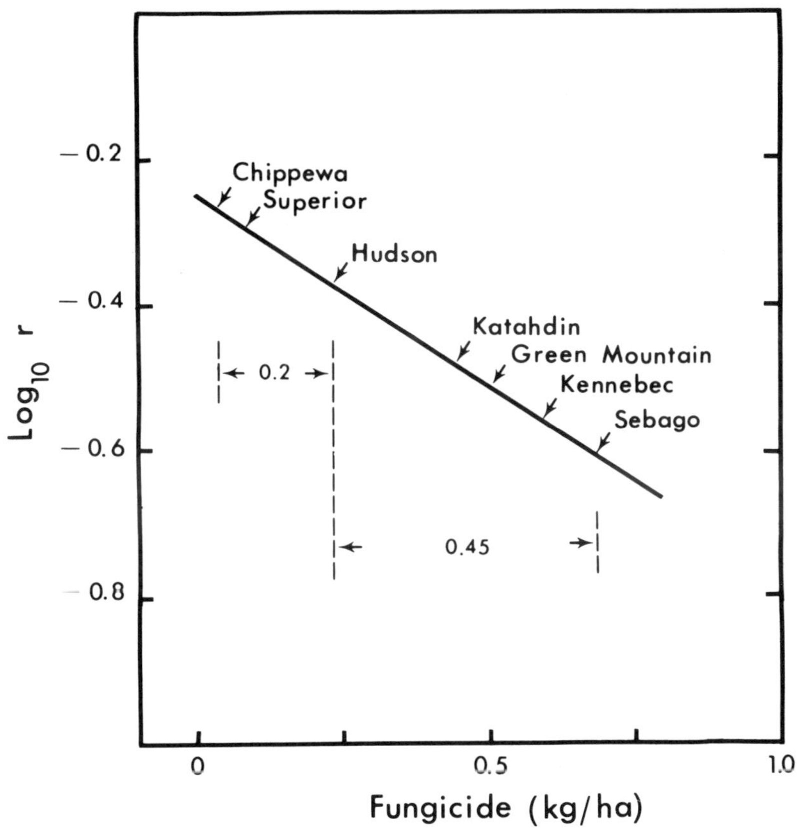

Figure 2
Relative foliage resistances of several potato cultivars. The regression line identifies the effect of weekly applications of fungicide (mancozeb) applied to the susceptible Russet Rural. The Y axis presents the log_{10} of the apparent infection rate (Van der Plank, 1963). The greater resistance in Sebago than in Hudson is approximately equivalent to the effect of 0.45 kg mancozeb/ha applied weekly (from Fry, 1975, by permission of The American Phytopathological Society).

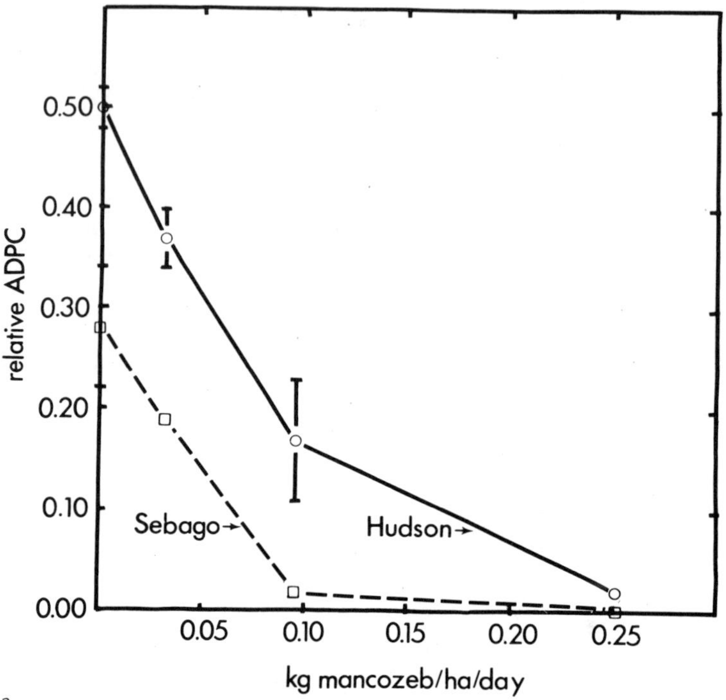

Figure 3
Effect of different amounts of fungicide (altered concentration) on potato late blight epidemics. Data are presented as relative areas under the disease progress curve (ADPC). Fungicide (mancozeb) dosage was adjusted by applying 0.22, 0.67, or 1.79 kg/ha every seven days (from Fry, 1978, by permission of The American Phytopathological Society).

progress curve) was similar to the effect achieved when fungicide dosage at each weekly application was adjusted (Figures 3 and 4). Weekly dosages ranged from 0.2 to 1.75 kg mancozeb per ha (Fry, 1978). The extent to which the interval between regular applications can be adjusted is limited. The limits are not well defined but because fungicide is physically and chemically removed from foliage and because new growth is not sprayed directly, one cannot extend intervals too far. With most protectant fungicides, regular intervals probably should not be extended beyond fourteen days.

Adjustment of fungicide dosage to complement cultivar resistance has been tested by growers to a limited extent in upstate New York. In 1978, two growers and in 1979 one grower applied less fungicide to one of their more resistant cultivars but maintained their regular dosage on their other more susceptible cultivars. The dosage applied to the resistant cultivar was calculated from the results of experiments in small

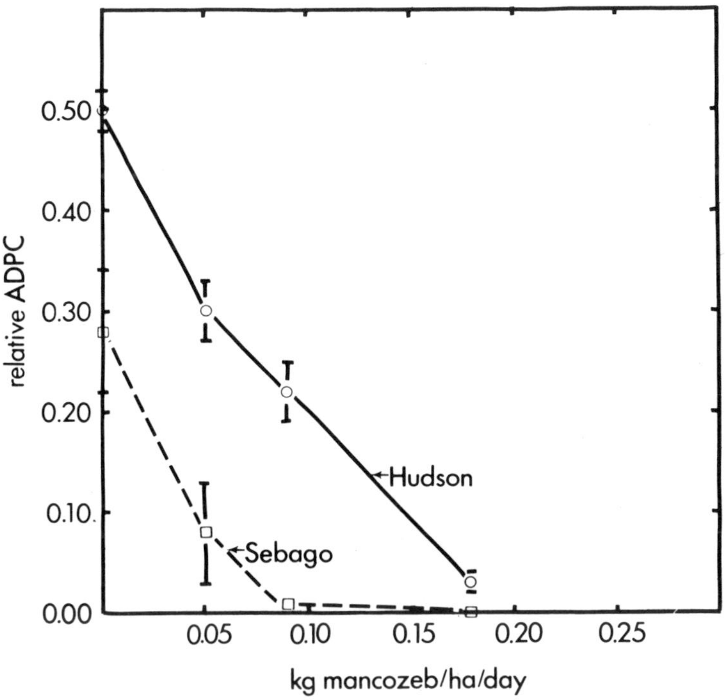

Figure 4
Effect of different amounts of fungicide (altered frequency) on late blight epidemics. Data are presented as relative areas under the disease progress curve (ADPC). Fungicide (mancozeb) dosage was altered by applying fungicide (0.36 kg/ha) every seven, four, or two days (from Fry, 1978, by permission of The American Phytopathological Society).

plots. None of these fields was affected by late blight, although late blight had occurred in other areas in upstate New York.

Pathogen Factors

The intensity of disease management should be greatest when the pathogen population is large and composed of individuals compatible with and aggressive toward the cultivars grown in the area. Although disease is likely to be more severe if infected cull potatoes are near, if infected volunteers occur in the field, or if infected seed tubers are planted, the influence of the pathogen from these sources has not yet been quantified. We do not yet know precisely how management activities should be adjusted if a source of inoculum is known to be present. Certainly the impact of initial inoculum is influenced by the weather, but the weather influence has not yet been quantified. On the other hand, if no source of inoculum exists, activities directed at suppressing late blight are unnecessary. Unfortunately our techniques have not yet been refined to enable us to know definitely that there is no source of inoculum. Detection of initial inoculum and quantification of its effect await further research before management programs can take these factors into account.

Weather and Other Environmental Factors

Late blight development is favored by abundant moisture at moderate temperatures. If relative humidities are low and leaves are not wet frequently, late blight is unlikely to develop. Germination of sporangia probably requires free moisture, and sporulation occurs only at relative humidities near 100 percent (deWeille, 1963). Very high temperatures (over 35°C) halt growth of *P. infestans*, and optimum mycelial growth occurs at about 23°C (Bruhn, 1979; Crosier, 1933). Optimum temperature for zoospore production from sporangia is 12°C (Crosier, 1933); however, because moisture is such an influential factor, one cannot describe pathogen development as a function of only temperature.

Because the proper combination of moisture and temperature are mandatory for rapid late blight development, techniques to identify these conditions, and thus predict late blight occurrence, have been developed in many potato production regions during the last thirty years. The use of forecasts to indicate the need for a fungicide application in disease management is probably analogous to the use of economic injury thresholds to time insecticide applications in insect management. For diseases such as late blight, weather monitoring of temperature, rainfall, and relative humidity replaces pest monitoring because in this case pathogen development is closely tied to the weather, and actual pathogen monitoring is technically difficult.

Forecasts can enhance the efficiency with which fungicides are used to suppress late blight. One of these, Blitecast, is a forecasting technique developed by Krause et al. (1975) at Pennsylvania State University. It is based on two older forecasting techniques (Hyre, 1954; Wallin, 1962). Basically when the weather has been especially favorable to *P. infestans*, Blitecast recommends that sprays be applied every five days. When the weather has been moderately favorable, the recommendation is for weekly application, and when the weather has been unfavorable, no spray is recommended (Table 3). The degree to which the weather is favorable depends on temperature and relative humidity (Wallin, 1962) or temperature and rainfall (Hyre, 1954).

The combination of these two techniques produced a reliable forecasting system. For example, from 1973 to

Table 3
Modification of Blitecast to Incorporate Host Resistance Effects

Cultivar	Recommended Frequency of Sprays Weather Influence		
	Unfavorable	Marginally Favorable	Favorable
Hudson (susceptible) (Blitecast)	No Spray	7-day	5-day
NY-59 (resistant)	No Spray	9-day	7-day

1975 in Ithaca, New York, potato plots sprayed according to Blitecast received fewer fungicide applications but were as well protected from late blight as were plots sprayed every week. During these three years, plots sprayed weekly had an average of ten applications per season, but plots sprayed according to Blitecast had an average of six applications per season. Blitecast is not necessarily appropriate for all locations, however. On Long Island, New York, weather is almost always favorable to late blight, and a five-day spray interval is recommended frequently. But generally the standard grower practice of a weekly application is sufficient to suppress late blight adequately, and Blitecast would have recommended more sprays than were needed.

Forecasting can be integrated with plant resistance to enhance the efficiency of fungicides. For example, fungicide dosage can be adjusted to complement cultivar resistance and be applied according to a forecast technique. In an experiment to test this integration, disease development on the resistant Sebago sprayed according to a forecast with a reduced fungicide dosage was compared to disease development on the susceptible Russet Rural sprayed weekly with the normal dosage. The combined effects of forecasting, resistance, and adjusted fungicide dosage provided disease control equivalent to that provided by standard applications to the susceptible cultivar but with about half as much fungicide (Fry, 1977).

It is possible also to integrate resistance, forecasts, and fungicides by adjusting the forecast to complement cultivar resistance. An experiment to test this forecast adjustment was done in 1979. Blitecast was modified so that applications were made less frequently to a resistant cultivar (NY-59) than to a susceptible cultivar (Hudson) (Table 3). Thus when the weather was especially favorable to *P. infestans*, sprays were applied every five days to Hudson (susceptible) but only every seven days to NY-59 (resistant), and when weather was moderately favorable to *P. infestans*, sprays were applied every seven and nine days to susceptible and resistant cultivars, respectively. These intervals were chosen based on the quantified effects of resistance in NY-59 (resistant) relative to Hudson (susceptible) (Fry, 1978). Late Blight developed so slowly in all plots that the dosage per application had to be reduced from 0.90 to 0.67 kg per ha. Consequently the fungicide effect was less and the resistance effect was greater than those originally planned (Table 4). The effect of resistance of NY-59 more than compensated for the fewer sprays applied to it. At the end of the season there was less than one-fifth as much foliar disease in plots of NY-59 sprayed according to a modified Blitecast (applications every seven or nine days) than occurred in plots of Hudson using the standard Blitecast recommendations (applications every five or seven days).

The results using a forecast adjusted to complement varietal resistance are consistent with the original predictions. The effect of resistance of NY-59 (Line L-521-7) was estimated to be equivalent to the effect of weekly applications of 0.56 kg mancozeb/ha (= 0.08 kg mancozeb per ha per day) applied to Hudson (Fry, 1978). Hudson treated according to the standard Blitecast received 7.41 kg per fifty-six days (= 0.132 kg per ha per day), and NY-59 treated according to the modified Blitecast received 5.84 kg mancozeb per ha (= 0.104 kg per ha per day) during the entire season. In order for disease development to have been equivalent on both varieties, NY 59 should have received only an average of about 0.05 kg mancozeb per ha per day (= 0.13 - 0.08) or a total application of 2.8 kg mancozeb.

Therapeutant Fungicides

The development of systemic fungicides with therapeutic properties will provide growers with an additional valuable tool for managing potato late blight. These chemicals are much more effective than protectant fungicides in suppressing established epidemics. Several such fungicides are in developmental stages. One of these chemicals, metalaxyl, has been investigated extensively.

Metalaxyl [CGA-48998, Ridomil®, N-(2,6-dimethylphenyl) -N-(methoxyacetyl)-alanine methyl ester] is systemic and is effective against P. infestans after infections in potato foliage have been established. This type of effect is unusual in terms of disease management (although it is the norm for insecticide efficacy). When applied after infections were established, metalaxyl suppressed further enlargement of late blight lesions and suppressed dramatically the degree to which P. infestans could sporulate from lesions. For example, when metalaxyl (at 25 μg a.i. per ml) was applied three days after inoculation, P. infestans was nearly unable to sporulate from treated lesions (Bruck et al., 1980). The combined effects of this compound (at 25 μg per ml) applied to foliage three days after inoculation were to suppress the number of viable sporangia to about 0.01 percent of those produced by P. infestans in untreated leaves. Thus metalaxyl should have an immediate effect on potato late blight epidemics. This expectation was fulfilled in a

Table 4
Effects on Potato Late Blight of Host Resistance and Fungicide, Integrated by Modification of Blitecast

Treatment	Final Disease Rating (%)	AUDPC[a]	Fungicide: Number of Applications	Fungicide: Total Amount (kg/ha)
Hudson - 7	30.2 AB	1.47 AB[d]	8	6.74[e]
Hudson - 5, 7 (B)[b]	18.1 B	0.85 B	9	7.41
Hudson - 7, 9 (B)[c]	32.3 A	1.80 A	7	5.84
NY-59 - 7	2.8 C	0.17 C	8	6.74
NY-59 - 5, 7 (B)[b]	1.5 C	0.10 C	9	7.41
NY-59 - 7, 9 (B)[c]	3.4 C	0.16 C	7	5.84
Untreated				
Hudson	98.3	16.5	-	-
NY-59	52.1	5.8	-	-

[a]Area under the disease progress curve in proportion-days.
[b]Fungicide (mancozeb) was not applied at all or on a seven- or five-day schedule as indicated by Blitecast (Krause et al., 1975).
[c]Fungicide (mancozeb) was not applied at all or on a seven- or five-day schedule as indicated by a modification of Blitecast.
[d]Numbers followed by the same letter are not significantly different as indicated by Duncan's new multiple range test.
[e]The fungicide dosage per application was 0.88 kg mancozeb/ha until August 20, 1979, after which it was 0.67 kg/ha. For the weekly spray, regular Blitecast and modified Blitecast treatments, there were six, six, and five applications, respectively, at the 0.88 kg/ha dosage.

series of tests. When applied to plots supporting epidemics of late blight, the suppressive effect of metalaxyl was observable within one to two days, and further development of the epidemic was greatly suppressed (Figure 5). As expected, however, the effect of a protectant fungicide was not detectable until about eight to ten days after its initial application. An additional benefit of metalaxyl is that it also suppressed tuber infections (Fry et al., 1979).

If fungicides such as metalaxyl become available for commercial use, they will reduce the perceived risk associated with enhancing the efficacy of fungicide use. Because growers would have available a method for suppressing established epidemics, they are likely to be more willing to enhance the efficiency of protectant fungicides by adjusting application amount and frequency to complement cultivar resistance and weather effects.

Because metalaxyl is relatively specific, some pathogens may become resistant to it (Ogawa et al., 1977). Several investigators have attempted to determine whether aggressive pathogen isolates can develop resistance to

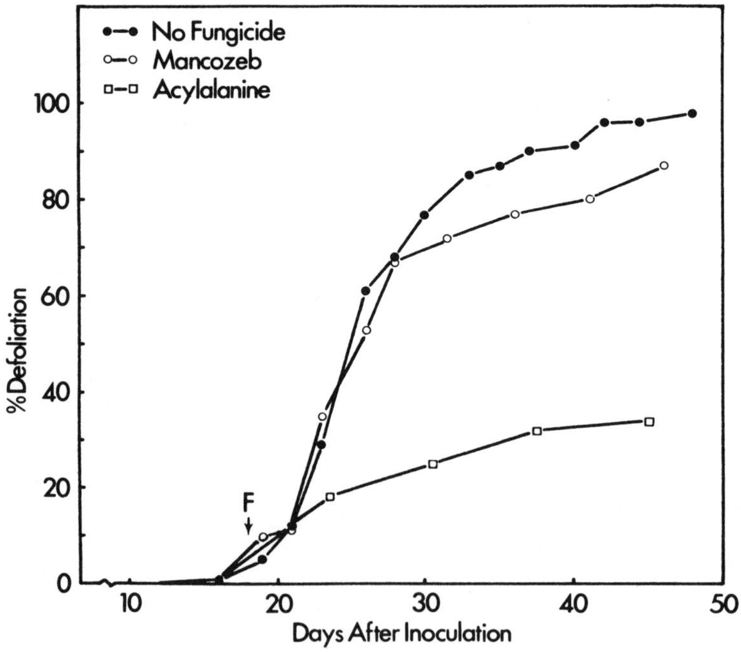

Figure 5
Effect of mancozeb ar acylalanine (matalaxyl) on established late blight epidemics. Fungicide applications (F) were initiated on day 18 at about 5 percent disease and were subsequently applied every seven and fourteen days for mancozeb (1.79 kg/ha) or metalaxyl (0.22 kg/ha), respectively (from Fry et al., 1979).

metalaxyl. Some isolates of *P. infestans* developed low levels of resistance to metalaxyl after repeated exposure to sublethal concentrations (Staub et al., 1979); however, isolates resistant in vitro were either nonpathogenic or sensitive when tested in vivo. When other *Phytophthora* species were tested, some resistant isolates retained pathogenicity. After exposure to a mutagen, some isolates of *P. megasperma* f. sp. *medicaginis* were resistant to high levels of metalaxyl and were aggressive pathogens on alfalfa (Davidse, 1980). Resistant isolates of *P. capsici* and *Pythium* spp. were obtained after culturing isolates on sublethal concentrations of Ridomil®. Control of damping off induced by the resistant *Pythium* isolates was achieved only by using very high levels of metalaxyl (Bruin and Edgington, 1979).

The usefulness of metalaxyl to suppress potato late blight epidemics will be maintained only if the pathogen population remains sensitive to it. It appears from the preliminary evidence that resistance to metalaxyl in *P. infestans* is at least a remote possibility. Consequently it seems appropriate to use the compound in such a way that selection pressure for resistant isolates will not be intense. One approach would be to restrict the application frequency on any field and to use other fungicides sequentially. It might even be wise to reserve metalaxyl for application to established epidemics because it is especially effective in this application. Regardless of the method, we need to use the compound judiciously so that resistant pathogen isolates do not become problematic.

PROPOSED PROGRAM FOR LATE BLIGHT MANAGEMENT

Efficient late blight management results from an integrated program, which incorporates knowledge of pathogen, host, and environment effects. In terms of pathogen effects, growers are urged to reduce the size of the initial pathogen population by removing volunteer and cull potatoes and by using seed tubers free of *P. infestans*. Should a source of inocuclum exist, we do not yet know q.antitatively how to adjust the intensity of other disease management activities. For example, a disease forecast should be modified if seed tubers were intected with *P. infestans*, and the degree of modification should relate to the incidence of infection; however, we have not defined the types of modifications that should occur.

As the environment becomes more favorable to *P. infestans*, disease management should become more intense. Empirically derived forecasts based on weather data have already proved useful in indicating the need for different intensities of management activity. However, the existing forecasts can be improved. For example, forecasts derived from a mechanistic understanding of late blight development might provide more accurate forecasts.

A computer simulator of potato late blight developed by Bruhn (1979) could serve as the basis for such a forecast. The influences of other factors in the environment need to be identified and incorporated into the disease management program. For example, some insecticides aid in suppressing late blight (Schumann and Thurston, 1976); when these insecticides are used, there may be less need for fungicides.

The intensity of disease management should be adjusted to complement host resistance. Preliminary guidelines for enhancing fungicide efficiency by adjusting dosage to complement host resistance have been developed and are being refined. When integrated with a forecasting technique, host resistance can further enhance fungicide efficiency.

We also need to determine the best approaches of integrating disease management with insect and weed management as components of the potato production system. Scientists working in pest management need to collaborate with systems scientists to identify an integrated, economically viable program for managing potato pests.

REFERENCES

Andrilenas, P. A., 1974, Farmers Use of Pesticides in 1971--Quantities. U.S. Dept. Agr., Econ. Res. Serv., Agr. Econ. Rept. No. 252, 56 p.

Andrilenas, P. A., 1975, Farmers' Use of Pesticides in 1971--Extent of Crop Use, Agr. Econ. Rept. No. 268, Econ. Res. Serv., U.S. Dept. Agr.

Anonymous, 1978, New York Certified Seed Potatoes: Crop Directory, N.Y. Seed Improvement Coop, Ithaca.

Bonde, R., and E. S. Schultz, 1943, Potato Refuse Piles as a Factor in the Dissemination of Late Blight, Bull. 416 Maine Agr. Exp. Stn., pp. 229-246.

Bruck, R. I., W. E. Fry, and A. E. Apple, 1980. Effect of Metalaxyl, an Acylalanine Fungicide on Developmental Stages of Phytophthora infestans, Phytopathology 70: (in press).

Bruhn, J. A., 1979, Simulation of the Potato Late Blight Management System, M.S. thesis, Cornell University, 238 p.

Bruin, G. C., and L. V. Edgington, 1979, Induced Resistance to Ridomil of Some Oomycetes, Presented at Northeastern Division of APS Mtgs., Springfield, Mass. Oct. 31-Nov. 2.

Crosier, W., 1933, Studies in the Biology of Phytophthora infestans (Mont.) de Bary, Memoir 155, Cornell University Agric. Exp. Stn.

Davidse, L. C., 1980, Resistance to Ridomil in Phytophthora megasperma f. sp. medicaginis, Acta Botanica Neerlandica (in press).

deWeille, G. A., 1963, Laboratory Results Regarding Potato Blight and Their Significance in the Epidemiology of Blight, Eur. Potato J. 6:121-130.

Ferris, V. R., 1955, Histological Study of a Pathogen-Suscept Relationship between *Phytophthora infestans* and Derivatives of *Solanum demissum*, *Phytopathology* 45:546-552.

Fry, W. E., 1975, Integrated Effects of Polygenic Resistance and a Protective Fungicide on Development of Potato Late Blight, *Phytopathology* 65:908-911.

Fry, W. E., 1977, Integrated Control of Potato Late Blight--Effects of Polygenic Resistance and Techniques of Timing Fungicide Applications, *Phytopathology* 67:415-420.

Fry, W. E., 1978, Quantification of General Resistance of Potato Cultivars and Fungicide Effects for Integrated Control of Potato Late Blight, *Phytopathology* 68:1650-1655.

Fry, W. E., R. I. Bruck, and C. C. Mundt, 1979*a* Retardation of Potato Late Blight Epidemics by Fungicides with Eradicant and Protectant Properties, *Plant Dis. Rep*. 63:970-974.

Fry, W. E., H. D. Thurston, K. Cathey, F. Holbrook, T. Lyons, J. B. Sieczka, D. Smith, and R. D. Sweet, 1979*b*, Potential for Intensified Integrated Management of Potato Pests in Northeastern USA, in *Pest Management Strategies*, vol. 2, Office of Technology Assessment, U.S. Congress, October.

Guzman-N, J., 1964, Nature of Partial Resistance of Certain Clones of Three *Solanum* Species to *Phytophthora infestans*, *Phytopathology* 54:1398-1404.

Hyre, R. A., 1954, Progress in Forecasting Late Blight of Potato and Tomato, *Plant Dis. Rep*. 38:245-253.

James, W. C., C. S. Shih, and L. C. Callbeck, 1973, Survey of Fungicide Spraying Practice for Potato Late Blight in Prince Edward Island 1972, *Can. Plant Dis. Surv*. 53:161-166.

Krause, R. A., L. B. Massie, and R. A. Hyre, 1975, Blitecast; A Computerized Forecast of Potato Late Blight, *Plant Dis. Rep*. 59:95-98.

Large, E. C., 1940, *The Advance of the Fungi*, Dover Publications, New York, 488 p.

Malcolmson, J. R., and W. Black, 1966, New R Genes in *Solanum demissum* Lindl. and Their Complementary Races of *Phytophthora infestans* (Mont.) de Bary, *Euphytica* 15:199-203.

Nelson, R. R., 1978, Genetics of Horizontal Resistance to Plant Diseases, *Annu. Rev. Phytopathol*. 16:359-378.

Niederhauser, J. S., and W. C. Cobb, 1959, The Late Blight of Potatoes, *Sci. Am*. Reprint May 1959.

Nugent, T. J. 1950, Three Years Experience Forecasting Late Blight in Tidewater, Virginia, In Plant Disease Forecasting: A Symposium, *Plant Dis. Rep. Supp.* 190:9-13.

Ogawa, J. M., J. D. Gilpatrick, and L. Chiarappa, 1977, Review of Plant Pathogens Resistant to Fungicides and Bactericides, *FAO Plant Protection Bull*. 25:97-111.

Peterson, L. C., 1947, The Overwintering of *Phytophthora infestans* (Mont.) de Bary under Long Island Conditions, *Am. Potato J*. 24:188-197.

Schumann, G., and H. D. Thurston, 1976, Foliar Infection of Potato by *Phytophthora infestans* Inhibited by Two Insecticides, *Plant Dis. Rep.* 60:734-735.
Staub, T., H. Dahmen, P. Urech, and F. Schwinn, 1979, Failure to Select for *in vivo* Resistance in *Phytophthora infestans* to Acylalanine Fungicides, *Plant Dis. Rep.* 63:385-389.
Thurston, H. D., 1971, Relationship of General Resistance: Late Blight of Potato, *Phytopathology* 61:620-626.
Van der Plank, J. E., 1963, *Plant Diseases: Epidemics and Control*, Academic Press, New York, 349 p.
Wallin, J. R., 1962, Summary of Recent Progress in Predicting Late Blight Epidemics in the United States and Canada, *Am. Potato J.* 39:306-312.

Automated Pest Management

D. R. MacKenzie and Z. Smilowitz

Integrated pest management (IPM) recently has become recognized as a means of contributing to a sophisticated control of pests, be they weeds, insects, or diseases. Many programs are making an increasing effort to develop integrated pest management strategies for many pests on many crops, and the potato has much of this attention, for a number of complex reasons. Most notable among these, no doubt, is the quantity of pesticide applied to the crop each year, particularly in the rain-fed Northeast where many potato pests have the potential for totally destroying a crop were it not for the common use of pesticides.

Many years ago it was recognized that the particular pest that causes potato late blight, *Phytophthora infestans*, could be a problem only when weather conditions were conducive to its epidemic development. Following up on twenty-five years of research by Wallin (1962), Wallin and Waggoner (1950), and Hyre (1954), a research team led by Ray Krause of the Pennsylvania State University organized a late blight forecasting service for potato growers of the northeast United States, named BLITECAST (Krause et al., 1975).

The essence of BLITECAST is a true IPM approach: the grower must use clean seed (sanitation), is markedly better off with a resistant variety (reduced apparent infection rate), and may use chemicals only when necessary (originally mancozeb). The BLITECAST system works in this way. Weather records on temperature, relative humidity, and rainfall from a particular potato field are recorded daily and interpreted weekly. Figure 1 is a graphic description of how the weekly weather is

---

Funds in support of part of this research came as a grant from the Pennsylvania Department of Agriculture.

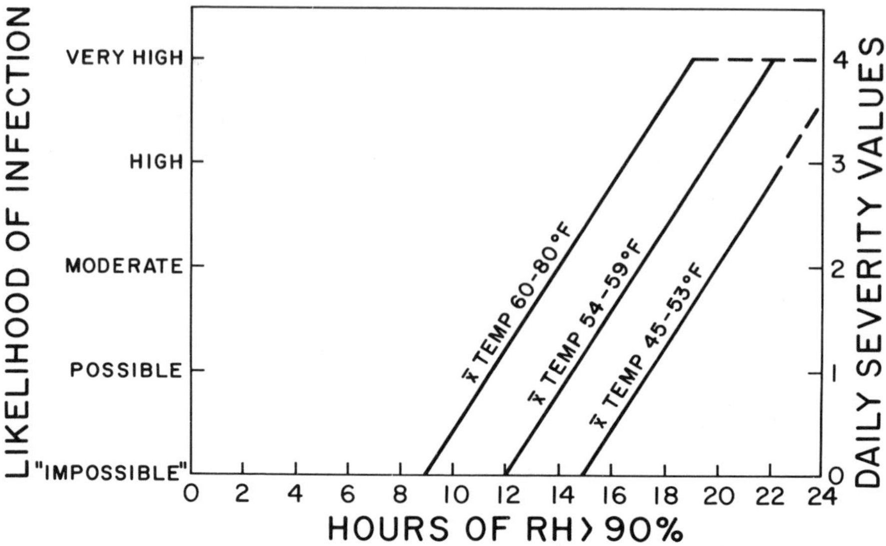

Figure 1
Relationship of relative humidity (RH) and temperature (TEMP) to the likelihood of infection of potato by Phytophthora infestans and assigned daily severity values subsequently used to recommend spray schedules for potato late blight control with BLITECAST.

summarized to describe the necessary relationships. The likelihood of infection is a subjective evaluation of the effect of temperature and extended periods of relative humidity above 90 percent. The interrelationship between temperature and hours of high relative humidity is used to assign daily severity values. These arbitrary numbers reflect the likelihood of late blight infection having taken place.

Spray recommendations are made from weekly summaries of seven-day totals of severity values. The interpretation of these totals depends upon rainfall for the past ten days (Table 1). The no-spray recommendation is given by BLITECAST following a week of no blight weather. An alert is issued when some blight weather was encountered but not enough to cause immediate action. Moderate spray schedules are as recommended on the fungicide label and correspond to normal procedures. A recommendation of heavy spray is given when excessive blight weather has been encountered and extra precautions are needed. More frequent spray applications are then required.

Spray recommendations are not given until eighteen severity values are accumulated if the cooperating grower has started early in the growing season at the green row stage. This bonus is a consequence of the use of disease-free seed and hence low initial levels of

Table 1
Spray Recommendation Table for BLITECAST

	Total Weekly Severity Values				
	0-2	3	4	5-6	7 or more
Low Rain	No Spray	No Spray	Alert	Moderate Spray	Heavy Spray
High Rain	No Spray	Alert	Moderate Spray	Heavy Spray	Heavy Spray

inoculum. This points out the reliance of BLITECAST on IPM concepts because even modest levels of inoculum on the seed will fool the forecasting system and threaten a potential late blight epidemic. One obvious implication of these underlying assumptions is that BLITECAST cooperators should not use seed from growers also using BLITECAST because the likelihood of higher levels of initial inoculum is greatly increased.

The BLITECAST system was tested during the 1970s and proved successful in an overwhelming majority of cases along the entire eastern coast of the United States. Nevertheless BLITECAST as a service soon became financially troubled when research funds were no longer available to extend it as a free service to growers. The subscription cost of $200 per station, reflecting the actual cost to the university, caused many growers to quit the system and return to weekly sprays for the control of late blight. At this point in our research program, we integrated BLITECAST with GPA-cast, an insect forecasting system, (Whalon and Smilowitz, 1979) to form the SPUDCAST system.

Previously it was common scientific procedure to select _a priori_ the most appropriate delivery system of several possible choices. Our first decision was to reject this methodology in favor of testing different types of delivery systems to understand better grower acceptance. Our first attempt to develop an alternative to the BLITECAST was an on-site microcomputer that would collect the necessary weather information, interpret the data, and issue a spray recommendation to potato growers (MacKenzie and Schimmelpfennig, 1978). This was our first introduction to semiconductor electronics, and it demonstrated the extensive possibilities for its application for agriculture. Our staff was comprised of one undergraduate computer science major and four electrical engineering undergraduate students who undertook the project with funding from our agricultural experiment station. The six month project was a success, and the design subsequently was made public. That microcomputer technology is now marketed by Campbell Scientific, Logan, Utah, for $1100. The unit, called the Blitecaster, is capable of collecting at periodic intervals all of the necessary weather information. It can operate the entire growing season on alkaline D batteries by virtue of its CMOS technology and its sleep cycle; the machine awakes periodically to collect its data and make calculations. To get a recommendation, all a potato grower needs to do is push a button.

We have been working very closely with entomologists on another application of electronics, testing the usage of hand-held calculators for the interpretation of weather data for late blight and green peach aphid forecasting (Lucas et al., 1979). We have discovered that with some very sophisticated programming, the original BLITECAST algorithm can be adapted to the hand-held calculator for direct use by potato growers. During the 1979 cropping season the BLITECAST and GPA-cast calculator programs were tested with several Pennsylvania potato growers for their evaluation of the system. The reaction of most of the growers was very positive, demonstrating that the calculator is an acceptable alternative to an on-site microcomputer (Lucas et al., 1979).

We are also investigating home-computing systems as an alternative to the larger IBM mainframe computer of the earlier BLITECAST system (MacKenzie, 1979). Our current evaluation of the potential on-farm market for home-computing systems suggest that some larger growers may choose to purchase machines similar to ours for their own use. In addition to pest forecasting, they would be able to find applications in farm business accouting, inventory, and other areas.

Particularly for small growers, we are now testing the application of a microcomputer system hooked up by a telephone line to participating growers. The computer, for example, might be located in a county agent's office or with a private consultant. Potato growers wishing to converse with the computer system would use a touch-tone pad with an acoustic coupler. The computer would be fitted with a voice synthesizer and the computer programmed to respond verbally to the input data, telling the grower the spray recommendation.

All of this technology exists. What is needed is a direct effort to test the application of these delivery systems for eastern growers.

Fundamental to all of these applications is an appreciation of the fact that no one delivery system will be suitable for all potato farmers. Some growers prefer to maintain close links with the University. Such individuals liked the original BLITECAST. Others want the independence offered with on-site microcomputers. Still others have multiple applications for electronic calculators. The more sophisticated growers are now purchasing home-computing systems for a variety of applications. Linkage by synthetic voice to a computer system appeals to small growers and can be offered without human supervision.

Throughout this project we have worked to integrate all of the potato pest (insect and disease) aspects to our management system. This, perhaps, makes the Pennsylvania State University system unique. Very rarely has IPM been truly integrated across disciplines.

In addition to the need for automated pest management we also recognize the need for far more research in the area of developing crop management systems acceptable to growers. For example, we lack information on crop loss, data necessary to a decision on when economic injury to a crop will occur. A crop loss model we are developing indicates that discount coefficients are needed to interpret our present single pest models.

Varietal differences are also extremely important in the consideration of IPM systems. This has become clear in the past four years in a sequence of studies that we are conducting on a farm near Hazleton, Pennsylvania. The residual varietal effects can be measured in subsequent years as lost yield. Much of this effect can be attributed to the increase of weeds due to the lack of competition by some varieties. The recent trend by potato breeders has been for selection of potatoes with smaller vines (and, perhaps, poorer competitive ability).

Observations on the differences in competitive abilities of potato varieties have caused me to examine the chemical umbrella applied to our potato crop in our efforts to produce potatoes commercially in the rain-fed Northeast. In preliminary investigations in an intense Colorado potato beetle area, we found that not all varieties suffer equally in the absence of chemicals from the attack of the Colorado potato beetle. In fact, some of the older varieties such as Kennebec gave a very respectable yield even without the application of any insecticide. Of equal interest to us was the observation that the old, large-vine-type potato varieties, such as Kennebec, tend to grow oversize and show growth cracks under our best chemical protection. Does this suggest that the abundance of vines produced by the old varieties compete with weeds and provide some cushion against pest attack? Perhaps the current trends toward varieties of small vine type may be moving us toward a greater dependence on the chemical umbrella.

In another study we investigated the effects of the rate of nitrogen fertilizer application on potato yield. Our purpose was not to repeat the usual study of agronomic experimentation but to address the question of why do growers persist in overfertilizing their potato crop. Potato growers prefer to keep their vines green as long as possible. They then use vine killers when they decide the crop has reached the proper size. Recommendations on nitrogen fertilization are not followed by potato growers who choose to use 30 or more kilograms per hectare more than recommended.

To study this practice we ran two parallel experiments. The first set of trials looked at the typical yield response curve to rate of nitrogen application in the absence of all pests. The curve we obtained was identical to that found by experiment station scientists for the past decades. Fitting the curve to the quadratic relationship and then determining its first derivative, we could then solve for the optimal amount of nitrogen fertilizer for maximum yield. This gave the standard recommendation of approximately 131 kilograms per hectare, precisely what the university community has been advising potato growers for many years.

In the second set of trials we allowed the common senescent-type pest of potato early blight disease (caused by _Alternaria solani_) to establish itself and progress. Our observation was that the amount of nitrogen fertilizer had a great effect on the development of the epidemic and on the final spread of disease. When higher rates of nitrogen were applied, early blight development slowed radically. Thus we found a very strong relationship between the final amount of disease and yield, and more importantly, the yield response curve to nitrogen rate was shifted dramatically to the right. Analyses of this response curve gave an optimum nitrogen application rate of 160 kilograms per hectare, the rate at which many growers are now fertilizing.

The lesson in this series of experiments was that there is a need to integrate research with pest management. As long as agricultural experiment stations are compartmentalized by academic disciplines, pest management will not receive the appropriate integrated research that is needed. When fertilizer trials are the domain of one department, disease research of another, and insect research of a third, the research priorities and the obvious team effort that is needed cannot contribute to the development of new IPM systems.

REFERENCES

Hyre, R. A., 1954, Progress in Forecasting Late Blight of Potato and Tomato, Plant Dis. Reptr. 38:245-253.

Krause, R. A., L. B. Massie, and R. A. Hyre, 1975, Blitecast, A Computerized Forecast of Potato Late Blight, Plant Dis. Reptr. 59:95-98.

Lucas, B. S., Z. Smilowitz, D. R. MacKenzie, and E. S. Noland, 1979, Use of the Hand-Held Programmable Calculator to Implement Two Potato Pest Forecasting Systems in Pennsylvania, Paper presented at the Entomological Society of America, National Meeting, Denver, Colorado.

MacKenzie, D. R., 1979, Delivery Systems for Pest Management, Conference Record of the 1979 International Micro and Mini Computer Conference, Houston, Texas, pp. 34-36.

MacKenzie, D. R., and H. Schimmelpfennig, 1978, Development of a Microcomputer Unit for Forecasting Potato Late Blight Using the Pennsylvania State University's Blitecast System, Am. Potato J. 55:384 (Abstr.).

Wallin, J. R., 1962, Summary of Recent Progress in Predicting Late Blight Epidemics in United States and Canada, Am. Potato J. 39:306-312.

Wallin, J. R., and P. E. Waggoner, 1950, The Influence of Climate on the Development and Spread of *Phytophthora infestans* in Artifically Inoculated Potato Plots, Plant Dis. Reptr. Suppl. 190:19-23.

Whalon, M. E., and Z. Smilowitz, 1979, GPA-Cast, A Computer Forecasting System for Predicting Populations and Implementing Control of the Green Peach Aphid on Potatoes, Environ. Entomol. 8:908-913.

Potential for Plant Resistance in Management of Arthropod Pests

Ward M. Tingey

The potato is subject to losses by arthropod pests virtually everywhere in the world it is grown. In recent years, the drawbacks associated with dependence on chemical insecticides for pest control have become increasingly apparent, and alternate control strategies have gained greater attention and appeal by consumers, farmers, and legislators alike. One possibility for controlling potato-infesting arthropods is to use a noninsecticidal strategy, genetic resistance.

GERM PLASM RESOURCES

Tuber-bearing species of the genus *Solanum* (Tourn.) have been extensively assessed for arthropod resistance over the past forty years, particularly in North America and Europe. A major portion of this effort has been directed toward identification of resistant germ plasm, with less emphasis placed on the nature and inheritance of resistance. The following discussion provides a summary of the current situation in breeding for resistance to three of the more thoroughly studied pests: green peach aphid, *Myzus persicae* (Sulzer), Colorado potato beetle, *Leptinotarsa decemlineata* (Say), and potato leafhopper, *Empoasca fabae* (Harris).

Myzus persicae

Clones and cultivars of Group Tuberosum (*Solanum tuberosum tuberosum* [L.], *S. t. andigena* Juz. & Buk.) vary in their response to aphid infestation, but levels of genetic resistance are low (Adams, 1946; Bradley and Ganong, 1951; Landis et al. 1972; Tingey and Plaisted, 1976) and the expression of resistance fluctuates with stage of plant growth (Taylor, 1962). The greatest genetic diversity for aphid resistance is in wild, tuber-bearing *Solanum* species (Table 1). Radcliffe and

Table 1
Sources of Resistance to the Potato Leafhopper and Green Peach Aphid in Tuber-Bearing *Solanum* Species

Species	Pest	Reference
Solanum berthaultii	*M. persicae*	Gibson, 1971
	E. fabae	Tingey and Gibson, 1978; Tingey et al., 1978
S. brachistotrichum	*M. persicae*	Radcliffe and Lauer, 1971b
S. bulbocastanum	*M. persicae*	Radcliffe and Lauer, 1971b
S. cardiophyllum	*M. persicae*	Radcliffe and Lauer, 1971b
S. chacoense	*E. fabae*	Sleesman, 1940; Tingey et al., 1978
S. commersonii	*E. fabae*	Sleesman, 1940
S. demissum	*E. fabae*	Sleesman, 1940
S. fendleri	*E. fabae*	Sleesman, 1940
S. hougasii	*E. fabae*	Radcliffe and Lauer, 1968; Tingey et al., 1978
S. kurtzianum	*E. fabae*	Radcliffe and Lauer, 1968
S. michoacanum	*M. persicae*	Radcliffe and Lauer, 1971b
S. pampasense	*E. fabae*	van de Klashorst and Tingey, 1979
S. pinnatisectum	*E. fabae*	Radcliffe and Lauer, 1968
S. polyadenium	*E. fabae*	Sleesman, 1940; Tingey et al., 1978; Tingey and Gibson, 1978
	M. persicae	Adams, 1946; Gibson, 1971
S. polytrichon	*E. fabae*	Radcliffe and Lauer, 1968
S. sanctae-rosae	*M. persicae*	Radcliffe and Lauer, 1971b
S. spegazzini	*E. fabae*	Radcliffe and Lauer, 1968
S. stenophyllidium	*M. persicae*	Radcliffe and Lauer, 1971b
S. tarijense	*M. persicae*	Gibson, 1971
S. tuberosum[a]	*E. fabae*	Sanford and Sleesman, 1970, 1974
S. t. andigena	*E. fabae*	Tingey and Plaisted, 1976
	M. persicae	Tingey and Plaisted, 1976

[a]Clones with *S. chacoense* or *S. demissum* ancestry.

coworkers (Radcliffe and Lauer, 1966, 1970, 1971a, 1971b; Radcliffe et al., 1974) have screened a major portion of the U.S. holdings of this germ plasm and reported excellent levels of resistance in accessions of various species, including *S. brachistotrichum* (Bitt.) Rydb., *S. bulbocastanum* Dun., *S. cardiophyllum* Lindl., *S. michoacanum* (Bitt.) Rydb., *S. sanctae-rosae* Hawkes, *S. stenophyllidium* Bitt., and *S. stoloniferum* Schlechtd. & Bché. Many of these species are difficult to exploit by traditional plant-breeding methods because of their genetic incompatibility with Gr. Tuberosum, but Sams et al. (1976) have reported progress in the transfer of aphid resistance to progeny of crosses between some wild species and *S. tuberosum*. The plant factors conferring aphid resistance in these species are unknown, whereas aphid resistance of the wild species, *S. berthaultii* Hawkes, *S. tarijense* Hawkes, and *S. polyadenium* Greenm., is conditioned by glandular pubescence.

Empoasca Fabae

Clones and cultivars of Gr. Tuberosum vary dramatically in infestation and yield response to the potato leafhopper (Sleesman and Bushnell, 1937; Sanford et al., 1972; Sanford and Sleesman, 1974; Tingey and Plaisted, 1976; Sanford and Webb, 1977), but plant age and clonal differences in time of tuberization or foliage maturity tend to confound the expression of genetic resistance (Sanford and Sleesman, 1970). The resistant cultivar Sequoia is the only named potato variety for which insect resistance is considered a major feature (Gardner et al., 1945; Peterson and Granovsky, 1950). Excellent levels of resistance to this pest are also found in accessions of wild species (Table 1) such as *S. berthaultii*, *S. chacoense* Bitt., *S. commersonii* Dun., *S. demissum* Lindl., *S. hougasii* Corr., *S. kurtzianum* Torr., *S. pinnatisectum* Dun., *S. polyadenium*, *S. polytrichon* Rydb., *S. spegazzini* Bitt. (Raman et al., 1979; Sleesman, 1940; Tingey and Gibson, 1978; Tingey et al., 1978), and in Gr. Tuberosum stocks with *S. chacoense* and *S. demissum* ancestry (Sanford and Sleesman, 1970; Sanford et al., 1972). At least two different mechanisms are associated with potato leafhopper resistance in wild *Solanum* species. First, resistance of *S. berthaultii* and *S. polyadenium* is conferred in part by glandular trichomes, secretions of which entrap adults and nymphs, or prevent feeding by occlusion of the labium (Tingey and Gibson, 1978). Second, resistance is associated with leaf glycoalkaloid content (Tingey et al., 1978).

Leptinotarsa decemlineata

Gr. Tuberosum cultivars are uniformly susceptible to attack by the Colorado potato beetle, although some clones differ in their ability to tolerate and compensate for defoliation (Schalk et al., 1975). The highest levels of resistance are present in accessions of wild species such as S. polyadenium, S. chacoense, and S. demissum (Swiniarski et al., 1958). Resistance in these species to the Colorado potato beetle is associated largely with levels and types of foliar glycoalkaloids (Prokoshev et al., 1952; Schreiber, 1957, 1958; Pierzchalski and Werner, 1958; Swiniarski et al., 1958). Individual glycoalkaloids differ in their activity against L. decemlineata; tomatine, chaconine, demissine, commersonine, and the water-soluble leptines appear to be the most toxic or repellent of the Solanum glycosides (Kuhn and Löw, 1955; S. L. Sinden, pers. comm.). Two other potential but largely unexplored defense mechanisms against the Colorado potato beetle are glandular trichomes (Gibson, 1976b) and proteinase inhibitors (Green and Ryan, 1972).

CHARACTERIZATION OF SPECIFIC DEFENSE MECHANISMS

Glycoalkaloids

Glycoalkaloids are a class of toxic nitrogen-containing steroidal glycosides found in foliage and tubers of potatoes. The first glycoalkaloid found in potatoes was α-solanine (Baup, 1826) (Figure 1). This was considered to be the only one present until the discovery of α-chaconine over a century and a quarter later (Kuhn and Löw, 1954). Although α-solanine and α-chaconine are, historically, the predominant potato glycoalkaloids, Schreiber (1968) has reported a wide range of individual glycoalkaloids in Solanum species, and there are indications that more remain to be characterized. Only recently a new glycoalkaloid, commersonine, was discovered in two Solanum species (Osman et al., 1976). Glycoalkaloids have no clear-cut metabolic role in solanaceous plants, but they have long been considered fungitoxic compounds (Agerburg et al., 1933; Arneson and Durbin, 1968; Sinden et al., 1973). These compounds have also been associated with gastrointestinal disturbances, nervous depression, and death following ingestion by humans (Willimot, 1933).

Potato glycoalkaloids have been implicated as resistance factors against the Colorado potato beetle (Kuhn and Löw, 1955; Schreiber, 1957; Buhr et al., 1958; Pierzchalski and Werner, 1958) and the potato leafhopper (Dahlman and Hibbs, 1967). More recently, we have demonstrated that foliar concentration of total glycoalkaloids in wild Solanum species is highly correlated ($r = -0.75$, $p \leq 0.01$) with field infestations of potato leafhopper nymphs (Tingey et al., 1978), suggesting a defensive role for these compounds. Subsequently (Raman et al., 1979), we performed feeding

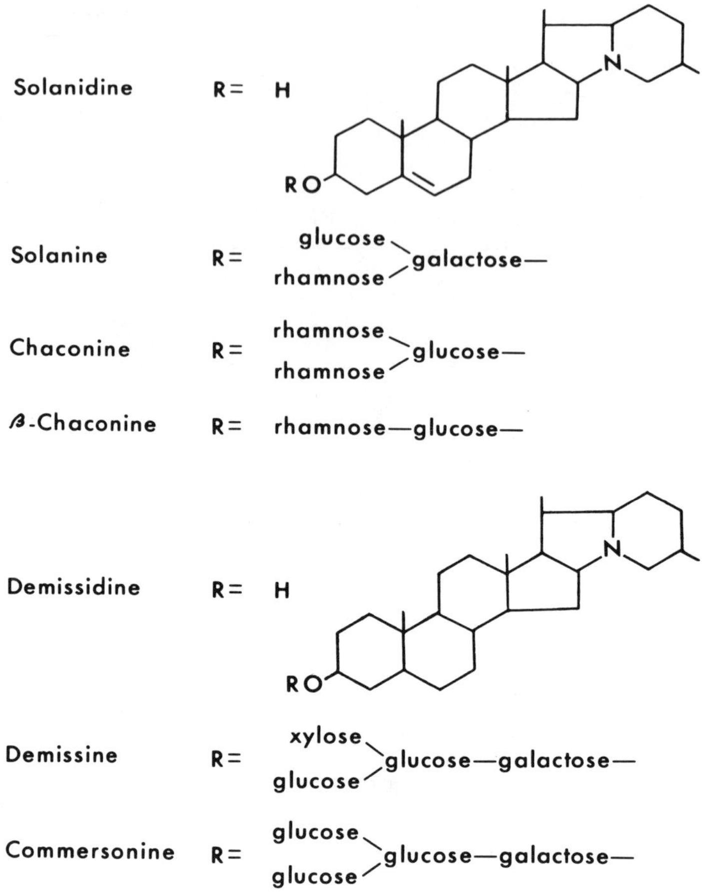

Figure 1
Structural formulas for the aglycones solanidine and demissidine and for five glycoalkaloids (from G. D. McCollum and S. L. Sinden, 1979, Inheritance Study of Tuber Glycoalkaloids in a Wild Potato, Solanum chacoense Bitter, Am. Potato J. 56:95-113, Copyright the Potato Association of America, reproduced by permission).

studies with total glycoalkaloid fractions isolated from the same accessions studied by Tingey et al. (1978). The effects of the total glycoalkaloid preparations on nymphal survival and on feeding behavior was assessed, and specific components of the feeding process affected by TGA were identified by use of an electronic recording system (Figure 2). We found that nymphal survival (Figure 3) and duration of settling, salivation-ingestion, and nonfeeding were significantly correlated with total glycoalkaloid concentration (r = -0.86, -0.79, -0.93, and 0.82 respectively), thus providing further evidence of a causal role for potato glycoalkaloids in leafhopper resistance.

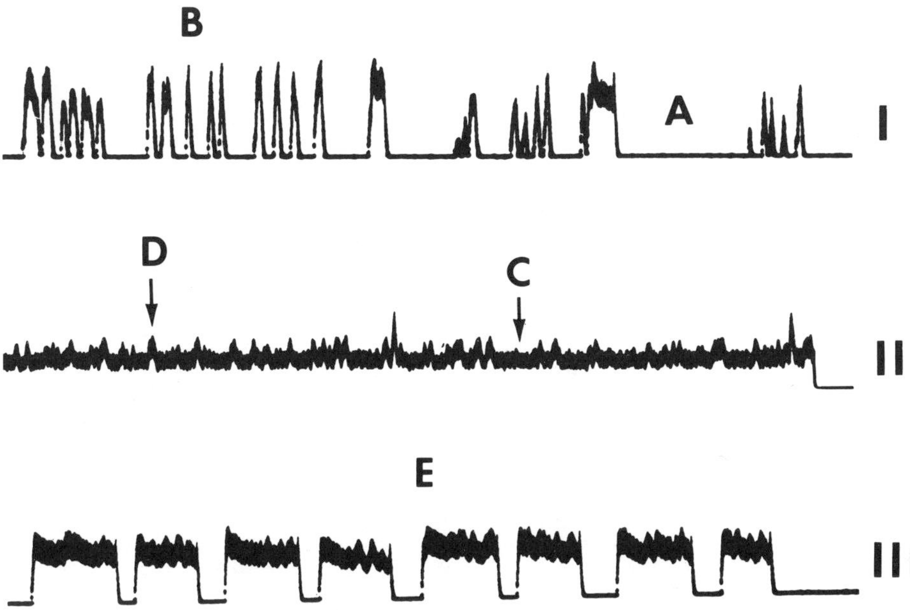

Figure 2
Electronic waveforms associated with feeding process of potato leaf-hopper nymphs on artificial diet. Direction of chart movement on horizontal (X) axis from right to left at 1.0 mm per sec. Vertical deflection away from the baseline indicates increase in voltage. Specific components of the feeding process include: (A) baseline, no feeding, (B) probing, brief insertion, and withdrawal of stylets, (C) ingestion, uptake of diet, (D) salivation. Specific waveforms include (I) initial probing of diet, (II) normal acceptance and sustained salivation-ingestion sequence on diet lacking glycoalkaloids, (III) interrupted salivation-ingestion process on diet containing total glycoalkaloid fraction from foliage of the resistant species, S. chacoense (PI 189220) (from R. V. Raman, W. M. Tingey, and P. Gregory, 1979, Potato Glycoalkaloids: Effect on Survival and Feeding Behavior of the Potato Leafhopper, J. Econ. Entomol. 72:337-341, Copyright Entomological Society of America, reproduced by permission).

Glandular Trichomes

The glandular trichomes of wild, tuber-bearing Solanum species defend against many potato-infesting pests (Gibson and Turner, 1977) and offer exceptional opportunity for exploitation in breeding resistant potato cultivars. Gibson (1971, 1974, 1976a) was among the first to associate glandular pubescence with insect resistance of the wild potato species S. berthaultii, S. polyadenium, and S. tarijense. These three species have high densities of four-lobe (type A) and/or simple glandular trichomes (type B) (Gibson 1976c) on leaves, stems, and petioles (Figure 4A); upon contact by an arthropod, a viscous exudate is discharged and

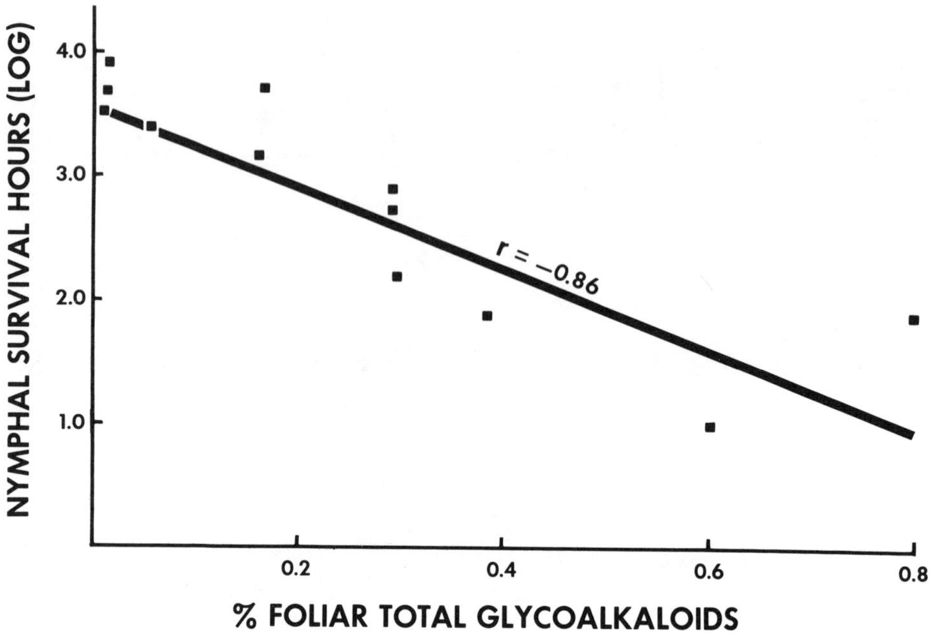

Figure 3
Relationship between nymphal survival and total glycoalkaloid concentration (TGA) for Empoasca fabae fed TGA extracts from twelve Solanum accessions (from K. V. Raman, W. M. Tingey, and P. Gregory, 1979, Potato Glycoalkaloids: Effect on Survival and Feeding Behavior of the Potato Leafhopper, J. Econ. Entomol. 72:337-341, Copyright Entomological Society of America, reproduced by permission).

accumulates on the arthropod's body. For the green peach aphid, potato aphid, Macrosiphum euphorbiae (Thomas), Colorado potato beetle, and two-spotted spider mite, Tetranychus urticae Koch, the exudate typically accumulates on tarsi, impeding the insect's movements or gluing the insect to the leaf (Gibson and Turner, 1977). This mechanical entrapment leads to considerable insect mortality within several days of confinement. Tingey and Gibson (1978) have demonstrated similar entrapment of Empoasca fabae by glandular trichomes of S. polyadenium and S. berthaultii and have also shown that trichome exudates accumulate on leafhopper mouth parts (Figure 5), preventing feeding and resulting in significant mortality (Table 2).

The biological activity, chemistry, and genetic exploitation of glandular trichomes have been under intensive study by our group at Cornell in recent years.

Biological Activity

Type B glandular trichomes are considerably more important in the expression of resistance than previously

Figure 4
Epidermal morphology of Solanum parents used in Cornell breeding program. (A) Type A and B glandular trichomes of the S. berthaultii clone, PI 310927-11 (X97.6), (B) predominantly nonglandular trichomes of the S. tuberosum cultivar Katahdin (X97.6).

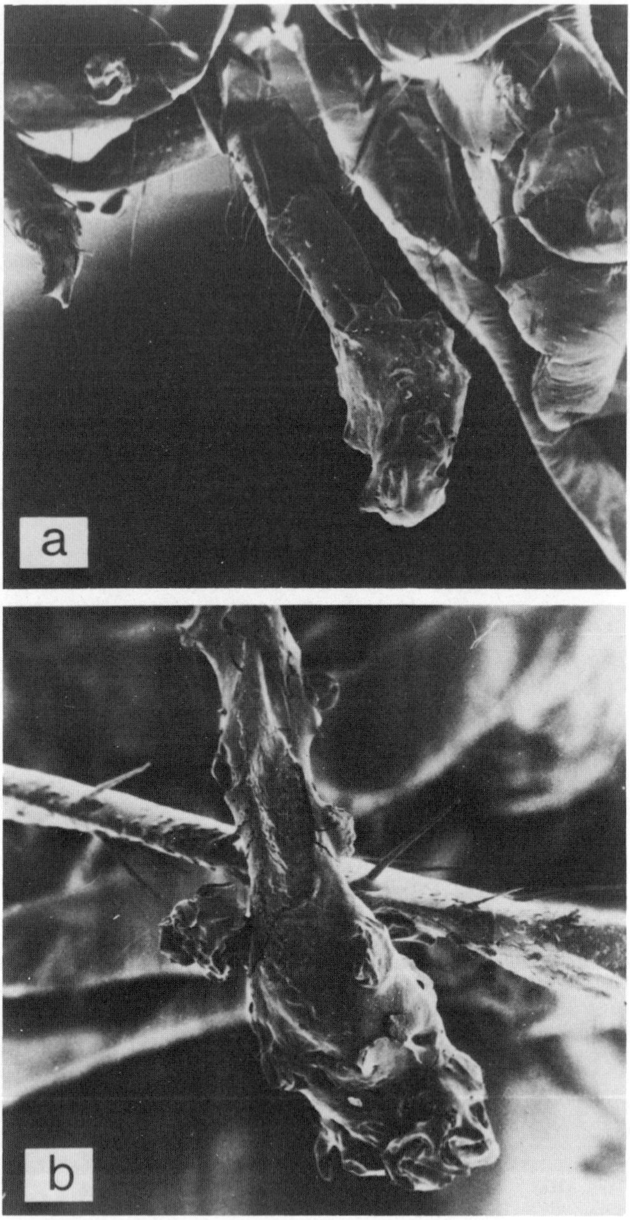

Figure 5
Glandular trichome exudate of Solanum polyadenium on (a) labium (X74.2) and (b) tarsus (X204.4) of Empoasca fabae nymph (from W. M. Tingey and R. W. Gibson, 1978, Feeding and Mobility of the Potato Leafhopper Impaired by Glandular Trichomes of Solanum berthaultii and S. polyadenium, J. Econ. Entomol. 71:856-858, Copyright Entomological Society of America, reproduced by permission).

Table 2
Relationship between Mortality and Trichome Exudate Accumulation on Bodies of *Empoasca fabae* Caged on Foliage of *Solanum polyadenium* and Katahdin.

Species	% Mortality		% Dead with Exudate on:[a]			
			Mouth Parts		Tarsi	
	Nymphs	Females	Nymphs	Females	Nymphs	Females
Solanum polyadenium[b]	78a	94a	75a	29a	83a	41a
S. tuberosum (Katahdin)[c]	11b	20b	0b	0b	0b	0b

Source: From Tingey and Gibson, 1978.

[a] Means within a column followed by the same letter are not significantly different at $p = 0.05$ by Duncan's multiple range test.

[b] A glandular-resistant species.

[c] A nonglandular-susceptible cultivar.

recognized. In field trials of S. berthaultii accessions, mean infestations of adult Empoasca fabae, apterous Myzus persicae, apterous Macrosiphum euphorbiae, and adult Epitrix cucumeris (Harris) were reduced 76 percent, 69 percent, 86 percent, and 77 percent, respectively, on accessions with both type A and type B trichomes in comparison to those with type A hairs alone, suggesting that type A and type B trichomes act in an additive or synergistic manner in plant defense against these pests. These results suggest a need to aggregate and stablize genes for type B trichomes in breeding populations. We have also found that variation within S. berthaultii for glandular trichome density and droplet size is far greater than generally recognized, thus highlighting the importance of selection of specific S. berthaultii parents for use in interspecific hybridization and variety development.

Levels of resistance, as measured by the immobilization of tarsi and labia by trichome exudate, increase with the pooled density of type A and type B trichomes for both the potato leafhopper and green peach aphid. In addition, the expression of resistance varies depending on the life stage. Immature stages are subject to greater mortality and immobilization of tarsi and labia by trichome exudate than are adults. For example, mortality of the green peach aphid ranged from 91 percent for second-stadium nymphs to 78 percent for alate adults. This phenomenon, perhaps resulting from different behavioral responses of immatures and adults or to differential body size and strength, is highly advantageous because it imposes a dramatic mortality effect on the target pest population at a prereproductive stage.

Chemistry of Glandular Trichomes

Although glandular trichomes represent a morphological barrier to insect attack, their main effect is a function of their chemistry. Consequently for maximum effectiveness, it is important to transfer genetic information not only for glandular trichome density but also for appropriate chemical composition. A browning reaction is responsible for entrapment of insects by glandular trichome exudates of Solanum berthaultii, S. polyadenium, and S. tarijense. Physical contact with the trichomes results in discharge of a clear substance that soon darkens and hardens as it accumulates on the insects' legs and mouth parts. Gibson (1971) suggested that the browning reaction might involve production of a polymeric phenol by action of polyphenol oxidase (PPO) because the browning reaction is inhibited in the absence of air; the copper-complexing agent, sodium diethylthiocarbamate, which inhibits polyphenol oxidase activity, inhibited the browning reaction; and use of Folin-Denis's reagent indicated the presence of phenols in the glandular trichomes. However, these conclusions were tentative because they were based entirely upon *in vivo* studies and not on direct biochemical evidence. Recent studies at Cornell have revealed the presence of phenolic oxidase activity (*in vitro*) in trichome secretions of *S. berthaultii* (P. Gregory et al., unpubl.). This oxidase activity appears consistent with that of PPO because treatment with two PPO inhibitors; diethyldithiocarbamate (DEDTC) and phenylthiourea (PTU), resulted in strong, concentration-dependent inhibition. In addition, several phenolic compounds have been found in trichome secretions of *S. berthaultii*, including protocatechuic acid, gallic acid, and catechol. Subsequent analyses by high-pressure liquid chromatography have confirmed these findings and shown protocatechuic acid to be a major phenolic component of *S. berthaultii* trichome secretions.

Genetic Exploitation

Interspecific tetraploid hybrids of *S. tuberosum* x *S. berthaultii* can be made relatively easily by use of the latter species as pollen parent. Such crosses are probably possible because of the tendency of some *S. berthaultii* clones to produce 2N male gametes. Excellent levels of resistance to the potato leafhopper and potato fleabeetle have been demonstrated at Cornell in F_2 and F_3 populations obtained through interspecific hybridization of these species. In 1979 field trials, for example, populations of nymphal and adult potato leafhoppers were reduced 98 percent and 97 percent respectively, on the most resistant hybrid (PCB4-103-2), compared to the cultivar Hudson. Populations of the potato fleabeetle were reduced 90 percent on the hybrid B197-20 compared to Hudson. A number of hybrid clones produced respectable tuber yields, considering that none of our populations have been selected for agronomic qualities, and our F_2 and F_3 populations are only two and three generations removed from a wild parent, *S. berthaultii*, which generally does not tuberize in the New York environment. Three of the highest yielding hybrids, B-227-59, B227-109, and PCB4-42-11, produced tuber yields of 4.9, 4.0, and 5.0 pounds (lb) per plant, respectively, compared to the susceptible cultivars, Superior (1.5 lb), Katahdin (2.3 lb), and Hudson (4.6 lb), and were nearly immune to infestation by the potato leafhopper and potato fleabeetle.

INHERITANCE OF RESISTANCE

Despite considerable progress in identifying sources of arthropod resistance in potatoes, relatively little is known about inheritance mechanisms. This may be due in part to the scarcity of information on specific resistance mechanisms, and in part to the complex genetics of the genus *Solanum*. Nevertheless significant progress toward an understanding of heritability has been made in several cases. Resistance to the green peach aphid is partially dominant and relatively well buffered against environmental influence in certain wild, tuber-bearing *Solanum* species and progenies of their interspecific crosses with Gr. Tuberosum (Sams et al., 1976). Heritability estimates of 50 to 70 percent were obtained for these populations. The heritability of resistance to the potato leafhopper, potato fleabeetle, *Epitrix cucumeris*, and tobacco fleabeetle, *Epitrix hirtipennis* (Melsheimer), in a population of tetraploid clones with *S. chacoense* and *S. demissum* ancestry, has also been demonstrated (Sanford and Sleesman, 1970; Sanford and Peel, 1970; Sanford et al., 1972). As for inheritance of specific defense mechanisms, potato total glycoalkaloid content is polygenically controlled (Sanford and Sinden 1972). In contrast, McCollum and Sinden (1979) have recently shown that the individual glycoalkaloids, α-solanine, α-chaconine, and commersonine, are under simple genetic control in *S. chacoense*. Simultaneous segregation of the three glycoalkaloids occurred, and simple segregation ratios of three-to-one in the F_2 and one-to-one in backcrosses were obtained for the presence or absence of these glycoalkaloids. With regard to glandular trichomes, Gibson (1979) concluded that a single dominant gene controls the presence of type B glandular trichomes on foliage of *S. berthaultii* and *S. tarijense*. In addition, at least one set of recessive genes is necessary for full secretion properties of type B hairs on interspecific hybrids of these species with *S. phureja* Juz. & Buk. and Gr. Tuberosum.

CONSIDERATIONS IN SELECTION AND EXPLOITATION OF RESISTANCE MECHANISMS

In most instances, the decision to implement breeding programs for insect resistance is made without knowledge of specific resistance mechanisms. From a historical perspective, many such programs have been highly successful in the development and release of resistant germ plasm. An understanding of resistance mechanisms can be particularly useful to the breeding program, however, in the analysis of inheritance patterns for resistance conferred by discrete chemical or physical factors. Moreover, such information may allow more discriminating decisions in the selection of resistance mechanisms for exploitation, particularly when faced with plant factors that confer undesirable horticultural or biological properties to the plant (Bergman and Tingey, 1979). Finally, knowledge of resistance mechanisms may aid in the selection of defensive attributes relatively well buffered against the genetic plasticity of the target pest and stable against environmental influence (Tingey and Singh 1980).

Leaf glycoalkaloid content can be manipulated by breeding to provide excellent levels of protection against the Colorado potato beetle and potato leafhopper. Unfortunately, however, levels of these broadly toxic steroidal glycosides in foliage and tubers are highly correlated (Schwarze, 1963). Although it may be possible to find clones segregating for pest-active levels in foliage and safe levels in tubers (Schwarze, 1962), such an approach would be feasible only with the availability of a rapid, inexpensive, and reliable screening method. Despite considerable research effort, screening for glycoalkaloid content currently is too problematical for routine adoption in potato breeding programs, and there appears to be no simple solution (MacKenzie and Gregory, 1979; Cadle et al., 1978). Until there are economical and streamlined analytical methods adaptable for use in screening large populations, breeding for glycoalkaloid-based resistance is not feasible. Although not as serious a limitation, glycoalkaloid levels of potato tissue vary depending on plant age, soil moisture, photoperiod, and light intensity (Pierzchalski, 1957; Pierzchalski and Werner, 1958; van de Klashorst and Tingey, 1979; Sinden and Webb. 1974). Finally, resistance conferred by glycoalkaloids is analogous in part to the protection provided by exogenous application of chemical insecticides and may be subject to similar failure following the appearance of host-specific biotypes.

The second major arthropod resistance mechanism in potato, glandular pubescence, currently appears to have fewer serious limitations as a pest management tool than do glycoalkaloids. Glandular trichomes are active against a wide range of small, soft-bodied pests (Gibson and Turner, 1977); the glandular species, S. berthaultii, can be readily hybridized with S. tuberosum (Figures 4, 6); many clones of the former species are relatively free of undesirably high levels of glycoalkaloids (Tingey et al., 1978; Raman et al., 1979); and pests are likely to counteradapt relatively slowly to the physical entrapment aspect of glandular trichome resistance. Another advantage of glandular-trichome-based resistance is the relative ease by which it can be assessed. At initial stages of screening, clones with superior densities of trichomes can be visually selected through use of simple microscopy. At later stages of selection, it will be desirable to assess the chemical activity of trichome secretions. Recent studies at Cornell (P. Gregory et al., unpubl.) suggest that this might be accomplished through use of a rapid and inexpensive measurement of phenolic oxidase levels.

FUTURE NEEDS AND OUTLOOK

In view of the importance of the potato as food for humans and animals, it is surprising that the development of varietal resistance to arthropods has lagged behind that for field and forage crops. In part, this may be due to the reluctance of potato breeders and entomologists to

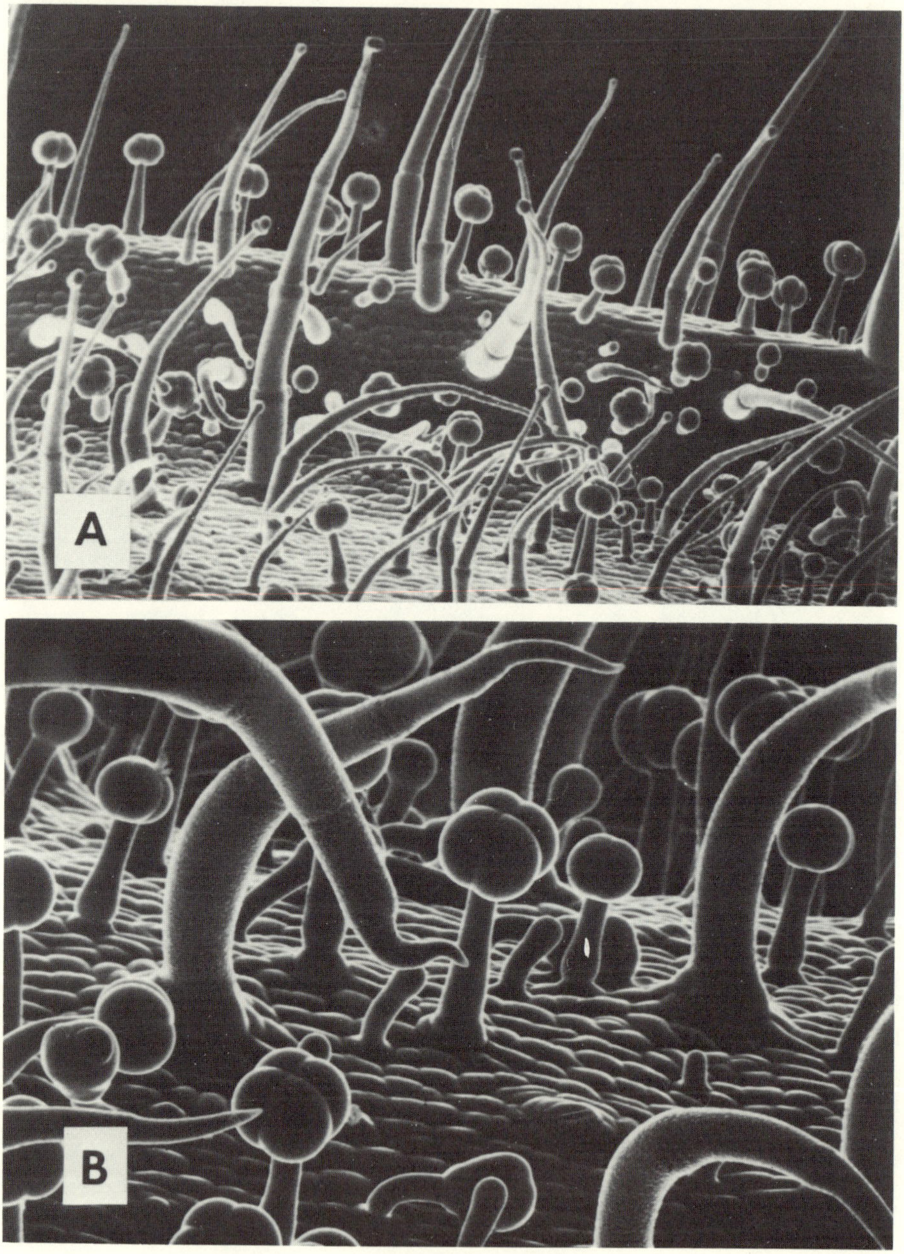

Figure 6
Epidermal morphology of S. tuberosum x S. berthaultii hybrids. (A) Type A and B glandular trichomes on foliage of F_3 clone, B203-38 (X97.6). (B) Type A glandular trichomes of F_1 clone, B264-7 (X244) (courtesy R. L. Plaisted and S. A. Mehlenbacher).

look beyond Gr. Tuberosum for sources of resistance. Far greater genetic diversity is available in wild, unadapted species than in the cultivated species of Solanum, but the use of wild, mostly diploid species poses a considerable challenge in breeding programs because of the complex genetics of the genus Solanum. Considerable progress has been made in recent years, however, in the development of methods for easing crossability barriers between normally incompatible species (Hermsen and Ramanna, 1973; Sams et al., 1977). Additional knowledge of Solanum genetics and of the inheritance of specific resistance mechanisms may also help breeders more effectively incorporate desirable traits in populations of interspecific hybrids while eliminating genes for undersirable characteristics.

The major effort to develop arthropod resistance in potato has been directed toward relatively few pests--such as Empoasca fabae, Leptinotarsa decemlineata, Macrosiphum euphorbiae, and Myzus persicae--although some information is available regarding resistance to Epitrix hirtipennis, E. cucumeris, wireworms, thrips, and mites (Table 3). Although many of these are major arthropod pests of potato in North American and Europe, more attention is needed in the development of genetic resistance for other key pests, of both regional and cosmopolitan importance, such as the potato tubermoth, Phthorimaea operculella (Zeller), the Andean weevil complex, and soil insects such as cutworms and white grubs.

Finally, more effort should be directed toward analysis of specific defense mechanisms of resistant germ plasm. It is remarkable, for example, that over thirty years passed before the significance of glandular trichomes in resistance of S. polyadenium was fully appreciated (Sleesman, 1940; Stringer, 1947; Gibson, 1971). Knowledge of specific defense mechanisms may help clarify the genetics of resistance, thus providing a more quantitative basis for selection and breeding.

REFERENCES

Agerburg, L. S., R. Schick, M. Schmidt, and R. V. Sengbusch, 1933, Die Bestimmung des Solaningehaltes von Pflanzen mit Hilfe von Cladosporium fulvum, Zuchter. 5:272-280.

Adams, J. B., 1946, Aphid Resistance in Potatoes, Am. Potato J. 23:1-2.

Arneson, P. A., and R. D. Durbin, 1968, Studies on the Mode of Action of Tomatine as a Fungitoxic Agent, Plant Physiol. 43:683-686.

Baup, S., 1826, [Letter to the editor concerning several new substances.], Ann. Chim. (Paris) 31:108-109 (in French).

Bergman, J. M., and W. M. Tingey, 1979, Aspects of Interaction between Plant Genotypes and Biological Control, Bull. Entomol. Soc. Am. 25:275-279.

Table 3
Sources of Resistance to Flea Beetles, Thrips, Mites, and Wireworms in Tuber-Bearing Species of *Solanum*

Pest	Source of Resistance	Reference
Epitrix cucumeris	*S. bulbocastanum*, *S. demissum*, *S. fendleri*, *S. polyadenium*	Sleesman, 1940
	S. tuberosum[a]	Sanford and Sleesman, 1970
E. harilana-rubia	*S. berthaultii*	Gibson, 1978
E. hirtipennis	*S. tuberosum*	Sanford and Peel, 1970
Limonius agronus	*S. tuberosum*	Rawlins, 1943
Polyphagotarsonemus latus	*S. berthaultii*, *S. polyadenium*, *S. tarijense*	Gibson and Valencia, 1978
Tetranychus urticae	*S. berthaultii*	Gibson, 1976c
Thrips tabaci	*S. berthaultii*	Gibson and Turner, 1977

[a]Clones with *S. chacoense* or *S. demissum* ancestry.

Bradley, R. H. E., and R. Y. Ganong, 1951, Aphid Infestations on Katahdin and on a Seedling Resistant to Myzus persicae (Sulz.), with Two Dates of Planting, Can. J. Zool. 29:329-338.

Buhr, H., R. Toball, and K. Schreiber, 1958, Die Wirkung von einigen pflanzenlichen Sonderstoffen, inbesondere von alkaloiden auf der Entwicklungen der Larven des Kartoffelkafers (Leptinotarsa decemlineata Say), Entomol. Exp. Appl. 1:209-224.

Cadle, L. S., D. A. Stelzig, K. L. Harper, and R. J. Young, 1978, Thin-Layer Chromatographic System for Identification and Quantification of Potato Tuber Glycoalkaloids, J. Agric. Food. Chem. 26:1453-1454.

Dahlman, D. L., and E. T. Hibbs, 1967, Responses of Empoasca fabae (Cicadellidae: Homoptera) to Tomatine, Solanine, Leptine I; Tomatidine, Solanidine, and Demissidine, Ann. Entomol. Soc. Am. 60:732-740.

Gardner, M. E., R. Schmidt, and F. J. Stevenson, 1945, The Sequoia Potato: A Recently-Introduced Insect-Resistant Variety. Am. Potato J. 22:97-103.

Gibson, R. W., 1971, Glandular Hairs Providing Resistance to Aphids in Certain Wild Potato Species, Ann. Appl. Biol. 68:113-119.

Gibson, R. W., 1974, Aphid-Trapping Glandular Hairs on Hybrids of Solanum tuberosum and S. berthaultii, Potato Res. 17:152-154.

Gibson, R. W., 1976a, Glandular Hairs are a Possible Means of Limiting Aphid Damage to the Potato Crops, Ann. Appl. Biol. 82:143-146.

Gibson, R. W., 1976b, Glandular Hairs on Solanum polyadenium Lessen Damage by the Colorado Beetle, Ann. Appl. Biol. 82:147-150.

Gibson, R. W., 1976c, Trapping of the Spider Mite Tetranychus urticae by Glandular Hairs on the Wild Potato Solanum berthaultii, Potato Res. 19:179-182.

Gibson, R. W., 1978, Resistance in Glandular-Haired Wild Potatoes to Flea Beetles, Am. Potato J. 55:595-598.

Gibson, R. W., 1979, The Geographical Distribution, Inheritance, and Pest-Resisting Properties of Sticky-tipped Foliar Hairs on Potato Species, Potato Res. 22:223-236.

Gibson, R. W., and R. H. Turner, 1977, Insect-Trapping Hairs on Potato Plants, P.A.N.S. 22:272-277.

Gibson, R. W., and L. Valencia, 1978, A survey of Potato Species for Resistance to the Mite Polyphagotarsonemus latus, with Particular Reference to the Protection of Solanum berthaultii and S. tarijense by Glandular Hairs, Potato Res. 21:217-223.

Green, T. R., and C. A. Ryan, 1972, Wound-Induced Proteinase Inhibitor in Plant Leaves: A Possible Defense Mechanism against Insects, Science 175:776-777.

Hermsen, J. G. Th., and M. S. Ramanna, 1973, Double-Bridge Hybrids of Solanum bulbocastanum and Cultivars of Solanum tuberosum, Euphytica 22:457-466.

van de Klashorst, G., and W. M. Tingey, 1979, Effect of Seedling Age, Environmental Temperature, and Foliar Total Glycoalkaloids on Resistance of Five *Solanum* Genotypes to the Potato Leafhopper, *Environ. Entomol*. 8:690-693.

Kuhn, R., and J. Löw, 1954, The Constitution of Solanine, *Angew Chem*. 66:639-640.

Kuhn, R., and R. Löw, 1955, Resistance Factors against *Leptinotarsa decemlineata* Say, Isolated from the Leaves of Wild *Solanum* Species, in M. G. Sevag, R. D. Reid, and O. E. Reynolds, *Origins of Resistance to Toxic Agents*, pp. 122-132, Academic Press, New York.

Landis, B. J., D. M. Powell, and L. Fox, 1972, Populations of Aphids on Named Varieties of Irish Potatoes, *Am. Potato J*. 49:63-69.

McCollum, G. D., and S. L. Sinden, 1979, Inheritance Study of Tuber Glycoalkaloids in a Wild Potato, *Solanum chacoense* Bitter, *Am. Potato J*. 56:95-113.

MacKenzie, J. D., and P. Gregory, 1979, Evaluation of a Comprehensive Method for Total Glycoalkaloid Determination, *Am. Potato J*. 56:27-33.

Osman, S. F., S. F. Herb, T. J. Fitzpatrick, and S. L. Sinden, 1976, Commersonine, a New Glycoalkaloid from Two *Solanum* Species, *Phytochem*. 15:1065-1067.

Peterson, A. G., and A. A. Granovsky, 1950, Feeding Effects of *Empoasca fabae* on a Resistant and Susceptible Variety of Potato, *Am. Potato J*. 27:366-371.

Pierzchalski, T., 1957, [Changes in glycoalkaloid contents in leaves of cultivated and wild potatoes and their hybrids during growth and their influence on the Colorado potato beetle], *Hodowla Rosl*., *Aklim*. *Nasienn*. 1:415-430 (in Polish).

Pierzchalski, T., and E. Werner, 1958, [Changes in glycoalkaloid contents in leaves of cultivated and wild potatoes and their hybrids during growth and their influence on the Colorado potato beetle (*Leptinotarsa decemlineata* Say)], *Hodowla. Rosl. Aklim, Nasienn*. 2:157-180 (in Polish).

Prokoshev, S. M., E. I. Petrochenko, and V. Z. Baranova, 1952, [Glycoalkaloids of tuber-bearing forms of *Solanum* in connection with resistance to Colorado potato beetle.], *Doklady Akad Nauk* (USSR)82:955-958 (in Russian).

Radcliffe, E. B., and F. I. Lauer, 1966, A Survey of Aphid Resistance in the Tuber-Bearing *Solanum* (Tourn.) L. Species, *Minn. Agric. Exp. Stn. Techn. Bull. 252*, 23 p.

Radcliffe, E. B., and F. I. Lauer, 1968, Resistance to *Myzus persicae* (Sulzer), *Macrosiphum euphorbiae* (Thomas), and *Empoasca Fabae* (Harris) in the Wild Tuber-Bearing *Solanum* (Tourn.) Species, *Minn. Agr. Exp. Stn. Tech. Bull. 259*, 27 p.

Radcliffe, E. B., and F. I. Lauer, 1970, Further Studies on Resistance to Green Peach Aphid and Potato Aphid in the Wild Tuber-Bearing *Solanum* Species, *J. Econ. Entomol*. 63:110-114.

Radcliffe, E. B., and F. I. Lauer, 1971a, An Appraisal of Aphid Resistant Tuber-Bearing Solanum Germ Plasm, Univ. Minn. Agric. Exp. Stn. Tech. Bull. 286, 22 p.

Radcliffe, E. B., and F. I. Lauer, 1971b, Resistance to Green Peach Aphid and Potato Aphid in Introductions of Wild Tuber-Bearing Solanum Species, J. Econ. Entomol. 64:1260-1266.

Radcliffe, E. B., F. I. Lauer, and R. E. Stucker, 1974, Stability of Green Peach Aphid Resistance in Tuber-Bearing Solanum Introductions and Its Effect on Screening Procedures, Environ. Entomol. 3:1022-1026.

Raman, K. V., W. M. Tingey, and P. Gregory, 1979, Potato Glycoalkaloids: Effect on Survival and Feeding Behavior of the Potato Leafhopper, J. Econ. Entomol. 72:337-341.

Rawlins, W. A., 1943, Some Varietal Differences in Wireworm Injury to Potatoes, Am. Potato J. 20:156-158.

Sams, D. W., F. I. Lauer, and E. B. Radcliffe, 1976, Breeding Behavior of Resistance to Green Peach Aphid in Tuber-Bearing Solanum Germ Plasm, Am. Potato J. 53:23-29.

Sams, D. W., P. D. Ascher, and F. I. Lauer, 1977, Crossability of Some Green-Peach-Aphid-Resistant Tuber-Bearing Solanum, Potential Bridging Species, and Solanum tuberosum ssp. tuberosum, Am. Potato J. 54:355-364.

Sanford, L. L., and S. L. Sinden, 1972, Inheritance of Potato Glycoalkaloids, Am. Potato J. 49:209-217.

Sanford, L. L., and J. P. Sleesman, 1970, Genetic Variation in a Population of Tetraploid Potatoes: Response to the Potato Leafhopper and the Potato Flea Beetle, Am. Potato J. 47:19-34.

Sanford, L. L., and J. P. Sleesman, 1974, Selection for Resistance to Potato Leafhopper in Potatoes, Am. Potato J. 51:44-50.

Sanford, L. L., and R. E. Webb, 1977, Loss of Yield in Potato Selections Infested with Potato Leafhoppers, Am. Potato J. 54:581-586.

Sanford, L. L., and R. D. Peel, 1970, Genetic Variation in a Population of Tetraploid Potatoes: Tuber Resistance to the Tobacco Flea Beetle, Am. Potato J. 47:169-175.

Sanford, L. O., O. V. Carlson, and E. T. Hibbs, 1972, Genetic Variation in a Population of Tetraploid Potatoes: Foliar Resistance to Oviposition of the Potato Leafhopper, Am. Potato J. 49:98-108.

Schalk, J. M., R. L. Plaisted, and L. L. Sanford, 1975, Progress Report: Resistance to the Colorado Potato Beetle and Potato Leafhopper in Solanum tuberosum subsp. andigena, Am. Potato J. 52:175-177.

Schreiber, K., 1957, [Naturally occurring plant resistance factors against the Colorado potato beetle (Leptinotarsa decemlineata) and their possible mode of action], Zuchter 27:289-299 (in German).

Schreiber, K., 1958, On Some Significant Substances of the Solanaceae and Their Importance for Potato Beetle Resistance, Entomol. Expt. App. 1:28-37.

Schreiber, K., 1968, Steroid Alkaloids: The *Solanum* Group, in R. H. F. Manske, ed., *The Alkaloids, Chemistry and Physiology*, pp. 1-192, Academic Press, New York.

Schwarze, P., 1962, [Methods for identification and determination of solanine in potato breeding material], *Zuchter*. 32:155-160 (in German).

Schwarze, P., 1963, [The glycoalkaloid content and the composition of the glycoalkaloid complex in progenies of the interspecific cross *S. tuberosum* x *S. chacoense*], *Zuchter*. 33:275-281(in German).

Sinden, S. L., R. W. Goth, and J. O'Brien, 1973, Effect of Potato Alkaloids on the Growth of *Alternaria solani* and Their Possible Role as Resistance Factors in Potatoes, *Phytopathol*. 63:303-307.

Sinden, S. L., and R. E. Webb, 1974, Effect of Environment on Glycoalkaloid Content of Six Potato Varieties at 39 Locations, *ARS-USDA Tech. Bull. 1472*, 30 p.

Sleesman, J. P., 1940, Resistance in Wild Potatoes to Attack by the Potato Leafhopper and the Potato Flea Beetle, *Am. Potato J*. 17:9-12.

Sleesman, J. P., and J. Bushnell, 1937, Variations in Nymphal Populations of the Potato Leafhopper on Different Varieties of Potatoes, *Am. Potato J*. 14:242-245.

Stringer, A., 1947, A Note on the Resistance of *Solanum polyadenium* to Aphids, *Rpt. Long Ashton Res. Stn. 1946*, pp. 88-89.

Swiniarski, E., E. Werner, and Z. Mierzwa, 1958, [Biochemical resistivity of wild potato species to Colorado potato beetle larvae (*Leptinotarsa decemlineata* Say)], *Hodowla Rosl Aklim Nasienn* 2:623-631 (in Polish).

Taylor, C. E., 1962, The Population Dynamics of Aphids Infesting the Potato Plant with Particular Reference to the Susceptibility of Certain Varieties to Infestation, *Eur. Potato J*. 5:204-219.

Tingey, W. M., and R. W. Gibson, 1978, Feeding and Mobility of the Potato Leafhopper Impaired by Glandular Trichomes of *Solanum berthaultii* and *S. polyadenium*, *J. Econ. Entomol*. 71:856-858.

Tingey, W. M., J. D. MacKenzie, and P. Gregory, 1978, Total Foliar Glycoalkaloids and Resistance of Potato Species to *Empoasca fabae* (Harris), *Am. Potato J*. 55:577-585.

Tingey, W. M., and R. L. Plaisted, 1976, Tetraploid Sources of Potato Resistance to *Myzus persicae*, *Macrosiphum euphorbiae*, and *Empoasca fabae*, *J. Econ. Entomol*. 69:673-676.

Tingey, W. M., and S. R. Singh, 1980, Environmental Factors Influencing the Magnitude and Expression of Resistance, in *Breeding Plants Resistant to Insects*, ed. F. G. Maxwell and P. R. Jennings, pp. 87-113, John Wiley & Sons, New York.

Willimot, S. C., 1933, An Investigation of Solanine Poisoning, *Analyst*. 58:431-439.

ABOUT THE EDITORS

JAMES H. LASHOMB has been an assistant professor in applied insect ecology at Cook College in Rutgers University since 1978. He received the B.S. degree in agriculture from Cornell University in 1970. The M.S. and Ph.D. degrees in entomology were conferred by the University of Maryland in 1973 and 1975, respectively. From 1975 to 1978, Dr. Lashomb was a research associate at Mississippi State University and worked on the Expanded Southern Pine Beetle Project in the areas of population ecology and toxicant evaluation. His present research interests are the population dynamics and natural enemies of the Colorado potato beetle.

RICHARD A. CASAGRANDE is an assistant professor of entomology at the University of Rhode Island. He received the M.S. and Ph.D. degrees in entomology from Michigan State University where he also completed a year of postdoctoral study in systems science in 1976. Dr. Casagrande presently directs a potato pest management program in Rhode Island and he conducts research on control alternatives for the Colorado potato beetle. His participation in this conference (Recent Advances in Potato Pest Management, Rutgers University, 1980) is an outgrowth of his interest in potato insect research and the close association he has maintained with Rutgers University since earning the B.S. degree in entomology there in 1969.